Mobile Ad Hoc Networks

From Wireless LANs to 4G Networks

George Aggelou

Prof., Institute of Technology, Crete, Greece
Director, G-Alpha Telecomms, Athens, Greece

McGraw-Hill

New York   Chicago   San Francisco   Lisbon   London   Madrid
Mexico City   Milan   New Delhi   San Juan   Seoul
Singapore   Sydney   Toronto

The McGraw·Hill Companies

Cataloging-in-Publication Data is on file with the Library of Congress.

1 2 3 4 5 6 7 8 9 0   DOC/DOC   0 1 0 9 8 7 6 5 4

ISBN 0-07-170074-9
ISBN 978-0-07-170074-0

The sponsoring editor for this book was Stephen S. Chapman and the production supervisor was Pamela A. Pelton. It was set in Century Schoolbook by International Typesetting and Composition. The art director for the cover was Anthony Landi.

This book was printed on recycled, acid-free paper containing a minimum of 50% recycled, de-inked fiber.

To my family
Vagelis, Stella, Vassilis, Alexander, Eleni, and
Eleni Neonaki

"*True knowledge exists in knowing that you know nothing. And in knowing that you know nothing, that makes you the smartest of all.*"

Socrates (470–399 B.C.)

"ἕν οἶδα ὅτι οὐδέν οἶδα"

Σωκράτης (470–399 π·χ·)

Contents

Preface

Mobile communications is the global success story of the late twentieth and early twenty-first centuries. The global system for mobile communication (GSM) is undoubtedly the largest technology worldwide. Second-generation telecommunication systems, such as the GSM, enabled voice traffic to go wireless the number of mobile phones already exceeds the number of landline phones and, according to the GSM MoU* Association, the mobile phone penetration exceeds 70 percent in countries with the most advanced wireless markets.

Meanwhile, as Internet users and content proliferated, the incredible penetration of personal communications systems as well as the increasing number of Internet connections have pushed the industry to come up with solutions to combine the multimedia capacities of the World Wide Web with the flexibility of wireless communications. The convergence of trends for next-generation mobile services, groundbreaking mobile techniques, and technological innovations made the vision of mobile multimedia a reality in a stronger way than anyone would have predicted. In this way, people using a lightweight and convenient pocket communicator will be able to experience audio and video clips, send electronic postcards to each other, and tap into entertainment and service databases personalized according to their preferences, location, or situation.

Facing the challenge of multimedia handling on-the-move, second-generation systems were soon proven limited. Mobile operators moved toward a new generation of technologies the third-generation (3G) mobile communication systems. 3G systems, or else *universal mobile telecommunication systems* (UMTS), are designed with the goal to complete the worldwide globalization process of mobile communications and provide the foundation for new services with high-rate and high-quality data. At the end of 2005, Mobile Network Operator (MNO)

---

*Memorandum of Understanding

professionals estimate that there will be 1.6 billion users, of which more than 1 billion will be 3G mobile Internet users. By 2010, the UMTS Forum foresees 28 percent of the world's 2.25 billion mobile cellular subscribers as subscribers of 3G systems.

Now that the first 3G systems are being built, it is natural to start asking what could be the next big step. Commentaries and predictions regarding wireless broadband communications are currently cultivating visions of the systems beyond the third generation, commonly know as fourth-generation (4G) systems. Maximum data rates up to 1 Gbps, the combination of several available, evolving, and emerging access technologies into a common platform, and their seamless interworking along with adaptive multimode terminals will be the key characteristics of beyond third generation mobile wireless systems.

The very high data rates envisioned for the 4G systems do not appear to be feasible with the conventional 2G and 3G cellular architectures. This is mainly attributed to two basic reasons. First, the transmission rates envisioned for 4G systems are as high as two orders of magnitude higher than those of the 3G systems, and it is known from theory that for a given transmit power level, the symbol (and thus bit) energy decreases linearly with the increasing transmission rate. Second, the spectrum that will be released for 4G systems will almost certainly be located well above the 2-GHz band used by the 3G systems. The radio propagation in these bands is significantly more vulnerable to non-line-of-sight conditions, which is the typical mode of operation in today's urban cellular communications.

The brute-force solution to this problem is to significantly increase the density of the base stations, resulting in considerably higher deployment costs, which would be feasible only if the number of subscribers also increased at the same rate. This seems unlikely, with the penetration of cellular phones already being high in the developed countries. On the other hand, the same number of subscribers will have a much higher demand in transmission rates, making the aggregate throughput rate the bottleneck in future wireless systems. Under the working assumption that subscribers would not be willing to pay the same amount per data bit as for voice bits, a drastic increase in the number of base stations does not seem economically justifiable.

With these factors in mind, it becomes obvious that more fundamental enhancements are necessary for the very ambitious throughput and coverage requirements of future systems. Toward this end in addition to advanced transmission techniques and collocated antenna technologies some major modifications in the wireless network architecture itself, which will enable effective distribution and collection of signals to and from wireless users, are required. The integration of multihop capability into conventional wireless networks is perhaps the most promising architectural upgrade to enhancing the area coverage without significant additional infrastructure cost. *Multihopping* is the ability of mobile radios to relay packets to one another without the use of base stations.

Multihop wireless networking traditionally has been studied in the context of ad hoc and peer-to-peer networks. A mobile ad hoc network (MANET) consists of wireless mobile nodes that can freely and dynamically self-organize into arbitrary and temporary network topologies, allowing people and devices to seamlessly internetwork in areas with no preexisting communication infrastructure. The salient feature of this breed of networks is that they can operate in different and differing propagation and network operating conditions, which cannot be predicted during the network design stage. Such a *self-organizing* and *rapidly deployable* network permits a new paradigm of wireless wearable devices that would immediately and easily enable instantaneous person-to-person, person-to-machine, or machine-to-person communications.

Such perceived versatility elicited immediate interest in the early days among military, police, and rescue agencies in the use of such networks, especially under disorganized or hostile conditions, including isolated scenes of natural disaster or armed conflict. Soldiers equipped with multimode mobile communicators can now communicate in an ad hoc manner, without the need for fixed wireless base stations. In addition, small vehicular devices equipped with audio sensors and cameras can be deployed at targeted regions to collect important location and environmental information that will be communicated back to a processing node via ad hoc mobile communications. Ship-to-ship ad hoc mobile communication is also desirable since it provides alternate communication paths without reliance on ground- or space-based communication infrastructures. Moreover, ad hoc mobile communication is particularly useful in relaying information (status, situation awareness, and the like) via data, video and/or voice from one rescue team member to another over a small handheld or wearable wireless device. Recently, home or small office networking and collaborative computing with laptop computers in a small area have emerged as alternative areas of application. People in a meeting or a conference can freely use their laptops, palmtops, or notebooks to have instant network formation in addition to file and information sharing without the presence of fixed base stations and systems administrators. A presenter can multicast slides and audio to intended recipients and attendees can ask questions and interact on a commonly shared white board.

Interestingly, the field of ad hoc mobile networks is rapidly growing and changing, and while there are still many challenges that need to be met, it is likely that such networks will see widespread use within the next few years. Over the past few years, we have witnessed an upsurge of interest in multihop-augmented infrastructure-based wireless networks in both industry and academia. Prominent examples include the "seed" concept in 3GPP, Mesh Networks in IEEE 802.16, and coverage extension of high performance local area network v.2 (HiperLAN/2) through relays or user-cooperative diversity mesh networks.

This book focuses on state-of-the-art in mobile ad hoc networks and highlights some of the emerging technologies, protocols, and approaches at different layers for realizing network services for users on-the-move in areas with possibly no pre-existing communication infrastructures. It also deals with 3G/UMTS standardization activities and illustrates key issues involved in the systems beyond 3G (B3G) pertinent to wireless ad hoc relaying.

The book is designed to be used both as a textbook and as a reference work. As a textbook, the material is suitable for courses on dynamic routing, multimedia services support over wireless mobile networks as well as third-generation wireless communication systems and beyond, ranging in level from advanced undergraduate to advanced graduate. As a reference work, it is intended primarily for ad hoc network technologists, engineers, and scientists in academia, industry, research labs, and the government, illustrating up-to-date technologies, experience reports, case studies, and descriptions of innovative ad hoc systems as well as latest major worldwide research findings in many aspects of wireless ad hoc networking. The reader is assumed to have some elementary knowledge of communications networks and how they function.

The book is organized as follows. Chapter 1 offers a general perspective on key features and operating principles of wireless mobile ad hoc networks, as well as an overview of the state of the art in current research on ad hoc relaying. It is followed by two chapters covering various technical challenges in wireless ad hoc networking with focus on dynamic routing, power/energy-efficient protocols, and quality-of-service support in highly dynamic wireless environments. In its general context, ad hoc networking is a multilayer problem. The absences of infrastructure and node mobility are the two basic causes of the dynamics in MANETs. Ensuring effective routing is one of the great challenges for ad hoc networking. Besides, the lack of central coordination facilities makes the control of channel resources difficult. Carefully designed distributed medium access techniques must be used for allowing fair access and semireliably transport data over the shared wireless links in the presence of rapid changes and hidden or exposed terminals. In addition, since the basic components of ad hoc wireless networks are mostly battery-operated portable devices, such as notebooks, personal digital assistants (PDAs), cellular phones, and microsensors, *power conservation* is one of the central issues of such networks. Power-conservative designs for ad hoc networks pose many challenges due to the lack of central coordination facilities.

Finally, Chapter 4 illustrates step-by-step the standardization procedures of 3G/UMTS and portrays its major evolutions from the second-generation schemes to true mobile multimedia wireless systems. It points out the principal features of systems beyond 3G (4G systems) and covers different approaches of exploiting the benefits of multihop communications in future-generation systems, such as solutions for radio range extension in mobile and wireless broadband cellular networks (trading range for capacity), and solutions to reduce infrastructure deployment costs. It finally draws out future research directions.

To a large extent, the material in this book is based on the author's widely recognized research and development work. With an abundance of in-depth material, this book covers the latest major worldwide research findings in many aspects of wireless ad hoc networking, providing all the theoretical and design knowledge and know-how needed to design with both present and future technology. I hope that this book will build your strength and knowledge in designing future mobile radio systems pertinent to wireless ad hoc relaying.

Acknowledgments

This book would have never been possible without the support and care of many wonderful people.

First and foremost I would like to express my deepest and sincere appreciation to the many people I have shared my career with, up to the present. I am especially indebted to Charlie Perkins (IBM T.J. Watson Research Center, NY), Pravin Bhagwat (IBM T.J. Watson Research Center, NY), Rahim Tafazolli (University of Surrey, UK), Ioannis Kriaras, (Lucent Research, UK), Laurent Lambrecht (CISCO Systems, UK), Nikos Pronios (INTRAKOM, Greece), and Manolis Antonidakis (Institute of Technology, Greece).

Special thanks go to Charlie Perkins who, back in 1996, invited me to join his group at IBM Research, NY. Charlie taught me how to do research and how to do good research. Through many spirited discussions on the topic of wireless routing, he constantly reminded me to be very careful about things that seemed obvious. Charlie is the person who stimulated my interests toward mobile wireless communications science. I will forever be indebted to him.

The team at McGraw-Hill Professional participating in the production of this book provided excellent support and worked hard to keep to the demanding schedule. I am extremely grateful to the editors of McGraw-Hill Professional for their many valuable comments that greatly improved the clarity and readability of this book.

Special thanks are also due to my students at the Institute of Technology, Greece, for their cooperation and understanding during the writing of this book, as well as to the receptionists from Irida Hotel (Crete) and Novotel Hotel (Brussels) for their patience and cooperation in providing printing facilities during late (often very late!) night sessions.

Finally, this endeavor would have been impossible without the love, support, encouragement, and blessings of my family Vagelis, Stella, Vassilis, and Eleni. Great many thanks go to Alex for his generous support on many aspects of this book. They all deserve very special thanks.

GEORGE AGGELOU
Institute of Technology, Greece

Wireless Ad Hoc Communications Technologies

Over the last several years the wireless world has seen significant developments with a new generation of radio frequency (RF) networking products. Advancements in wireless communications technologies together with the availability of wireless communications devices with increased processing capabilities have enabled wireless connectivity of mobile users to the global Internet. At the same time, the proliferation of mobile computing devices (e.g., laptops, handheld digital devices, personal digital assistants (PDAs), and wearable computers) has brought about a revolution in the computing world. Not only general-purpose computers but also handheld terminals designed for specific tasks (store management, hospital data loggers, and so forth) are available at modest prices. Users have improved their working methods by taking advantage of the new computer capabilities. Technology trends have thus evolved rapidly from the personal computer (PC) age (i.e., one computing device per person) to the ubiquitous computing age, in which individual users simultaneously use several electronic platforms through which they can access all the required information whenever and wherever they may be, thus providing a mobile and ubiquitous connection to Internet information services [BRUNO02].

Of late, there has been an increasing development in local area networking (LAN) technology. LANs, in general, allow for the sharing of information in cooperative work where they offer a fast and reliable connection to desktop computers. Most LANs are based on transmission media consisting of wire, coaxial cable, twisted pair, or optical fibers. See Fig. 1.1.

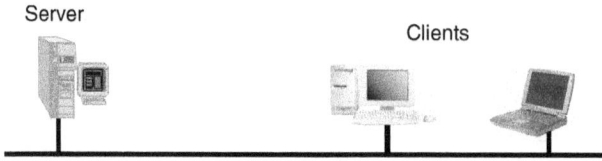

Figure 1.1 Traditional wired LAN.

The investment costs for a new LAN installation, however, can be as high as 40 percent of the whole installation, including the cost of software, hardware, and cabling, as problems may arise when the network is reconfigured [PAHL98]. All the equipment and software can be reused, but the cables are fixed. Therefore, the cost needed to move them may turn out to be more or less the same as a new installation. This economic projection has caused a growing interest in wireless local area networks (W-LANs). W-LANs can offer, in principle, portability and lower installation costs (Fig. 1.2). New generations of wireless networks have thus evolved to enable personal high-speed interconnectivity with wide area networks through the infrastructure of a wireless carrier.

Wireless systems can be installed in different environments such as offices, manufacturing floors, research laboratories, hospitals, or universities offering a set of applications that include communication between terminals and connections to the telephone network (wireless PBX or pocket phones). The nature of ubiquitous devices makes wireless networks the easiest solution for their interconnection. This has led to a rapid growth in the use of wireless technologies for the LAN environment. Portability is one direct result of this evolution, which enables users to take their primary information working tools with them.

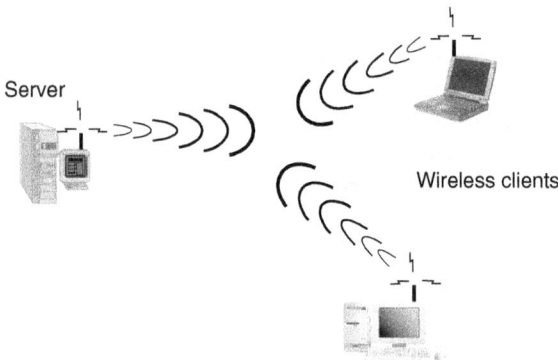

Figure 1.2 Wireless LAN.

For this purpose, a family of networking protocols, such as the IEEE 802.11 standard (see Sec. 1.7.1), was developed to support wireless high-speed connectivity of LANs (such as corporate or university network infrastructures) that interconnect with wide area network (WAN) services. The W-LAN technology has already been making inroads into universities as a high-speed campus network, and it is also being tested in other markets such as coffee shops and airports to provide wireless access to networked services.

To address the public demand for tetherless communications, in the late 1990s, a new family of short-range technologies appeared in the W-LAN landscape with the goal to make cable connectivity among various communications devices redundant. Bluetooth is one of the most popular short-range wireless technologies. It is a specification for short-range, low-cost communication that provides a cable replacement solution for consumer devices enabling user-friendly connectivity among computing devices. This technology can be spontaneously set up to serve a specific function and collapsed when no longer needed. No longer would you need to wire up PDAs, notebooks, mobile phones, headsets, MP3 players, digital cameras, and even computers but you could effortlessly achieve interdevice communication by using radio—so long as they are less than 10 m apart.

Bluetooth has the potential for impacting many areas—including applications that would have been inconceivable a few years ago—such as a dishwasher wirelessly signaling a WebCam to the user that its cycle has finished and that it now needs more salt for its water softener! However, one particular area where Bluetooth professionals expect to have a significant impact in the future is its interworking with wide-area wireless delivery mechanisms such as cellular telephony. Examples include enabling wireless connectivity between a computer and a cellular phone, a personal digital assistant (PDA) and a computer, or a PDA and a cellular phone. When Bluetooth is used to interconnect a cellular phone to another computing device, the cellular phone can, in effect, operate as a personal wireless gateway to send and receive information over a local or wide area network. These scenarios can become real using Bluetooth wireless networking without the requirement for line-of-sight communications whereas in the case of the cellular phone, it can be located within a closed briefcase and can easily interconnect with a PDA or laptop computer. Bluetooth thus brings a new level of interconnectivity to the personal dimension, which aims at providing an air interface that is universal, low cost, user friendly, and capable of replacing the variety of proprietary cables that consumers need to carry and use to connect their personal devices   [BISDIK01].

With the increasing proliferation of short-range wireless LANs and portable/mobile terminals, soon the need for standardization activities became evident so as to allow technology interoperability for an increasingly mobile workforce. In this regard, wireless carriers are taking a closer look at how wireless local area network developments can augment their aspirations for the next-generation wireless network implementations. Both technologies, IEEE 802.11 and Bluetooth, compete in the same frequency band (both operate in the industrial/scientific/ medical (ISM) 2.4 GHz band) and in the same market. The Bluetooth Special Interest Group (SIG) has thus been working on efforts that can lead to coexistence developments between W-LAN standards in the ISM band. Coexistence is seen as a key trend in Bluetooth and IEEE 802.11 communications so as to create possibilities of bridging devices that could link Bluetooth to 2.4 GHz W-LANs. One can easily go with the following scenario regarding coexistence: a traveler at an airport might use a Bluetooth-enabled device such as a PDA or cellular phone via associated access points to check flight schedules and to book hotel reservations. Then, the traveler could well use a W-LAN to surf the Web and to view video e-mails via other W-LAN access points. That is, a user might want to insert an IEEE 802.11 PC card into a notebook that already contains a Bluetooth module. Considering this case, several developments are already underway that can significantly enhance this communications experience.

In an effort to define the respective roles that the IEEE 802.11 family of specifications and Bluetooth-like technologies may play in the networking realm, specific niche areas have emerged for each; namely the traditional W-LAN market for IEEE 802.11 and the wireless personal area networks (W-PAN) and body area networks (W-BAN) market for Bluetooth wireless technology. A taxonomy of W-LANs is depicted in Fig. 1.3.

A W-PAN is a new class of wireless networks that allows the proximal devices to dynamically share information with minimum power consumption [ZIMMER96a], compared to IEEE 802.11 (W-LAN) products.

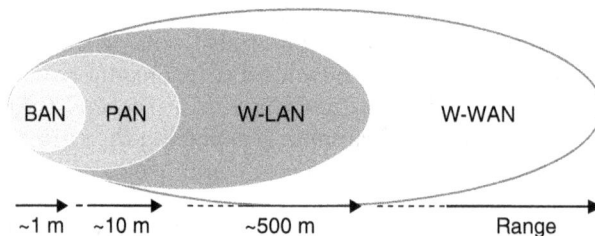

Figure 1.3 A taxonomy of W-LANs based on their operating range.

W-LANs and PANs, however, do not meet all the networking requirements of ubiquitous computing. Situations exist where carrying and
holding a computer are not practical (e.g., assembly line work). A *wearable* computer solves these problems by distributing computer components (e.g., head-mounted displays, microphones, earphones, processors,
and mass storage) on the body [ZIMMER96a]. Users can thus receive
job-critical information and maintain control of their devices while
their hands remain free for other work. A network with a transmission
range of a human body is commonly referred to as a *body area network*
(BAN) and constitutes the best solution for connecting wearable
devices.

1.1   Wireless Local Area Networks (W-LANs)

Wireless LANs (W-LAN) were originally introduced to solve four nag-
ging problems—mobility, mobile-to-mobile networking, relocation, and
an alternative to locations that are difficult to wire. A W-LAN is a flexible data communication system implemented as an extension to, or as
an alternative for, a wired LAN within a building or campus. Like wired
LANs, a wireless LAN (Fig. 1.2) has a communication range typical of
a single building or a cluster of buildings, i.e., 100 to 500 m. Using electromagnetic waves, W-LANs transmit and receive data over the air,
minimizing the need for wired connections. Thus W-LANs combine data
connectivity with user mobility, and through simplified configuration,
enable movable LANs.

For radio frequency wireless LANs, the availability of unlicensed
spectrum is a significant enabler. In the United States, it was the
Federal Communications Commission's rule change in 1985 (first published in 1985, rules modified in 1990) allowing unlicensed spread spectrum use of the three ISM frequency bands, that encouraged the
development of a number of wireless technologies. Today, unlicensed
wireless LAN products are available in all three of the ISM bands at 902
to 928 MHz, 2.400 to 2.4835 GHz, and 5.725 to 5.850 GHz.

To summarize, W-LANs allow users to gain access to shared information without looking for a place to plug in and network managers to
set up or augment networks without installing or moving wires and
cables. In general terms, the advantages of W-LANs over traditional
wired networks can be categorized into the following five broad categories:

- *Installation speed and simplicity.*   W-LAN technology eliminates the
  need to pull cable through the walls and ceilings.

- *Installation flexibility.*   Wireless technology allows networks to go
  where wired networks cannot go.

- *Scalability.* W-LANs systems can be configured in a variety of topologies to meet the needs of specific applications and installations. Configurations can be easily changed and ranged from independent networks suitable for a small number of users to full infrastructure networks of thousands of users.

- *Improved productivity and service.* It allows people to access shared resources anywhere in their organization.

- *Reduced cost-of-ownership.* Although the initial investment required for W-LAN hardware can be higher than the cost of wired LAN hardware, overall installation expenses and life-cycle costs can be significantly lower.

1.2   Wireless LAN Services

Over the last decade, W-LANs have gained strong popularity in a number of vertical markets, including the health-care, retail, manufacturing, warehousing, and academic arenas. These industries have profited from the productivity gains of using handheld terminals and notebook computers to transmit real-time information to a centralized host for processing. Today, W-LANs are becoming more widely recognized as a general-purpose alternative for a broad range of business customers. Examples of these types of applications include inventory control in store and warehouse environments, point of sale terminals, and rental car check-in. Mobile professionals are thus now able to work at their convenience in hot spot locations where access is affordable and convenient. Wireless LANs are also increasingly being used in the hospital and university environments where users are highly mobile and bandwidth requirements are moderate.

Besides, a practical scenario that exposes the applicability of W-LANs comes from real-life—the home. Imagine that you have a broadband line installed so that you can work from home from time to time, except that your partner needs to check emails and do some online shopping, your daughter needs to use the Internet for her exam project, and your son wants to get online to "shoot" his friends in the latest Internet-based adventure game. Anything that would enable you to share the broadband connection without trailing wires all round the house would be a good thing, right? Well, the response from the consumer market seems to be a unanimous *yes*. In a recent report released by Infonetics Research on the broadband gateway market (routers enabling multiple users to simultaneously utilize a single broadband connection), 7.5 million such units were shipped worldwide during 2002, a total market worth US$645 million. This market is booming, driven by increasing broadband availability, the rising popularity of telecommuting, cheap and simple consumer technology, and an alarmingly simple cost-benefit.

Market analysts forecast that this market will grow to 16.3 million units by 2006, an increase of 117 percent, reaching extremely high penetration levels within broadband/multiple PC homes.

Success in the consumer market is driving wireless LAN technology into the enterprise too. W-LAN has leaked into the enterprise environment with individuals setting up ad hoc wireless LAN access point devices to enjoy the same benefits of wire-free mobility that they enjoy at home. These "rogue" access points are the bane of the network manager's life as they are usually deployed inside the corporate firewall and leave huge gaps in network security.

In an enterprise environment the IT organization maps corporate policies onto the computing and network infrastructure. Users are granted access to resources based on their identity/role in the organization and other related factors. In the wired LAN these policies are applied to physical ports in the switching fabric—the ports where specific users connect. So while network managers can "trust the wire" in terms of knowing who is on the network and where they are stationed, this model is completely flawed for wireless LANs where radio signals bleed through walls, shared access points connect multiple users, and mobility implies a mix of users at any access point.

1.3 The W-LAN Market

Technological developments are conspiring to make public access W-LAN rollout on a nationwide scale a comparatively cheap and easy-to-install proposition. Today's customer base of 400 to 500 million fixed-line Internet users, combined with increasing laptop sales, could boost the global W-LAN market size to as much as a hundred million users by the second half of this decade (Fig. 1.4).

PC cards for the IEEE 802.11 interface, for instance, are a commodity item and are available with the new notebook computers. Airports, for example, may offer W-LAN access as a free value-added service to their business class customers in much the same way as they provide free drinks and snacks in lounges. Hotels too could offer this as a value addition to attract more customers, perhaps for hosting conferences when there will be a higher demand for wireless Internet access [3GSM01].

Besides, a W-LAN market entry of telephone companies or cable service providers who already have widely distributed infrastructure that could support W-LANs is a possible scenario. A new and lucrative revenue stream could be seen in installing and maintaining enterprise-wide secure W-LAN solutions and integrating a possible wireless VoIP (Voice over IP) in existing telephone infrastructure. This would save companies long-distance telephone costs as well as increase efficiency through true wireless flexibility. Table 1.1 contains a partial list of W-LAN commercial products.

TABLE 1.1 Partial List of W-LAN Commercial Products

Product company location	Freq. (MHz)	Link rate	User rate	Protocol(s)	Access	No. of chan. or spread factor	Mod./coding	Power	Network topology
Altair Plus II Motorola Arlington Hts. IL	18–19 GHz	15 Mb/s	5.7 Mb/s	Ethernet			4 level FSK	25 mW peak	Eight devices/radio; radio to base to Ethernet
WaveLAN NCR/AT&T Dayton, OH	902–928	2 Mb/s	1.6 Mb/s	Ethernet-like	DS SS		DQPSK	250 mW	Peer to peer
AirLAN Selected San Diego, CA	902–928		2 Mb/s	Ethernet	DS SS		DQPSK	250 mW	PCMCIA w/ant; radio to hub
Freeport Windata Inc. Northboro, MA	902–928	16 Mb/s	5.7 Mb/s	Ethernet	DS SS	32 chips/bit	16 PSK trellis coding	650 mW	Hub
Intersect Persoft Inc. Madison, WI	902–928		2 Mb/s	Ethernet, token-ring	DS SS		DQPSK	250 mW	Hub
LAWN O'Neill Comm. Horshern, PA	902–928		35.8 Mb/s	AX.25	SS	20 users/chan.; max 4 chan	"unconventional"	20 mW	Peer to peer
WiLAN Wi-LAN INC. Calgary, Alta.	902–928	20 Mb/s	1.5 Mb/s/chan.	Ethernet, token-ring	CDMA/TDMA	3 chan. 10–15 links each		30 mW	Peer to peer

RadioPort ALPS Electric, United States	902–928		242 kb/s	Ethernet	SS	7/3 channels	100 mW	Peer to peer
ArLAN 600 Telesys. SLW Don Mills, Ont.	902–928; 2.4 GHz		1.35 Mb/s	Ethernet	SS		1 W max	Pos with ant.; radio to hub
RadioLink Cal. Microwave Sunnyvale, CA	902–928; 2.4 GHz	250 kb/s	64 kb/s		FH SS	250 ms/hop 500 kHz space		Hub
Range Lan Proxim Inc. Mountain View, CA	902–928		242 kb/s	Ethernet, token-ring	DS SS	3 chan.	100 mW	
RangeLan2 Proxim Inc. Mountain View, CA	2.4 GHz	1.6 Mb/s	50 kb/s max.	Ethernet, token-ring	FH SS	10 chan. 5 kb/s; 15 subch. each	100 mW	Peer to peer bridge
Netwave Xircom Calabersas, CA	2.4 GHz	1 Mb/s		Ethernet, token-ring	FH SS	82 1-MHz chan. Or "hops"		Hub
Freelink Cabletron Sys. Rochester, NH	2.4 and 5.8 GHz		5.7 Mb/s	Ethernet	DS SS	32 chips/bit	100 mW 16 PSK trellis coding	Hub

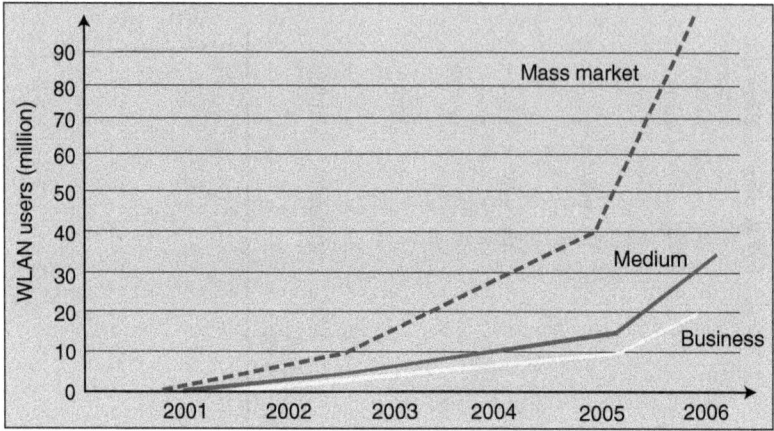

Source: Siemens

Figure 1.4 W-LAN users worldwide.

Current estimates and future projections of the W-LAN market growth show the W-LAN market to be a profitable venture for cellular operators to pursue despite the conservative approach taken in estimating cost and revenue projections [HENKEL02]. Emerging community networks and recently announced business models based on revenue sharing among private W-LAN owners indicate a promising future of wireless LANs [TEL-INTER02v36]. See Fig. 1.5

Although this is exciting, it is worth remembering that despite wireless LAN being a significant technology with a huge variety of applications in the home, enterprise, and public areas, there is much still to be

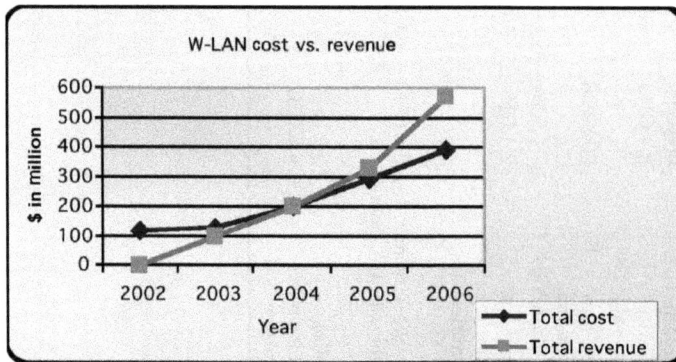

Figure 1.5 Cost-benefit analysis. [HENKEL02]

resolved; making money from the W-LAN market will be no walk-in-the-park for operators eyeing this technology as a comparatively cheap route to offering high-speed Internet access to business users. There's already too much hype about what wireless LAN can and will do, and it is not clear (and will not be for a while) which wireless LAN business models will work.

The failure of Metricom in the United States (in late 2001), for example, serves as a reminder that it will not be easy for start-ups to make money from providing W-LAN-only services. Among the issues to be resolved is the technology itself—the network coverage can be poor, roaming between different *mobile network operators* (MNOs) is not always supported, security has been virtually nonexistent, and IEEE 802.11 currently cannot support voice calls to any satisfactory level.

W-LAN roaming, especially, is of significant importance—optimal roaming model involves agreements between MNOs, owners of hot spots, and wireline service providers. Lots of hot spots in urban environments will be needed to meet this need. British Telecom, for example, plans to have 4000 in the United Kingdom by June 2005 [3GSM01, TEL-INTER02v36], a move that is likely to force the United Kingdom's Radio Communications Agency to remove the current ban on the use of this spectrum for public telecommunications services.

The inherent complexity of W-LAN roaming can be resolved by brokers, i.e., independent companies that function as clearing houses, setting up agreements with the relevant parties and splitting the income. In order to go nationwide and be able to offer a single account, brokers will need to set up partnerships with similar outfits. And in order to become an international service, major league MNOs and ISPs would have to come on board. We take 2G (second generation) roaming of wireless voice service for granted, but it took several years to establish all the one-on-one agreements. Besides, a key issue for MNOs and W-LAN hot spot players to make this service profitable is to provide mobile professionals with the data equivalent of their 2G voice service, including payment at the end of the month, and on the same bill, for voice and national and international access.

1.4 W-LAN Network Configurations

Two different approaches can be followed in the implementation of a W-LAN network (see Fig. 1.6)—an *infrastructure-based* and an *infrastructure-free*, or ad hoc [IEEE-STDRD] network configuration. An infrastructure-based architecture is a centrally coordinated and controlled network. The centralized controller, named *access point* (see Fig. 1.6 (*b*)), is connected to the wired network, thus providing Internet access to mobile devices.

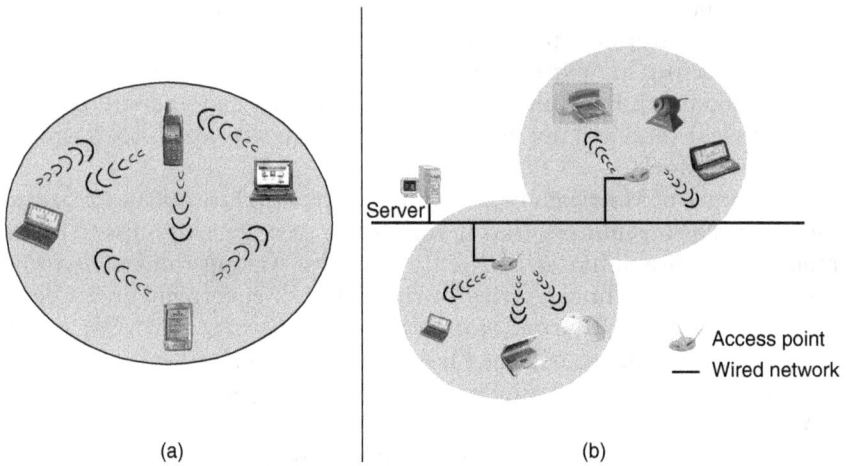

Figure 1.6   W-LAN configurations: (*a*) ad hoc networking and (*b*) infrastructure-based.

In contrast, an ad hoc wireless network architecture is a self-organizing and distributed controlled network formed by a set of stations that can freely and dynamically self-configure and organize themselves to set up a temporary wireless network [see Fig. 1.6 (*a*)]. In the ad hoc configuration, no fixed controller is thus required. Instead, a controller is dynamically elected from among all the stations participating in the communication.

The main drawback of W-LAN technology, when a wireless network needs to be quickly deployed, is its dependency on preconfigured hardware modules such as network access points and cabling, and on administrative tasks, which consume a considerable amount of manpower for a W-LAN of hundreds or thousands of stations.

The design objectives of an ad hoc wireless network include the speed of connection setup, the ease of removal of services and users, and the any-time, any-place network service access. In this regard, mobile professionals consider the introduction of ad hoc networking technologies into local mobility scenarios as the buzzword for tomorrow's mobile wireless communications systems. There are a few factors that turn the wireless ad hoc networking technology to a special case for adapting to specific network and application peculiarities as the technology evolves, is developed, and deployed. These include:

- Wireless ad hoc networks are particularly tailored for operation in situations where the *wireless communication coverage* of a single base station is not sufficient to cover a given area of mobile terminals.

There indeed exist places where any single wireless communication platform would fail to provide successful communication coverage. These places, called *dead spots*, include subway train platforms, indoor environments, and basements.

In addition to radio coverage limitations, a malfunction of the serving base station would impair the communication of all mobiles in a given area. A malfunction of base station, however, can be easily overcome through network self-configuration and rerouting of ongoing calls to neighbor base station(s) via mobile-to-mobile, or *multihop*, relaying.

Wireless ad hoc, or *relaying,* is a technology that could be employed to improve coverage and robustness against radio link failures, such that mobile terminals are capable of relaying radio signals of other mobile terminals.

- *Area coverage enhancement for high data-rate applications.* Generally, data transmission requires transmitting power almost in proportion to its bit rate. Given that the communication devices are portable terminals with lightweight batteries, to provide sufficient battery life, power must be reasonably conserved. Power efficiency in turn imposes limitations on the transmission range, data rate, communication activity (both transmitting and receiving), and processing speed of these devices.

  Moreover, wide area coverage for high bit-rate data transmission leads to high base-station density and increase in infrastructure cost. A common solution to support high bit-rate services is that the area coverage for high bit-rate data transmission is limited to the vicinity of the base stations or the place where line-of-sight propagation is available, while low bit-rate data transmission service can cover an entire cell.

  Relaying is a technology that could be employed to offset the difficulties of high data transmission rate, and associated intercell interference, over significant distances with negligible increase in the mobile station's complexity or cost.

- Ad hoc networks are easily set up in a wireless environment. If a company moves to a new location, the wireless system is much easier to move than ripping up all the cables that a wired system would have snaked throughout the building.

- There are many *applications* that benefit from ad hoc networks, especially where base stations cannot be deployed, or are destroyed, or are malfunctioning. Examples include emergency relief activities needed in a devastated area affected by a storm or attack, military operations into unknown territory, and locations where it is hard to deploy a base station to provide coverage (like in a mine shaft or in a canyon).

To conclude, the adoption of ad hoc networking principles enables engineers to build wireless local (as well as wide) area networks using modest expenditures (such as the cost of a computer, radio, and antenna) while keeping network administration costs at a minimum.

1.5   Physical Media for W-LANs

Three different technologies can be used at the physical layer for transmission over wireless LANs—infrared, radio frequency, and microwave. In 1985 the United States released the ISM frequency bands. These bands, are the 902 to 928 MHz, 2.4 to 2.4853 GHz, and 5.725 to 5.85 GHz, and do not require licensing by the Federal Communications Commission (FCC). This prompted most of the wireless LAN products to operate within the ISM bands although several restrictions are applied on these.

1.5.1   Infrared

Infrared LANs (or IR LANs) use the wavelength band, which is between the visible spectrum and microwaves (780 and 950 nm); hence the name *infrared.*

One attractive feature of IR systems is that they are not bandwidth limited and thus can achieve transmission speeds greater than other media systems. In addition, IR LANs offer more resistance to electromagnetic noise than spread spectrum systems (see Sec. 5.5.4 for details on spread spectrum systems), which makes them admirably suitable for use in very noisy environments such as factory buildings.

The major deficiencies of IR LAN systems include:

- Infrared transmission operates in the light spectrum. The transmission spectrum is thus shared with the sun and other things such as fluorescent lights. If there is enough interference from other sources it can render the LAN useless. Directional transceivers with special filters can, however, offset this problem.

- IR systems require an unobstructed line of sight (LOS).

- IR signals cannot penetrate opaque objects. This means that walls, dividers, curtains, or even fog can obstruct the signal.

There are two conventional ways to set up an IR LAN:

1. *To transmit omnidirectionally (nondirected IR LANs).* Signals bounce off of everything and in every direction (omni transmissions). This technique implies a loss of energy, which reduces the communication coverage to 30 to 60 ft.

2. *To transmit directionally (directed IR LANs).* Directed transmissions induce a good range of a couple of kilometers. It is thus recommended for outdoors usage. This system however is generally less robust than omnidirectional IR LANs, since it becomes more difficult to block all the light reflected from large surface areas.

Current implementations of IR LANs yield performances that can match (or even exceed in some cases) the data rate of wired-based LANs. Transmission systems allowing 50 Mb/s or 100 Mb/s have been demonstrated in controlled environments. In general conditions, slower data transfers are observed. Specifically, three different data transmission rates have been identified by the IrDA (Infrared Data Association)—an industry consortium developing infrared products for computer connectivity—115 kb/s, 1.15 Mb/s, and 4 Mb/s.

1.5.2 Microwave

A microwave is a short radio wave that varies from 1 mm to 30 cm in length. In contrast to longer radio waves, which are either blocked or reflected, microwaves can pass through the ionosphere. This makes microwave technology more suitable for long-range applications such as communications with satellites.

Two microwave configurations dominate the field today:

1. Point-to-point systems—designed for low- and medium-density communications.

2. Point-to-multipoint systems—provide communications between a central node and remote data units. They provide backbone links, enabling less populated areas to be covered on a more economical basis.

Microwave W-LANs usually operate at less than 500 mW of power in compliance with FCC regulations. They use narrow-band transmissions with single-frequency modulation and are set up mostly in the 5.8 GHz band.

Recent regulations impel vendors of microwave systems to obtain a license for use in their frequency spectrum. Although this justifies why microwave LANs are by far the most expensive W-LANs on the market, on the other hand, since the licensed company has only the right to use this frequency, this makes Microwave W-LANs almost interference-free to surrounding noise.

1.5.3 Radio frequency

In order to implement a radio frequency (RF) link, two methods can be used—narrowband and spread-spectrum techniques. Narrowband

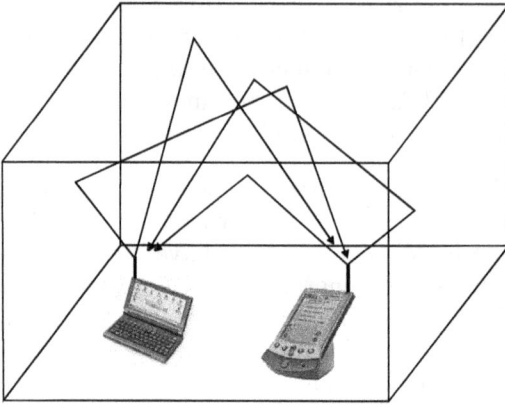

Figure 1.7   Example of multipath fading.

modulation schemes have problems with multipath transmission (Fig. 1.7) as they are very sensitive to interference. Spread-spectrum technology (SST), on the other hand, is highly resistant to interference, which nominates it as one of the most popular and preferred W-LAN technologies of today.

First commercially used in the mid-1980s, spread spectrum radio is highly secure, can operate under intense frequency jamming, and provides good signal integrity. The name spread spectrum comes from the technique used to modulate the radio signal, which "spreads" the signal over a portion of the radio spectrum. The spreading technique permits an increase in the number of bits transmitted at the expense of the available bandwidth. The receiver must use the same spreading code as the sender to be able to correlate and reassemble all data received into the original message.

By using a nonspread signal (such as a narrowband microwave), power is not concentrated in a single narrow frequency band. This makes the signal sound like noise. As a result, interception is a difficult practice in spread spectrum systems. Besides, with this spread of the signal's power over a wide band of frequencies, not only is the signal more difficult to intercept, but more resistant to electromagnetic interferences (surrounding noise) as well. Noises in general include:

- *Interference.*   Disruption by external sources, such as the electromagnetic emissions of electronic devices, or internal sources, such as crosstalk.

- *Jamming.*   Disruption caused by a stronger signal, which overwhelms the weaker signal.

- *Multipath.* The message is reflected by solid objects, which causes a distortion in the signal.

Spread spectrum technology is further divided into two subcategories—*direct sequence spread spectrum* (DSSS) and *frequency-hopping spread spectrum* (FHSS).

In the following section we provide a synopsis of the internal mechanics of SS techniques. We will look at SS techniques in more depth in Chap. 3.

Direct sequence spread spectrum (DSSS). DSSS avoids excessive power concentration by spreading the signal over a wider frequency band. At the transmitter, each bit of data is mapped into a pattern of "chips." The higher modulation rate is achieved by multiplying the digital signal with a chip sequence. The ratio of chips per bit is called the *spreading ratio*. In general, a high spreading ratio increases the resistance of the signal to interference, whereas a low spreading ratio increases the net bandwidth available to a user.

In practice, DSSS spreading ratios are quite small. The Federal Communications Commission (FCC) in particular requires that the spreading ratio must be greater than 10. Virtually all manufacturers of 2.4 GHz products offer a spreading ratio of less than 20. The proposed IEEE 802.11 standard, for example, specifies a spreading ratio of 11. At the receiver, the chips are mapped back into a bit, recreating the original data. To assure proper operation, the transmitter and the receiver must then be synchronized.

The rationale behind spread-spectrum technique is the realization that two different signals, each spread-coded with a unique spread code, cannot have the same spectral characteristics, thus enabling several networks to operate at the same location. By assigning unique (orthogonal) spreading codes all transmissions can use the same frequency band without interfering with each other. Signals with different spreading codes look like noise and are filtered out at the receiver end.

Frequency-hopping Spread Spectrum (FHSS). FHSS spreads the signal by transmitting a short burst on one frequency, "hopping" to another frequency for another short burst and so on. The source and destination of a transmission must be synchronized so as to operate on the same frequency at the same time.

The hopping pattern (frequencies and order in which they are used) as well as the dwell time (time for reception or transmission at each frequency) are determined according to a pseudorandom code sequence. These are restricted by most regulatory agencies. The FCC, for example, requires that the band is split into at least 75 frequencies (subchannels) and that the dwell time is no longer than 400 ms. If interference occurs at one frequency, then the data are retransmitted on a subsequent hop on another frequency.

More than one FHSS LANs can be colocated given that an orthogonal hopping sequence is used. Comparing FHSS and DSSS LANs, the number of colocated LANs can be greater with FHSS systems, which is mainly attributed to the fact that the subchannels are smaller than in DSSS. In addition, because there are a large number of possible sequences in the 2.4 GHz band, FHSS allows many nonoverlapping channels to be deployed.

Frequency-hopping systems are also less susceptible to interception because the frequency is constantly shifting. This attribute gives them a high degree of security. In order to jam a frequency-hopping system, the whole band must be jammed. These features are very attractive to agencies involved with law enforcement or the military.

To summarize, since a frequency-hopping system can offer a larger number of channels (i.e., frequency-hopping patterns) than a direct sequence system, a frequency-hopping system may be more useful for dense environments in which cells have overlapped with many adjacent cells.

Moreover, frequency-hopping and direct sequence systems have somewhat different types of resiliences to narrowband interference. Frequency-hopping systems experience the interference only for a fraction of time, whereas direct sequence systems experience a fraction of the interference power all the time. Thus, frequency-hopping systems have the performance advantage if the interference is high and direct sequence systems have the advantage if the interference is low.

Finally, both of these types of radio systems aim at transmitting at power levels of 100 mW or less, which enables them to achieve ranges of up to 100 m indoors, depending on the data rate and the building geometry and composition.

1.6   State of the Art in Mobile Ad Hoc Networks

The history of ad hoc networks can be traced back to 1972 and the Department of Defense (DoD) sponsored *packet radio network* (PRNET), which evolved into the *survivable adaptive radio networks* (SURAN) program in the early 1980s [FREEBER01]. The goal of these programs was to provide packet-switched networking to mobile battlefield elements in an infrastructureless, hostile environment (soldiers, tanks, aircraft, and the like forming the nodes in the network).

The PRNET used a combination of ALOHA and *carrier sense medium access* (CSMA) approaches for medium access, and a kind of distance-vector routing. It was totally asynchronous and was based on a completely distributed architecture. PRNET handled datagram traffic reasonably well, but did not offer efficient multimedia support. SURAN significantly improved upon radios (making them smaller, cheaper, and power-thrifty), scalability of algorithms, and resilience to electronic attacks.

In the early 1990s, a spate of new developments signaled a new phase in ad hoc networking. Notebook computers became popular, as did open-source software, and viable communications equipment based on RF and infrared. The idea of an infrastructureless collection of mobile hosts was the topic of many conference papers (such as [PERK94, JOHN94]), and the IEEE 802.11 subcommittee adopted the term "ad hoc networks". At around the same time, the DoD continued from where it left off— funding programs such as the *global mobile information systems* (GloMo), and the *near-term digital radio* (NTDR). The goal of GloMo was to provide office-environment ethernet-type multimedia connectivity anytime, anywhere, in handheld devices. Channel access approaches were now in the CSMA/CA (*collision avoidance*) and TDMA molds, and several novel routing and topology control schemes were developed. The NTDR used clustering and link-state routing and self-organized into a two-tier ad hoc network. As of late 2002, NTDR was used by the U.S. army, and was the only "real" (nonprototypical) ad hoc network in use then.

A *mobile ad hoc network* (MANET) is a system of wireless mobile nodes that can freely and dynamically self-organize in arbitrary and temporary network topologies without the need of a wired backbone or a centralized administration. People and devices can be seamlessly internetworked in areas without any preexisting communication infrastructure or when the use of such infrastructure requires wireless extension [CORS00]. With these in mind, mobile ad hoc networking offers unique benefits and versatility for certain environments and applications. First, since they have no fixed infrastructure including base stations as prerequisites, they can be created and used any time, anywhere. Second, such networks can be intrinsically fault resilient, for they do not operate under the limitations of a fixed topology. Indeed, since all nodes are allowed to be mobile, the composition of such networks is necessarily time varying. Addition and deletion of nodes occurs only by interactions with other nodes; no other agency is involved.

Such perceived advantages elicited immediate interest in the early days among military, police, and rescue agencies in the use of such networks, especially under disorganized or hostile conditions, including isolated scenes of natural disaster or armed conflict. Soldiers equipped with multimode mobile communicators can now communicate in an ad hoc manner without the need for fixed wireless base stations. In addition, small vehicular devices equipped with audio sensors and cameras can be deployed at targeted regions to collect important location and environmental information which can be communicated back to a processing node via ad hoc mobile communications. Ship-to-ship ad hoc mobile communication is also desirable since it provides alternate communication paths without reliance on ground- or space-based communication infrastructures. Moreover, ad hoc mobile communication is particularly useful in relaying information (such as status and situation awareness) via data, video, and/or voice from one rescue team member to another over a small handheld or wearable wireless device. In recent days, home or small office networking and collaborative computing with laptop computers in a small area (such as a conference or classroom and convention centers) have emerged as other major areas of application. People in a meeting or conference can freely use their laptops, palmtops, and notebooks to have instant network formation—in addition to file and information sharing—without the presence of fixed base stations and systems administrators. Presenters can multicast slides and audio to intended recipients. Attendees can ask questions and interact on a commonly-shared white board.

While many challenges remain to be resolved before large scale MANETs can be widely deployed (discussions follow), small-scale mobile ad hoc (*single-hop*) networks already appear in the public market [BRUNO01]. Network cards for single-hop ad hoc wireless networks constitute the building blocks to construct small-scale ad hoc (*multihop*) networks that extend the range of single-hop wireless technologies over a few radio hops [CONTI02, CORS00].

Wide area ad hoc networks, on the other hand, are *mobile multihop* wireless networks. A multihop wireless network uses no wired backbone or centralized administration. This is to be sure that the network will not collapse if one of the mobile nodes moves out of transmitter range of the others. Nodes should be able to enter/leave the network as they wish.

Laptop computers and personal digital assistants that communicate directly with each other are some examples of nodes in a multihop wireless network. Figure 1.8 depicts the peer-level multihop representation of such networks. Mobile devices communicate with each other directly (such as PDA and cellular phone) whenever a radio channel with adequate propagation characteristics is available between them. Otherwise, multihop communication is necessary, in which one or more intermediate nodes act as a relay (router) between the communicating nodes. For example, there is no direct radio channel (shown by the lines in Fig. 1.8)

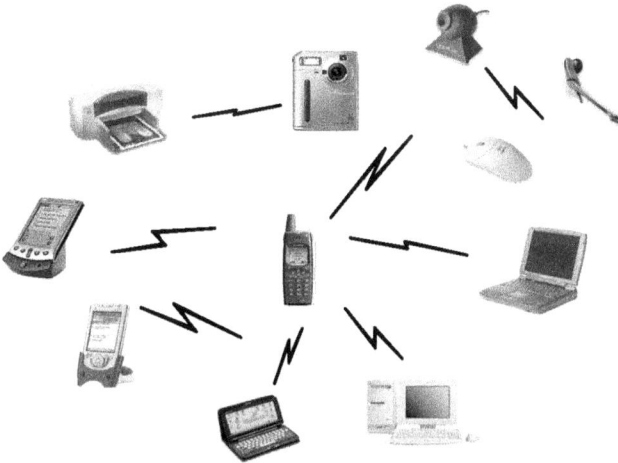

Figure 1.8 A mobile ad hoc network.

between the PDA and the WebCam. In this case, terminals with relaying capabilities (such as the cellular phone, the portable printer, and the laptop.) could serve as intermediate routers and help the PDA and the WebCam to establish their communication.

The limitations on power consumption, imposed by portable wireless radios, result in a node transmission range that is typically small relative to the span of the network. Consequently, in mobile ad hoc networks the mobile terminals do not always have direct radio links to all the radio terminals in the network, resulting in a distributed multihop network with a time-varying topology. This, coupled with the fact that the communication infrastructure does not rely on the assistance of base stations, implies that terminals must communicate with each other either directly or indirectly, using relaying stations via intermediate mobile hosts. Therefore, nodes also act as routers (also called *mobile routers* [CHAMBER02]) and dynamically establish communications among themselves to form an infrastructureless wireless network.

Spurred by the growing interest in ad hoc networking, a number of standards activities, and commercial standards evolved in the mid-to-late nineties. Within the IETF, the MANET working group was born, and sought to standardize routing protocols for ad hoc networks. The 802.11 subcommittee standardized a medium access protocol that was based on collision avoidance and tolerated hidden terminals, making it usable for building mobile ad hoc network prototypes out of notebooks and 802.11 PCMCIA cards. HiperLAN and Bluetooth were some other standards that addressed and benefited ad hoc networking. In recent years, there has been an upsurge of interest in multihop-augmented infrastructure-based networks in both industry and academia, such as the "seed" concept in 3GPP, mesh networks in IEEE 802.16, coverage extension of HiperLAN/2 through relays, or user-cooperative diversity mesh networks.

The following sections describe briefly the IEEE 802.11, ETSI HiperLAN, and Bluetooth standards.

1.7   Technologies for Ad Hoc Networks

Currently, three main communication standards with ad hoc capabilities are completed, each addressing a specific range of commercial applications. These standards include the IEEE 802.11 family of protocols [IEEE-STDRD], the high-performance LAN (HiperLAN) 2 protocol, and the Bluetooth specifications [BISDIK01, BLUETH-SPEC, MILLER00]. Examples of commercial applications are the *wireless body* and *personal area networks* (W-BANs and W-PANs) for short-range communication of small user devices, *wireless local area networks* (W-LAN) mostly for user and data communication and *in-house digital networks* (IHDN) for audio, video, and data exchange. The IEEE 802.11 family of protocols is mainly used for W-LANs, the HiperLAN for W-LAN and IHDN, and the Bluetooth specifications for short-range wireless communications (W-BANs and W-PANs).

1.7.1   The IEEE 802.11 Protocol specifications

In 1997, the Institute of Electrical and Electronics Engineers (IEEE) adopted the first digital wireless data transmitting standard, named IEEE 802.11, with data rates up to 2 Mb/s [IEEE-STDRD]. Originally, IEEE 802.11 was conceived as part of the IEEE 802.4 token bus standard with a given name of 802.4L. In 1990, the 802.4L group was renamed as the IEEE 802.11 W-LAN Project Committee, which created an independent 802 standard tasked with defining three physical (PHY) layer specifications and one common *medium access control* (MAC) layer for the lower portion of the data-link layer for W-LANs. The purpose of the IEEE 802.11 standard was to foster industry product compatibility between W-LAN product vendors (802.3 standard (Ethernet) [METCAL76]) which, consequently, led to the approval of the IEEE 802.11 standard on June 27, 1997 [IEEE-STDRD].

Since then, two IEEE standards have been ratified to extend the data rate of W-LANs by enhancing the physical layer specifications. These specifications are the IEEE 802.11a and the IEEE 802.11b; both ratified in 1999. The IEEE 802.11a task group created a standard for W-LAN operations in the 5 GHz UNII (unlicensed national information infra-structure) band, with data rates up to 54 Mb/s. The IEEE 802.11b task group produced a standard for W-LAN operations in the 2.4 GHz band, with data rates up to 11 Mb/s. Both the standards—IEEE 802.11a and b—share the same MAC specifications with the original IEEE 802.11 standard [IEEE-STDRD]. The differences are evident in newer PHY specifications, where IEEE 802.11a uses *orthogonal frequency-division multiplexing* (OFDM) while IEEE 802.11b uses *complementary code keying* (CCK).

Apart from these two standards, several other task groups (designated by letters) have been created to extend the IEEE 802.11 standard [IEEE-STDRD]. Among these the IEEE 802.11e task group targets at supporting voice and video over IEEE 802.11 networks using an enhanced MAC layer with QoS features, and the IEEE 802.11g task group, which is working to develop a higher-speed extension to IEEE 802.11b while retaining compatibility, i.e., it uses the 2.4 GHz frequency band. Figure 1.9 illustrates this evolutionary path of IEEE 802.11 standards.

The key features of IEEE 802.11 family of specifications are summarized in Table 1.2.

IEEE 802.11 Architecture. An IEEE 802.11 W-LAN can be implemented either with infrastructure or without infrastructure support (i.e., wireless ad hoc communications). In an infrastructure-based network, there is a centralized controller for each cell, often referred to as *access point*.

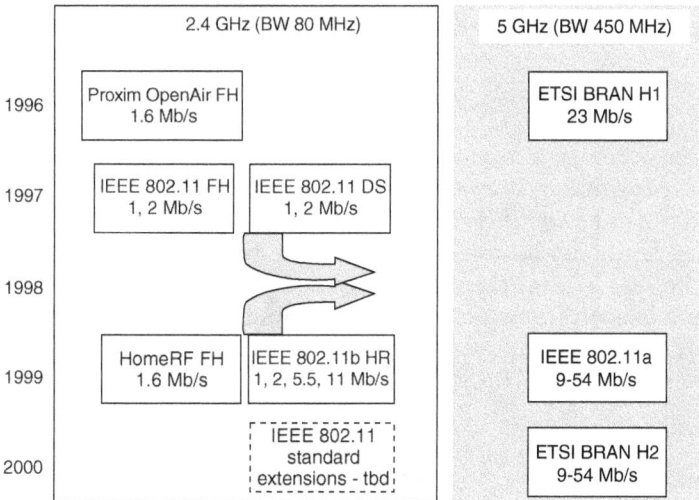

Figure 1.9 W-LAN standards' evolution.

TABLE 1.1   Key Features of IEEE 802.11 Standards

Industry standards	Roaming support	Supported PHY technology	Data rate (Mb/s)	ISM band (GHz)	UNII band (GHz)	Network classification
IEEE 802.11	Yes	DSSS, FHSS, Diffuse Ir	1, 2	2.4–2.48	N/A	W-LAN
IEEE 802.11a	Yes	OFDM	6, 9, 12, 18, 24, 36, 48, 54	N/A	5.15–5.25 5.25–5.35 5.72–5.87	W-LAN
IEEE 802.11b	Yes	DSSS	1, 2, 5.5, 11	2.4–2.48	N/A	W-LAN

The access point is normally connected to the wired backbone[1] network (distribution system in IEEE 802.11 nomenclature) thus providing Internet access to mobile devices. All traffic goes through the access point, even when this is sent to a destination that belongs to the same cell. It may also act as a point coordinator to provide contention-free services to the associated stations [MATTHIA01].

In the infrastructure-free network topology, a group of stations communicate directly with each other in an ad hoc fashion, independent of any infrastructure or base stations.

When two or more stations come together to communicate with each other they form a *basic service set* (BSS). The minimum BSS consists of two stations. A BSS which stands alone and is not connected to a base is called an *independent basic service set* (IBSS). The IBSS addresses the mobile peer-to-peer (ad hoc) configuration mode. Stations in IBSS-mode periodically broadcast beacons. The first station to instantiate an IBSS sets the *beacon interval*, which is broadcast with every beacon or *probe response frame*. Every station in the IBSS is therefore aware of the beacon intervals of the IBSS. For a station prior to transmitting its beacon, it shall first calculate a random *beacon backoff* interval. If the station has not received a beacon before its backoff interval expires, it proceeds and transmits its beacon frame. During the time of this beacon contention, data and *announcement traffic indication message*[2] (ATIM) transmissions are being halted. The algorithm makes sure that there will always be one beacon transmitted in the IBSS. The station last transmitting a beacon frame is elected to respond to probe requests. A new station in an IBSS does not transmit any beacon or probe response until it hears one from its IBSS.

---

[1]This backbone network is typically wired, but can also be wireless. For the case of a wireless backbone, the IEEE 802.11 standard makes use of a special frame format that effectively tunnels the original frame over the IEEE 802.11 wireless network.

[2]ATIMs are used for power save mode-see "IEEE 802.11 Architecture" in Sec. 1.7.1

Furthermore, in contrast to centralized polling-based mode, where the AP provides its own clock as common clock and handles the timing synchronization among the mobiles, synchronization among the IBSS stations is achieved by means of a distributed algorithm. The IEEE 802.11 uses two main functions for the synchronization of the stations in an IBSS:

1. *Synchronization acquisition.* This functionality is necessary for joining an existing IBSS. The discovery of existing IBSSs is the result of a scanning procedure of the wireless medium. During the scanning, the station receiver is tuned to different radio frequencies, searching for particular control frames. The station may initialize a new IBSS only if the scanning procedure does not result in finding any IBSS.

2. *Synchronization maintenance.* The system maintenance is implemented via a distributed algorithm, which is based on the transmission of beacon frames at a known nominal rate. The station that initialized the IBSS decides the beacon interval.

Creating large and complex networks using a number of BSSs leads us to the next level of hierarchy, the *extended service set* or ESS. The ESS shown in Fig. 1.10 consists of a series of overlapping BSSs (each containing

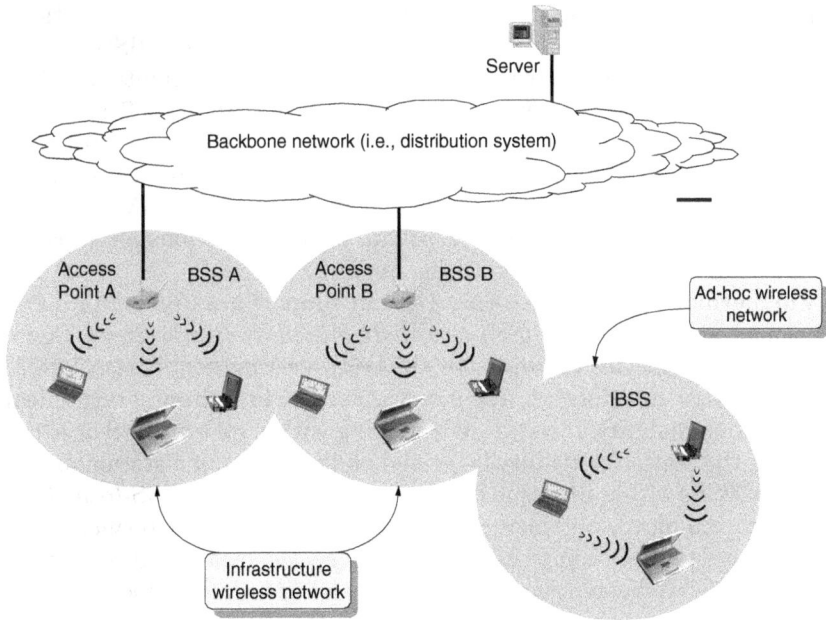

Figure 1.10 Infrastructure and ad hoc IEEE 802.11 W-LANs.

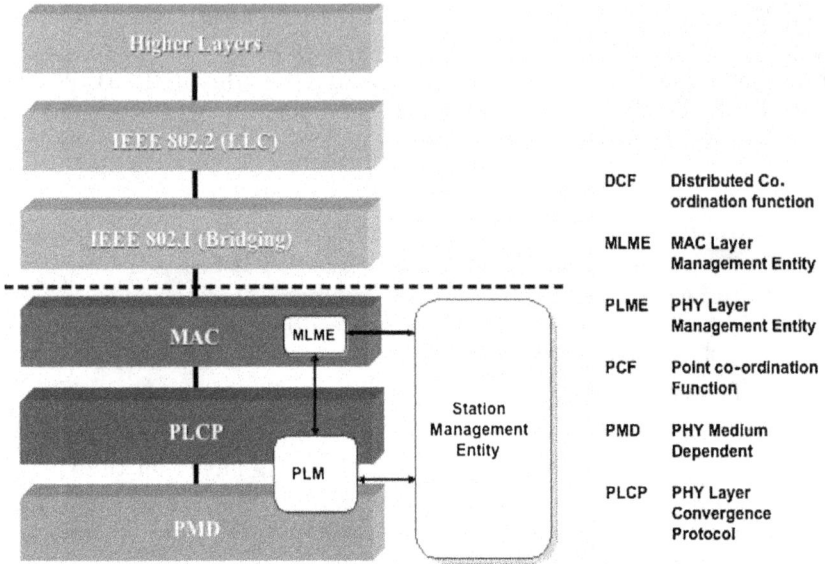

DCF	Distributed Co. ordination function
MLME	MAC Layer Management Entity
PLME	PHY Layer Management Entity
PCF	Point co-ordination Function
PMD	PHY Medium Dependent
PLCP	PHY Layer Convergence Protocol

Figure 1.11   IEEE 802.11 reference model.

an AP) connected together by means of the distribution system (DS). The beauty of the ESS is that the entire network looks like an independent basic service set to the link control layer. This means that stations within the ESS can communicate or even move between BSSs transparently to the link control layer. One of the requirements of IEEE 802.11 is to be used with existing wired networks. IEEE 802.11 solved this challenge with the use of a *portal*. A portal is the logical integration between wired LANs and IEEE 802.11. It also can serve as the access point to the DS. All data going to an IEEE 802.11 LAN from an 802.X LAN must pass through a portal. It thus functions as a bridge between the wired and the wireless. See Fig. 1.11.

Furthermore, the services that DS can support are divided into two sections—*station services* (SS) and *distribution system services* (DSS). There are five services provided by the DSS—*association, reassociation, disassociation, distribution,* and *integration.* The first three services deal with station mobility. If a station is moving within its own BSS or is not moving, the station's mobility is termed *no-transition.* If a station moves between BSSs within the same ESS, its mobility is termed *BSS-transition.* If the station moves between BSSs of differing ESSs, it is termed *ESS-transition.* For a station to use the LAN services, it must affiliate itself with the BSS infrastructure. This is done by associating itself with an access point. Associations are dynamic in nature because stations move, turn on, or turn off. A station can only be associated with one AP. This ensures that the DS always knows where the station is. Association supports no-transition mobility but is not enough to support BSS-transition. BSS-transition is achieved through the reassociation service, which allows the station to switch its association from one AP to another. Both association and reassociation are initiated by the station. Disassociation is when the association between the station and the AP is terminated. This can be initiated by either party. A disassociated station cannot send or receive data. Notice that ESS-transition has not been mentioned so far. That is because ESS-transition is not supported. A station can move to a new ESS but will have to reinitiate connections.

Distribution and integration are the remaining DS services. The distribution service is simply forwarding data from the sender to the intended receiver. The message is sent to the local AP (input AP), then distributed through the DS to the AP (output AP) that the recipient is associated with. If the sender and receiver are in the same BSS, the input and out APs are the same. So the distribution service is logically invoked whether the data are going through the DS or not. Integration is when the output AP is a portal. Thus 802.x LANs are integrated into the IEEE 802.11 DS.

Station services (SS) are *authentication, deauthentication, privacy,* and *MAC service data unit (MSDU) delivery.* With a wireless system, the medium is not exactly bound as with a wired system. In order to control access to the network stations before they are allowed to converse, they must first pass a series of tests to ensure that they are legitimate users. That is really what authentication is all about. Once a station is authenticated, it may then associate itself. The authentication relationship may be between two stations inside an IBSS or to the AP of the BSS. Authentication outside of the BSS does not take place. There are two types of authentication services offered by IEEE 802.11. The first is *open system authentication.* This means that anyone who attempts to authenticate will receive authentication. The second type is *shared key authentication.* In order to become authenticated the users must be in possession of a shared secret. The shared secret is implemented with the use of a data encryption algorithm called the *wired equivalent privacy* (WEP)[3] algorithm to protect authorized stations from eavesdroppers. The shared secret is delivered to all stations ahead of time in some secure method (such as someone walking around and loading the secret onto each station). Deauthentication occurs when either the station or AP wishes to terminate a station's authentication. When this happens the station is automatically disassociated. If WEP is not used then stations are "in the

---

[3] WEP algorithm is based on the RC4 PRNG algorithm developed by RSA Data Security, Inc.

clear" or "in the red" status, meaning that their traffic is not encrypted. Data transmitted in the clear are called *plaintext*, whereas encrypted data are called *ciphertext*. All stations start in the red until they are authenticated.

Finally, with regard to addressing in IEEE 802.11, the authors of the IEEE 802.11 standard allowed for the possibility that the wireless media, the distribution system, and the wired LAN infrastructure would all use different address spaces. IEEE 802.11 only specifies addressing over the wireless medium, though it was intended specifically to facilitate integration with IEEE 802.3 wired Ethernet LANs. IEEE802 48-bit addressing scheme was therefore adopted for IEEE 802.11, thereby maintaining address compatibility with the entire family of IEEE 802 standards. That is, each device owns a unique 48-bit station address. Further, each subnet has its unique identifier, which is the access point's MAC address (48 bit) in infrastructure networks or a random 48-bit IEEE 802 locally administered address (two bits + 46-bit random number) in infrastructureless networks. In the vast majority of installations, the distribution system is an IEEE 802 wired LAN and all three logical addressing spaces are identical.

Key IEEE 802.11 medium access control layer features.   The IEEE 802.11 MAC layer provides a basic access mechanism that supports several characteristics such as clear channel assessment, both asynchronous and time critical data delivery, link setup, encryption and authentication, power management, and channel synchronization [BRAY00]. Further, link control and management, fragmentation and defragmentation (also known as fragmentation and reassembly), and roaming are considered to be MAC layer entities [MATTHIA01]. Roaming, a key feature of W-LANs, allows mobile users to roam about BSSs. Figure 1.10 illustrates how two access points are interconnected with the wired backbone infrastructure while a mobile user seamlessly moves between two BSSs. This roaming capability is a key feature of the IEEE 802.11 family of specifications that other short-range wireless technologies, such as Bluetooth, do not support.

IEEE 802.11 MAC layer provides IEEE 802.11 users both contention-based and contention-free access control services for IEEE 802.11 W-LANs. The basic access method for peer-to-peer ad hoc networking in the IEEE 802.11 MAC protocol is the *distributed coordination function* (DCF) that incorporates a CSMA/CA protocol as the basic medium access mechanism. On top of the DCF resides the *point coordination function* (PCF) that provides contention-free medium access in infrastructure-based network configurations [MATTHIA01]. See Fig. 1.12. The PCF operates similarly to a polling system [CONTI97], where a

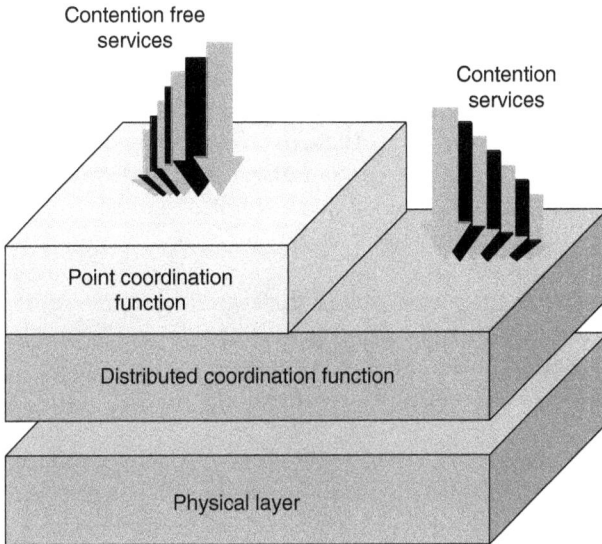

Figure 1.12 IEEE 802.11 architecture.

point coordinator provides the transmission rights at a single station through a polling mechanism.

IEEE 802.11 MAC management entities.   Both the specified layers in IEEE 802.11—DCF and PCF—contain their own management entities called *management information bases* (MIBs) that contain layer-specific information [MATTHIA01]. Specifically, there are three management entities—*the physical layer management entity (PLME), the MAC layer management entity (MLME),* and the *station management entity (SME).*

The PLME provides services to set the physical transmission channel, such as the hopping pattern in the FHSS system.

The *MAC sublayer managment entity (MLME)* is responsible for connection setup and maintenance, and power management. It provides the following services—power management, scanning, synchronization, authentication, deauthentication, association, reassociation, disassociation, reset, and start. Scanning for stations is used to find a station in the communication range; the standard defines active or passive scanning. To synchronize all timers in the stations in a BSS and an IBSS, the MLME provides the synchronization service. The synchronization information is either provided by the access point in a BSS (i.e., centralized) or via a distributed algorithm in an IBSS. In the centralized configuration the AP sends a beacon frame containing

BSS properties and the timing synchronization information at regular intervals.

Furthermore, a station must first authenticate itself before it associates itself with a BSS. Associated stations may use the distribution system (e.g., communicate with other stations in other subnets, requesting point coordination services), whereas nonassociated stations must not. The reset-service resets the MAC entity, and the start-service is used to start either a BSS or an IBSS, if the station has not synchronized itself with a BSS.

The *SME* services are not fully specified in the standard. It is a layer-independent entity and would typically get and set values from the MLME and the PLME, to pass them to higher-level management functions. It can be considered as residing in a control plane.

The IEEE 802.11 physical layers. The PHY in IEEE 802.11 consists of two sublayers—the physical-medium-dependant sublayer and the physical layer convergence sublayer on top. The physical-medium-dependant sublayer specifies how to send and receive data over the wireless medium, whereas the convergence sublayer maps the MAC frames to the physical-medium-dependant functions [MATTHIA01]. The original IEEE 802.11 standard specifies the use of three different PHY layers, any of which can utilize the same MAC layer. These PHY layers include two spread spectrum techniques at 2.4 GHz in the ISM band—FHSS and DSSS. The third PHY layer is an optical technique using *diffuse infrared* (DFIR). The specifics of each technology are discussed in Sec. 1.5.

In Europe, the same 2.4 GHz band (as the U.S. ISM band) has been allocated to allow wireless LAN operation (see Table 1.3), whereas in Japan only the frequencies from 2.471 to 2.497 GHz have been allocated, requiring, therefore, special provisions in the IEEE 802.11 draft standard. The IEEE 802.11 committee allowed the definition of multiple PHY layers, in part, because the members of the committee had some interest in each of the previously mentioned PHY layers and hence they sought to accommodate all of them. The benefit of this approach is that the various advantages of each of the physical layers can be exploited by users who want an IEEE 802.11-compliant wireless LAN (e.g., see [GOLDB95]).

TABLE 1.3   Global Spectrum Allocation at 2.4 GHz

Region	Allocated spectrum
United States	2.400–2.4835 GHz
Europe	2.400–2.4835 GHz
Japan	2.471–2.497 GHz
France	2.4465–2.4835 GHz
Spain	2.445–2.475 GHz

The disadvantage of this approach is that in order to permit interoperability between them, the users need to specify additionally the type and data rate of their wireless LAN system.

All IEEE 802.11 devices using one of these three technologies are required to operate at 1 Mb/s data rate, with 2 Mb/s as an option. The maximal size of an MSDU is 2312 bytes, excluding the MAC header and the physical preamble. There are several physical-layer dependent parameters that are relevant to the design of the MAC protocol. On one hand, there is the Rx/Tx (receiver/transmitter) turnaround time that varies as:

0 ms (infrared)

10 ms (direct sequence)

19 ms (frequency-hopping)

This time is contained in both the length of the interframe spaces and the length of the backoff slots in the contention window. The backoff slot time for the three layers is defined as:

6 ms for infrared

20 ms for direct-sequence

50 ms for frequency-hopping

In addition, each physical layer introduces a physical preamble with different length, which is added to each packet. Infrared adds 92 to 112 timeslots of 250 ns + 32 bits, direct sequence 192 bit, and frequency-hopping adds 122 bits.

Given, therefore, a specific physical layer technique with different backoff slot times and physical preamble lengths, significantly different performance measures are observed at the MAC layer.

The IEEE 802.11 DCF method for medium access in infrastructureless wireless networks.   The DCF specifies the use of the CSMA protocol with CA. The CSMA used in wireless networks is similar to the CSMA scheme used in wired LANs. However, the *collision detection* (CD) technique for wired LANs cannot be used effectively for wireless LANs since nodes cannot detect over-the-air collisions when they occur. The absence of detection is caused by the strong signals present at the transmitters that also serve to drown out other communicating signals [WICKEL96].

The CA property of the CSMA-based W-LANs helps them to reduce the number of over-the-air collisions. In short, the basic CSMA/CA medium access function allows for options that can minimize collisions by using *request-to-send* (RTS), *clear-to-send* (CTS), data, and *acknowledge* (ACK)

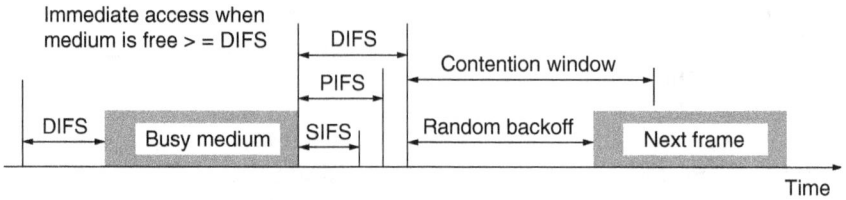

Figure 1.13 Primary access mechanism.

transmission frames, in a sequential fashion. Communication is established when one of the wireless nodes sends a short message RTS
frame. The RTS frame includes the destination and the length of message. The message duration is known as the *network allocation vector*
(NAV). The NAV alerts all others in the medium to back off for the
duration of the transmission. The receiving station issues a CTS frame
which echoes the sender's address and the NAV. If the CTS frame is not
received, it is assumed that a collision occurred and the RTS process starts
over. After the data frame is received, an ACK frame is sent back verify-
ing a successful data transmission.

Before delving into the details of DCF, however, it is sensible to introduce the interframe space concept of IEEE 802.11 that is used to provide priority medium access. In general, this can be seen as a set of
medium access timings for the different frame types—the random backoff procedure, the frame transfer procedures, the acknowledgement procedures, and the RTS/CTS procedures. More specifically, each timer
expires after a certain time period. Depending on the timer that expires,
certain frame types may or may not be sent, subject to their priority.
These interframe spaces are summarized in Fig. 1.13.

To this end, when using the DCF access method (summarized in
Fig. 1.14), before a station initiates a transmission, it senses the channel to determine whether another station is transmitting. If the medium
is found to be idle for an interval that exceeds the *distributed interframe
space* (DIFS), the station continues with its transmission.[4] The transmitted packet contains the projected length of the transmission. Each
active station stores this information in a local variable, named the
NAV. The NAV, therefore, conveys information about how long the
medium will remain busy [see Fig. 1.14 (a)]. This mechanism prevents
a station from listening to the channel during transmissions thus
enabling the implementation of power-saving policies.

---

[4]To guarantee fair access to the shared medium, a station that has just transmitted a
packet and has another packet ready for transmission must perform the backoff procedure before initiating the second transmission.

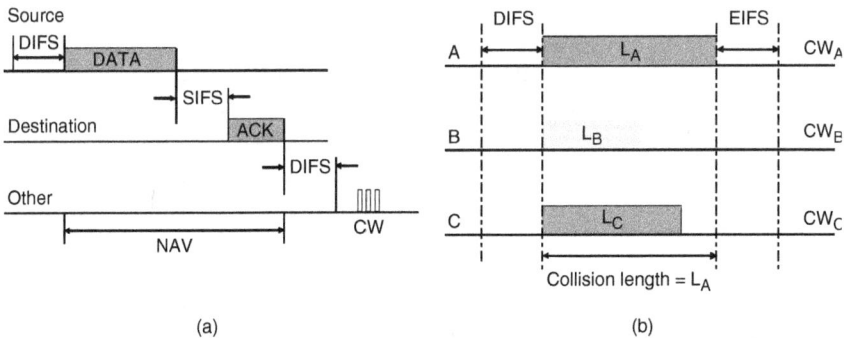

Figure 1.14   EEE 802.11 DCF: (*a*) a successful transmission; (*b*) a collision.

The CSMA/CA protocol does not rely on the capability of the stations to detect a collision by hearing their own transmissions. Hence, immediate positive acknowledgments are employed to ascertain the successful reception of each packet transmission. Specifically, the receiver, after the reception of the data frame, waits for a time interval called the *short interframe space* (SIFS), which is less than the DIFS, and then initiates the transmission of an ACK frame. The ACK is not transmitted if the packet is corrupted or lost due to collisions. A cyclic redundancy check (CRC) algorithm is adopted to discover transmission errors. Collisions among stations occur when two or more stations start transmitting at the same time [see Fig. 1.14 (*b*)]. If an acknowledgment is not received, the data frame is presumed to have been lost, and a retransmission is scheduled.

After an erroneous frame is detected (due to collisions or transmission errors), the channel must remain idle for at least an *extended interframe space* (EIFS) interval before the stations reactivate the backoff algorithm to schedule their transmissions [see Fig. 1.14 (*b*)].

The CA scheme in CSMA/CA further provides a random backoff delay feature before a new transmission attempt is executed, which guarantees a time spreading of the transmissions. This random delay helps avoid collisions from simultaneous multiuser transmissions, since other wireless nodes could also be waiting to send data over the network [WICKEL96]. When a station *S*, with a packet ready for transmission, observes a busy channel, it defers the transmission until the end of the ongoing transmission. At the end of the channel busy period, the station *S* initializes a counter (called the *backoff timer*) by selecting a random interval (*backoff interval*) for scheduling its transmission attempt. The backoff timer is decreased for as long as the channel is sensed as idle, stopped when a transmission is detected on the channel, and reactivated when the channel is sensed as idle again for more than a DIFS. The station transmits when the backoff timer reaches zero.

The hidden and exposed terminal problem.   A common limitation associated with wireless LAN systems is the *hidden and exposed terminal problem* [KLEIN75, KLEIN75a, LEIN87]. In particular, hidden and exposed nodes are a phenomenon of carrier sensing and packet sensing MAC protocols. Packet sensing is performed at the MAC layer and carrier sensing at the physical layer. Packet sensing ensures that received packets are decoded only if their energy level is above a certain threshold. Carrier sensing, on the other hand, mandates that a station that has data to send but hears a level of energy higher than a given threshold, defer and try to access the channel after some random time.

Nodes, therefore, defer attempting to gain access when they sense another node using the channel. A "hidden node" is a node that cannot hear the transmission that another node is receiving and thus does not defer attempting to gain access. The reason for not hearing the transmission can be distance, an obstacle in the propagation path, being in the null of an antenna, or, in the case of DSSS, the use of a different code. When this station thus tries to gain access, it interferes with the reception at the receiving node.

The exposed node problem is very different. An exposed node is a node that hears multiple disjoint sections of a network and never gets the opportunity to contend since it is always deferring to someone else.

The example in Fig. 1.15 demonstrates the hidden/exposed terminal problem. Let us assume that the stations A and B are hidden from each other and that both wish to transmit to a third station, named *Receiver*. When A is transmitting to *Receiver*, the carrier sensing of B cannot capture any transmission event, and thus B can immediately start a transmission to *Receiver* as well. Therefore, both stations A and B would transmit at the same time to *Receiver*.

Now consider the case when A is transmitting to C. Since *Receiver* can "hear" A, *Receiver* cannot risk initiating a transmission to B for fear of causing a collision at A. In Fig. 1.15 *Receiver* is "exposed" to A.

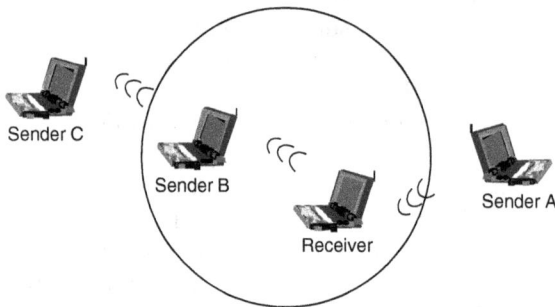

Figure 1.15 Illustration of the hidden and exposed terminal problem.

Approaches to mitigate the hidden and exposed terminal problem. The hidden and exposed stations phenomenon may occur in both infrastructure-based and ad hoc networks. Notably, the problem becomes more severe in ad hoc networks where almost no coordination exists among the stations.

The hidden terminal problem can significantly reduce the amount of traffic carried by the system. Some of this lost capacity can be regained through special mechanisms that allow receivers to control access to the channel. A few techniques are briefly discussed in the following paragraphs.

Tackling the problem at the MAC layer.   In general, the shared wireless medium of mobile ad hoc networks requires the use of appropriate MAC protocols to mitigate the medium contention issues as well as to allow for an efficient use of the limited bandwidth. Specialized MAC protocols could also help to alleviate the hidden and exposed terminal problem. For instance, to avoid the hidden and exposed terminal problem, the IEEE 802.11 basic CSMA access mechanism is extended with a virtual carrier sensing mechanism, called RTS/CTS. As pointed out earlier, in the RTS/CTS mechanism after access to the medium is gained and before transmission of a data packet begins, a short control packet, called RTS, is sent to the receiving station announcing the upcoming transmission. This message contains the destination address and the duration of the transmission. The receiver replies to this with a CTS packet to indicate readiness to receive the data. CTS packets also contain the projected length of the transmission. This information is stored by each active station in its NAV, the value of which becomes equal to the end of the channel busy period. Therefore, all stations within the range of at least one of the two stations (receiver and transmitter) know how long the channel will be used for this data transmission (see Fig. 1.16).

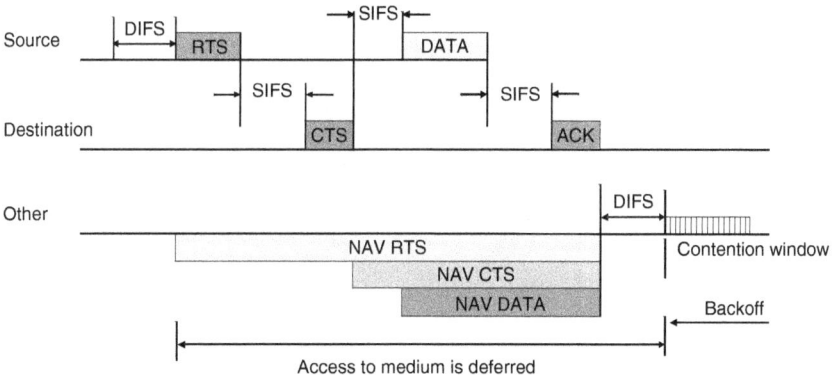

Figure 1.16 The RTS/CTS mechanism.

Referring to the hidden and exposed terminal example of Fig. 1.15, when B wishes to transmit to *Receiver*, it first sends an RTS message to *Receiver*. In response, Receiver broadcasts a CTS message that is received by both A and B. Since B has received the CTS message unsolicited, A knows that *Receiver* is granting permission to send to a hidden terminal and hence refrains from transmitting. Upon receiving the CTS message from *Receiver* in response to its RTS message, B transmits its own message.

Not only does the previously mentioned procedure solve the hidden terminal problem but also solves the exposed terminal problem. Upon receiving an unsolicited CTS message, A refrains from transmitting. After an appropriate interval, determined by the attributes of the channel (i.e., duration of a time slot, and the like.), A can send its own RTS message to C as the prelude to a message transmission.

The RTS/CTS mechanism, for the majority of the carrier-sensing protocols, resumes with an acknowledgement sent for each data packet. If an acknowledgement is not received, the MAC layer retransmits the data. This entire sequence is called the *4-way handshake* as shown by Fig. 1.17.

RTS/CTS collisions and loss of state information.   Even though the RTS/CTS mechanism solves the hidden station problem *during the transmission of user data*, collisions may still occur *during the transmissions of control packets* (the small RTS and CTS packets), thus resulting in loss of RTS/CTS packets. In addition, the RTS-CTS solution proposed for solving the hidden terminal problem in various MACs, such as IEEE 802.11 and MACA, assumes bidirectional links. Bidirectionality assures that the transmission of an RTS is always followed from a CTS exchange. In most wireless environments, however,

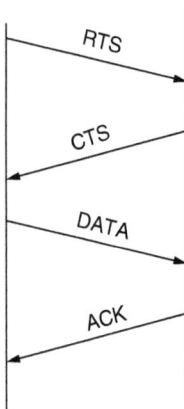

Figure 1.17 The 4-way handshake.

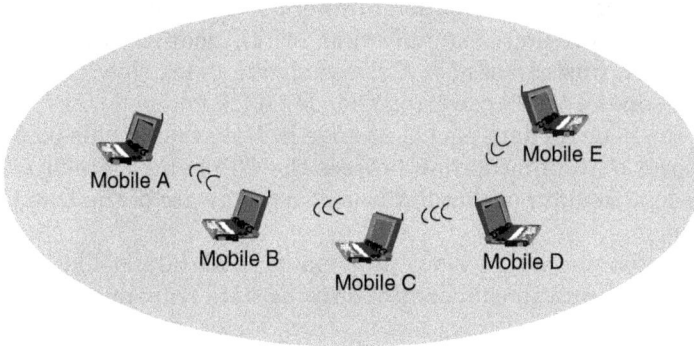

Figure 1.18 Illustration of RTS-CTS loss.

due to asymmetrical propagation conditions, links are often unidirectional. To this avail, and if unidirectional links are assumed, the RTS/CTS exchange may indeed fail and MAC cannot do much about the exposed and hidden terminal problem. In this case, packet loss has to be handled at a different layer.

To illustrate how RTS/CTS information may be lost, consider the following example in Fig. 1.18 (for single channel operation). Assume that node D transmits a data packet to node E (using the RTS/CTS handshake). When the packet is in transit, node A decides to transmit to node B, also using the RTS/CTS handshake. Let us see now what node C experiences:

1. Node C receives node D's RTS.

2. Node C receives node D's data. (It becomes "jammed" by this transmission)

3. Node C receives node B's CTS (sent in response of node A's RTS). Since node C is jammed by the data sent from D to E, it cannot successfully decode node B's CTS. Therefore, node C will not learn that node B is about to receive a packet.

4. After node D completes its transmission, node C is free to transmit. Node C does not know that B is receiving a packet and assumes that the channel is free. Let's say that node C decides to transmit to node D.

5. Node C sends an RTS to D, receives a CTS back and then starts sending data. At this point, node C's data (or even the RTS) collides (at node B) with node A's data.

This example illustrates how a node can lose an RTS or CTS (or, typically, *state* information). Moreover, a node that loses state information may cause the loss of state information to other nodes as well. For example, assume in the previous example that in (4), node D decides to transmit to node C (instead of node C transmitting to D). Now, (5) will be that node C sends a CTS answering node D's RTS, and node C's CTS collides with the node A's data packet at node B. Note that at this point node B also loses state information (misses the CTS) and therefore it does not only lose A's data packet but it is also not aware of the D to C communication.

State information loss is a serious problem in IEEE 802.11-like protocols as this may result in collisions of ongoing data transmissions of other stations.

Similar situations (although less often) can occur with multichannel systems, where the trigger of "loss of state" is not jamming a node due to simultaneous communication but due to the fact that during normal reception (in a single transceiver system) a node can only listen to one channel at a time. Thus, a node that is listening to the data channel is not aware of the activities in the control channel, potentially causing it to lose state information.

Tackling the problem at the physical layer.   When the density of nodes is very high, the usage of control signals (RTS/CTS) could cause congestion problems in the network. The wireless spectrum is admittedly a scarce resource and mobiles must consume bandwidth judiciously. With this in mind, the hidden and exposed node problems can also be tackled at the physical layer by using multiple codes (e.g., the transmitter-directed codes).

Using multiple codes implies that each node has a unique PN code in the network and transmits using its own PN code. The transmitter senses the channel for the known PN codes and if the receiver is idle (i.e., there are no data for this node), the node is free to transmit. Since the PN codes are unique in the network, the hidden node problem is addressed. Also since the node should not be idle until the completion of the ongoing data transfer of neighboring nodes, the exposed node problem is also addressed.

Another configuration of using a set of codes is for the sender to include in its RTS that is sent over a common code, a preferred code (unused at that time), and for the receiver to send a CTS in response, if the code is acceptable. Then data-ACK takes over the selected code.

A downside of using completely orthogonal multiple spreading codes is the requirement of a synchronous system. It is, however, very difficult to achieve perfect synchronization in a pure ad hoc system without the aid of sophisticated clocks or advanced and complex synchronization mechanisms.

Tackling the problem with smart antennas.   From the preceding discussions on the hidden/exposed terminal problem, one can easily

understand that the main problem comes from the propagation properties of omnidirectionally transmitted signals. In this regard, smart
directional antennas may well be the solution. Smart antennas can provide much greater capacity and performance benefits than standard
antennas. This is attributed to their ability to fine-tune their coverage
patterns to match the traffic conditions as well as to adjust to the changing RF conditions.

Briefly, smart antennas can be realized by an antenna array at the
base station and sophisticated baseband signal processing. Adaptive
directional reception can be achieved on the uplink and adaptive directional transmission on the downlink. Thereby, an increased antenna
gain and an increased diversity gain are realized toward the desired
user, whereas, at the same time, less interference is encountered from
the other directions on the uplink or transmitted in the other directions on the downlink. Adaptive antennas are described in more detail
in Chap. 3.

In mobile ad hoc networks, however, if both the transmitter and the
receiver employ directional antennas, a lot of issues need working out,
including pragmatic ones. In general, using smaller antennas implies a
higher operating frequency which in turn implies poorer propagation
characteristics. With smart antennas, one needs (typically) a half-wave-
length spacing between elements. At 2.4 GHz, a cylindrical 8-element
array would have a radius of about 8 cm—making it quite unwieldy to
carry on a PDA or laptop. If the directional beams are not very narrow,
using directional antennas does not make the problem go away altogether, it just makes it less severe. So, raising the operating frequency
to 2.4 GHz ISM band, you get a mere 0.8 cm radius. But then line-of-
sight operation is required.

Additionally, if the option to use beams exists, the implicit gain of the
focused beam would imply that two nodes can communicate reliably
over much greater distances than in the omnidirectional mode. For
example, if both antennas show a 6 dB gain, the effective signal will be
16 times as strong, so the effective range could be two to four times as
great. If that is the case, then MANET protocols that seek to locate a
node precisely in order to set up a temporary point-to-point link will be
needed, instead of a multihop path. Of course, if the range is too wide,
there will also be a need to set up multihop paths, but each link could
be in the form of vectors, rather than nodes. This may be worth it, but
it sounds more complex.

On the other hand, even if one could construct an antenna capable of
synthesizing perfectly power-controlled pencil beams and these could be
used for communications, there would still be a need for omnidirectional
broadcasting. Beacons, broadcasts, and initial signal acquisition, all use
omnidirectional coverage. Therefore, at least when starting up the link,

the receiver can be expected to have an omnireceiving antenna. Once a link is established it should do its best to "act like a piece of wire," or in RF parlance, a motion tracking power-controlled pencil beam.

In addition, there is at least one other wrinkle in using the omnimode of operation. With omnidirectional broadcasts, it is not always necessary to send an ACK. If A sends a packet to B for forwarding, as B rebroadcasts the packet, A hears it and knows all is well. An explicit ACK in this case is not needed. Directional antennas, on the other hand, do foster the need for an explicit ACK in almost all cases.

Without loss of generality, these observations make us note that any antenna system in MANETs might eventually be required to function as omnidirectional as well as directional.

The IEEE 802.11 point coordination function (PCF).   In order to support time-bounded services, the IEEE 802.11 standard specifies the optional use of the PCF, in which a point coordinator (or PCF station) has priority control over the medium. The point coordinator (PC) resides in the access point. It schedules a contention-free period, which is announced by sending a beacon frame after the SIFS timer expires. That is, when the PCF is active, the PCF station allows only a single station in each cell to have priority access to the medium at any particular time.

This is implemented through the use of the previously mentioned PIFS and a beacon frame (see Fig. 1.13) that alerts all the stations in the cell to defer transmitting for the length of the *contention-free period* (CFP). If one of the stations does not hear the expected beacon, it sets its NAV to a known maximum value for the length of the CFP. The length of the CFP can vary within each CFP repetition interval according to the system load. Having silenced all the stations, the PCF station can allow a given station to have contention-free access through the use of an (optional) polling frame that is sent by the PCF station. A station interprets a poll as a resource grant to transmit. Quality of service (QoS) guarantees could be foreseen using this mechanism.

A typical wireless LAN installation would use different channels for adjacent cells so as to prevent two PCF stations (i.e., access points) from using the same channel during the contention-free period. This would allow coexistence with an ad hoc network that is using DCF only, even on the same channel (see Fig. 1.12).

Power saving.   Since most IEEE 802.11 devices are wireless mobiles (e.g., small handhelds and personal digital assistants), power consumption optimization is a critical matter. There are two power saving modes included in the MAC protocol—awake and doze. In the awake mode, stations are fully powered all the time. In the doze mode, nodes must "wake up" periodically to listen for beacons which indicate that the

AP has queued messages. Nodes must inform their associated AP before entering doze.

A station, in the power saving mode (i.e., the doze state), cannot transmit or receive frames. The IEEE 802.11 standard defines power management procedures for cases with and without infrastructure. In the presence of infrastructure, a dozing station periodically wakes up and listens to selected time-stamped beacons that are sent by the access point. If the beacon indicates that the access point has queued data for that station, the station sends a special poll frame that tells the access point to send the data.

In the absence of infrastructure, the policy adopted within an IBSS is completely distributed for preserving the self-organizing behavior. The power conserving stations in the ad hoc cell wake up for only short predefined periods of time, which are announced in the ATIM window, to hear if they should remain awake to receive a frame. All requests for transmissions are placed within the ATIM window. A station in sleep mode that receives one, wakes up for an interval which equals at least the next beacon interval. The ATIM window is also determined by the first station to initialize the IBSS. If the time ATIM window is zero, power-save mode is not used.

IEEE 802.11 MAC framing.   There are different types of MAC-level messages. These are classified as data, management, and control. Data frames contain user data, management frames support the different MAC services (e.g., authentication frame), and control frames are meant to support delivery control of data and management frames. According to the IEEE 802.11 standard, an IEEE 802.11 MAC frame consists of a MAC header, a frame body, and a CRC-32 *frame check sequence* (FCS). The basic structure of a MAC frame is depicted in Fig. 1.19 [IEEE-STDRD].

The MAC header consists of seven fields and is 30 bytes long. These fields are frame control, duration, address 1, address 2, address 3, sequence control, and address 4.

The *duration/ID* field is 2 bytes long. It contains data on the duration value for each field and for control frames it carries the associated identity of the transmitting station. The *address* fields identify the basic service set, the destination address, the source address, and the receiver

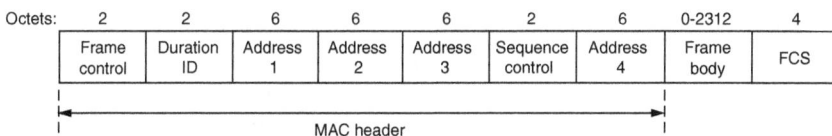

Octets:	2	2	6	6	6	2	6	0-2312	4
	Frame control	Duration ID	Address 1	Address 2	Address 3	Sequence control	Address 4	Frame body	FCS

MAC header

Figure 1.19 IEEE 802.11 frame.

Octets: 2	2	4	1	1	1	1	1	1	1	1
Protocol version	Type	Subtype	To DS	From DS	More flag	Retry	Power mngnt	More flag	WEP	Order

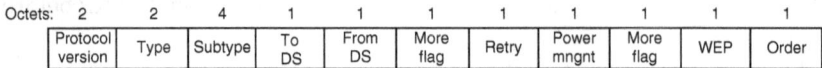

Figure 1.20   IEEE 802.11 MAC header.

and transmitter addresses. Each address field is 6 bytes long. The *sequence control* field is 2 bytes and is split into 2 subfields—fragment number and sequence number. The fragment number is 4 bits and tells how many fragments the MSDU is broken into. The sequence number field is 12 bits and indicates the sequence number of the MSDU. The *frame body* is a variable length field from 0 to 2312 octets (bytes) long. This is the payload. The FCS is a 32-bit cyclic redundancy check which ensures that there are no errors in the frame. For the standard generator polynomial see App. B, Sec. B.2.

The *frame control* field shown in Fig. 1.20 is 2 bytes long and is composed of the following fields—Protocol Version, Type, Subtype, To DS, From DS, More Flag, Retry, Power Management, More Flag, More Data, WEP, and Order.

Specifically, the *Protocol Version* field is 2 bits in length and will carry the version of the IEEE 802.11 standard. *Type* and *Subtype* fields are 2 and 4 bits, respectively. They work together hierarchically to determine the function of the frame. The remaining eight fields are all 1 bit in length. The *To DS* field is set to 1 if the frame is destined for the distribution system. The *From DS* field is set to 1 when frames exit the distribution system. Frames that stay within their basic service set have both of these fields set to 0. The *More Flag* field is set to 1 if there is a following fragment of the current MSDU. *Retry* is set to 1 if this frame is a retransmission. *Power Management* field indicates if a station is in power save mode (set to 1) or active (set to 0). *More Data* field is set to 1 if any MSDUs are buffered for that station. The *WEP* field is set to 1 if the information in the frame body was processed with the WEP algorithm. The *Order* field is set to 1 if the frames must be strictly ordered.

IEEE 802.11 as ad hoc network.   As previously mentioned, the IEEE 802.11 concept allows for infrastructureless networks. The IEEE 802.11 standard is a good platform to implement a single-hop local ad hoc network mainly because of its extreme simplicity. Multihop networks covering areas of several square kilometers could also be built by exploiting the IEEE 802.11 technology.

When referring to an ad hoc configuration, IEEE 802.11 refers to the IBSS in which all stations are in mutual communication range and communicate directly. The IBSS has no access point and thus cannot use the distribution system. Since there is no access point available, stations cannot associate to any BSS and hence only frames which do not use the

DS are allowed[5]. Therefore the stations make use of just the SS (authentication, MSDU delivery, privacy). The major changes occur in the MAC layer, in particular the MLME, which provides contention services (DCF services), authentication, and privacy to the higher layers.

As with the globally managed *basic service set identifier* (BSSID) (in infrastructure mode), this is now locally assigned. It is an IEEE 802 locally administered address with the individual/group bit set to 0 and the universal/local bit set to 1. The remaining 46-bit number is chosen at random. The random algorithm should, for similar seeds (adjacent IBSS for example, if local time is used as seed), return different numbers.

1.7.1   ETSI HiperLAN: layers, services, and entities

High-performance radio LAN is a part of a high-speed radio access framework developed by ETSI BRAN, the European Telecommunications Standard Institute responsible for broadband radio access networks. The goal was to produce a wireless LAN that would be indistinguishable in performance from wired LANs, such as Ethernet, as well as to support both isochronous and asynchronous services. Operating in the 5-GHz band, HiperLAN is designed to carry multimedia data and support QoS as it provides basic data transport functions with bit rates from 6 to 54 Mbit/s. Unlike the IEEE 802.11 standard, this committee was not driven by existing products or regulations. Instead, a set of functional requirements was defined and a committee set out to satisfy these requirements. The standards work was confined to the lowest two OSI layers [TANENB96] whereas a draft standard [HLAN-SPEC] was released in July 1995 for imminent ratification.

There are four different types of HiperLAN networks defined.

- *HiperLAN type 1.*   The transmission has a range of about 50 m at a rate of 23 Mb/s and uses the 5-GHz radio frequency.

- *HiperLAN type 2.*   The transmission has a range of about 200 m for wireless ATM networks at a rate of 23 Mb/s and uses the 5-GHz radio frequency.

- *HiperLAN type 3.*   The transmission has a range of about 5000 m for wireless ATM networks at a rate of 20 Mb/s and uses the 5-GHz radio frequency. This network can be used for WLL (wireless local loop) communications.

---

[5]These frames are called Class 1 and Class 2 frames in the standard.

- *HiperLAN type 4.*   The transmission has a range of about 200 m for wireless ATM connections at a rate of 155 Mb/s and uses the 17-GHz radio frequency.

Scenarios for usage and the choices considered for different aspects of the standard are described in Refs. [WILKIN95, HALLS95]. In brief, the standard allows for a radio LAN system operating at 23.529 Mb/s with support for multihop routing, time bounded services, and power saving.

Similar to IEEE 802.11, HiperLAN organizes mobile terminals in cells. Specifically, two modes of operation are possible:

- In a base-station oriented mode the network is organized like a traditional cellular radio network, in which a so-called access point acts as the base station and gateway to the core (wired) network.

- In the ad hoc mode no core network is present. The network is self-organizing such that a station is dynamically chosen to act as an AP, called *central controller* (CC). The advantage of this organization is that the same centralized MAC protocol can be applied in both modes of operation.

The HiperLAN project has defined a system architecture as shown in Fig. 1.21. The HiperLAN reference model is depicted in Fig. 1.22. As illustrated, on top of the physical layer specification, a separate sublayer has been integrated, containing the channel access mechanism. This mechanism is used by the different functional entities, offering a multitude of services with different requirements. These requirements include:

- Short-range—10 to 100 m at the most—communication

- To work with and without infrastructure (i.e., two stations can communicate directly, if needed)

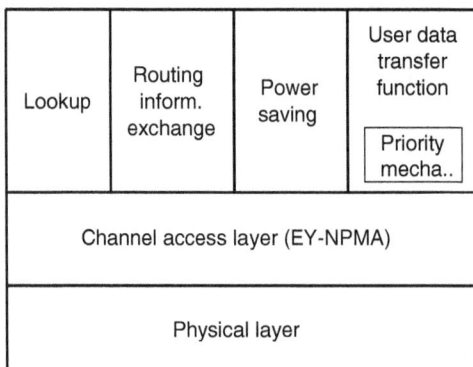

Lookup	Routing inform. exchange	Power saving	User data transfer function
			Priority mecha..
Channel access layer (EY-NPMA)			
Physical layer			

Figure 1.21   Layer architecture of the HiperLAN reference model.

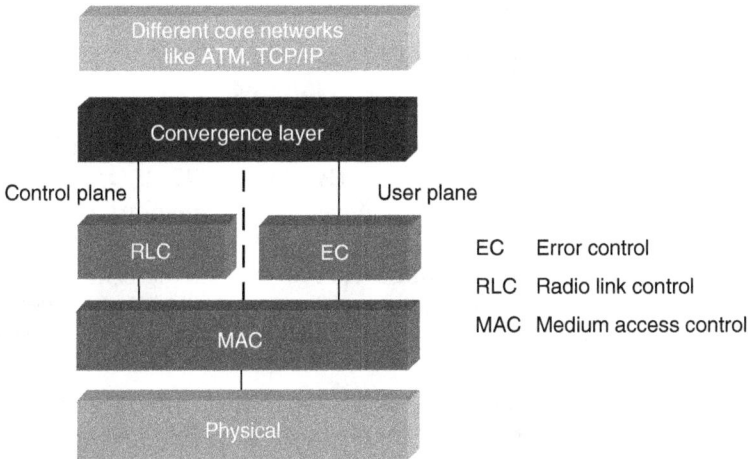

Figure 1.22   HiperLAN/2 reference model.

- To allow for low mobility, with a user's speed of 1.4 m/s at the most
- To support isochron ous traffic—audio traffic, at 32 kb/s with a 10 ns latency, and video traffic, at 2 Mb/s with a 100 ns latency
- To support asynchronous traffic—data traffic at 10 Mb/s, with immediate access

To support forwarding of packets to stations outside the radio range of the sender, routing information exchange functionality is present. A lookup functionality is added to enable a colocated operation of distinct W-LANs. Optional encryption/decryption may be used. The standard stays away from defining the particular encryption method used, but defines methods to inform the receiver which of a particular set of encryption keys is used to encrypt the packet.

Physical layer.   The high data rate together with the need for a number of channels requires a reasonably large amount of spectrum, on the order of 150 MHz or more. The HiperLAN committee initially identified two bands, from 5.15 to 5.30 GHz and 17.1 to 17.2 GHz. Currently, the standard addresses mainly the 5-GHz band, which was ratified[6] for HiperLAN use by the Conference of European Posts and Telecommunications Administration (CEPT). The band is divided into five channels, with the lower three channels available in Pan-European countries and the

---

[6]HiperLAN does not have exclusive use of either the 5-GHz or 17-GHz band and can use these bands on a nonprotected basis.

upper two channels only available in some countries. The channel center frequencies start at 5.176468 GHz and the center frequencies are separated by 23.5294 MHz.

The PHY layer is based on OFDM (orthogonal frequency division multiplexing) with 52 subcarriers. Each subcarrier can be modulated with four different modulation schemes (BPSK, QPSK, 16QAM, and 64QAM). Forward error correction is achieved with a convolutional code with code rate 1/2 and constraint length 7. Different code rates (1/2, 9/16, and 3/4) can be achieved by the application of puncturing schemes. A combination of a modulation scheme and code rate is called a PHY-mode. With the highest PHY-mode (64QAM3/4), a data rate of 54 Mbit/s can be achieved.

The chosen modulation method trades off data rate and reduction of adjacent channel interference. The goal is to attain packet error rates below 10^{-3}. The coding scheme offers protection, with respect to error-correction per block, from at least two random errors and burst errors less than 32-bits long.

HiperLAN medium access mechanism.   EY-NPMA (elimination yield-nonpreemptive priority multiple access) is the MAC protocol for HiperLAN. EY-NPMA is based on a carrier sensing mechanism, but is quite different in its details from that used in the IEEE 802.3 (Ethernet) or the IEEE 802.11 standard. It offers a mechanism that requires a minimal number of reception/transmission turnarounds, while still resulting in a single winning station with high probability (97.8 percent).

EY-NPMA specific features include:

- No preemption by frames with higher priority after the priority resolution (see the following text).
- Fair contention resolution of frames with the same priority (see the following text).
- Hierarchical independence of performance.

The access mechanism is split into three phases: *priority resolution, elimination,* and *yield.*

In case the medium is sensed to be free for a sufficient length of time (1700-bit times, to be specific), immediate transmission is allowed. If not, the three medium access phases are adopted (Fig. 1.23).

The *priority resolution* phase is aimed at allowing only those nodes that have packets of the highest available priority to contend for channel access. This phase consists of a number of slots according to the number of priorities. A station seeking access to the media transmits a burst for the *priority assertion period* (PAD) in the fitting priority slot. In this

Prioritization phase	Elimination phase	Yield phase
Has 1-5 slots	Has n+1 slots, n ≤ 12	Has m slots, m ≤ 15
Slot is 256 bits long	Slot is 256 bits long	Slot is 64 bits long

Transmission ends, followed by one synchronization slot of 256 bits

For each contending user, the probability of transmitting a burst of l slots, i ≤ 12, is $0.5^{(j+1)}$

Each contending user defers from transmission for j slots, j < 14, with a probability of 0.1×0.9^{j}

Transmission ends

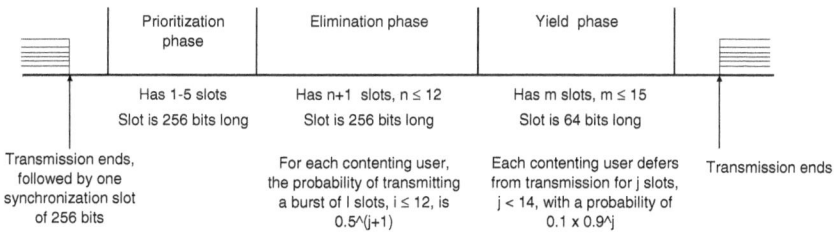

Figure 1.23 Channel access for HiperLAN.

sense, a node having a packet with priority p is transmitting a burst in slot $(p + 1)$, if no burst with higher priority was heard before. At the end of the first burst on the channel, at least one station survives this phase. The prioritization phase resumes and the elimination phase begins.

In the *elimination* phase, all surviving stations from the priority resolution phase transmit bursts and then listen to the channel for an *elimination survival verification period* (ESVP). The burst length is different for every station, bounded and defined by a certain geometrical probability distribution. A station that sends a signal in the ESVP stands back from transmission and listens to the channel for one time slot. If another burst is heard while listening to the channel, the node stops contending for the channel. Thus, only the node(s) with the longest burst will, in the absence of the hidden node problem, be allowed to further contend for the channel. Immediately after the longest burst and listening period of the elimination phase, at least one station survives. The elimination phase resumes and the yield phase starts.

In the yield phase, all surviving stations from the elimination phase defer transmitting for a geometrically distributed number of slots, while listening to the channel for a time equal to the *yield period* (YP). The YP length is different for every station, again bounded by a discrete probability distribution. A station that hears a signal stands back from transmission; otherwise it transmits the data frame immediately after the yield period.

To summarize, the purpose of the elimination phase is to decrease the number of contenders, and then the yield phase attempts to ensure that only one node eventually transmits. The chances of actual collisions for data are negligibly small (less than 3 percent). Furthermore, the residual lifetime of a packet together with its priority is used to determine its channel access priority and the support of different QoS levels for packet delivery. The priority of a packet is distinguished as high and normal, and the packet lifetime is measured in integral milliseconds with a range of 0 to 32767 ms (default value is 500 ms).

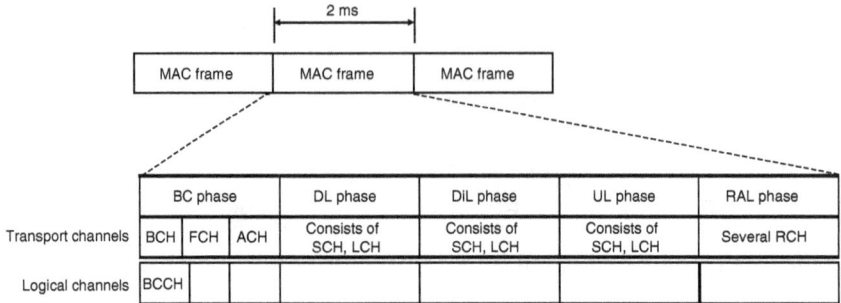

Figure 1.24 The structure of a HiperLAN MAC frame.

HiperLAN MAC channeling.   The MAC sublayer provides logical channels to the upper sublayers based on a dynamic TDMA/TDD scheme. The MAC frames are of length of 2 ms, i.e., 500 OFDM symbols and are centrally scheduled by the AP/CC.

As illustrated in Fig. 1.24, MAC frames contain the transport channels that describe the basic building blocks to form a *packet data unit* (PDU) train. These channels are mapped onto different phases:

- *Broadcast (BC) phase.*   Carries the broadcast control channel (BCCH) and the frame control channel (FCCH). The BCCH contains general announcements and some status bits announcing the appearance of more detailed broadcast information in the downlink phase (DL). The FCCH carries the information about the structure of the ongoing frame, containing the exact position of all following emissions, their usage, and the content type. The messages in the FCCH are called *resource grants* (RG).

- *Downlink (DL) phase.*   Carries user specific control information and user data transmitted from AP/CC to mobiles. Additionally, the DL phase may contain further broadcast information which does not fit in the fixed BCCH field.

- *Uplink (UL) phase.*   Carries control and user data from the mobiles to the AP/CC. The *mobile terminals* (MTs) have to request capacity for one of the following frames in order to get resources. Resources are granted from the AP/CC.

- *Direct Link (DiL) phase.*   Carries user data traffic between mobiles without direct involvement of the AP/CC. For control traffic, however, the AP/CC is indirectly involved by receiving *resource requests* (RR) from MTs and transmitting RGs in the FCCH. In ad hoc networks this phase is mandatory, whereas it is not used in the centralized configuration.

- *Random access (RA) phase.* Carriers a number of *random access channels* (RCH). MTs that have been granted no capacity in the UL phase use this phase for the transmission of control information. Nonassociated MTs use RCHs for the initial signaling exchange with an AP/CC. This phase is also used during a handover so that active connections are switched over to a new AP/CC.

The boundaries between the different phases are adjusted, according to data contained in the BCCH and the FCCH (logical) channels. The access mechanism foresees that terminals request resources within the so-called *short channels* (SCH) that are transmitted piggyback to one or several data packets. Data are segmented and transmitted in packets of 48-byte length that fit into the so-called *long channels* (LCH). The AP/CC collects all RRs that are received during a frame and allocates resources in the next MAC frame taking into account the perconnection QoS requirements. Resource grants of the AP/CC are announced in the broadcast channel at the beginning of each MAC frame.

The radio link control and error control sublayers. The *radio link control* (RLC) *sublayer* of MAC layer is responsible for link setup and maintenance. It, therefore, appears in the control plane of the reference model (Fig. 1.22). There are three entities providing different functions. First, the *association control* (AC) functions provide such services as link capabilities negotiation, encryption, authentication provision, association with an AP/CC, and information on the AP/CC. These are broadcast in the beacon with every MAC frame. Second, the *radio resource control* (RRC) functions provide dynamic frequency selection so that the most suitable channel is selected for transmission, based on measurements of both the AP/CC and the MTs. Further, transmit power control and radio presence maintenance (power save modes, alive queries, absence for scanning other channels, and the like.) are implemented. Handovers are also handled by the radio resource control functions. Third, the *DLC connection control* (DCC) functions carry out the connection setup, release, and modifications. In addition, multicast join and leave functions are provided by the DCC entity.

In the user plane, the *error control* (EC) entity supports three different modes of error control. These can be acknowledged mode with retransmissions, repetition mode with repeating the LCHs so as to provide a better link quality, or unacknowledged (i.e., unreliable) mode.

As the uppermost layer in the HiperLAN specification, a set of *convergence layers* (CL) is specified, which adapt the HiperLAN transport functions to the different core networks such as ATM and IP.

HiperLAN lookup. The HiperLAN lookup functionality is designed to explore the HiperLAN communication environment. The function allows the retrieval of the HiperLAN identifier for a specific HiperLAN, declared by name. A new HiperLAN can be created by choosing an unused HiperLAN identifier. An already existing HiperLAN is joined by merely using its identifier. A HiperLAN is destroyed when no entity makes use of its identifier.

Priority classes.   Although the HiperLAN draft standard does not define different priority classes for the specific traffic classes, like multimedia or file transfer, it however supports time bounded delivery of packets. This task is performed by assigning channel access priorities dynamically to the packets. A channel access priority depends on the *normalized residual MSDU lifetime* (NRMT) and the user-assigned priority. The NRMT, which is assigned from the MAC layer, is the ratio of residual lifetime and the distance between source and destination (measured in hops). Thus the priority of each packet increases while its lifetime expires. In each access cycle, only packets with the same access priority compete for the channel. This is because the access mechanism guarantees hierarchical independence of performance between packets with different channel access priorities. Table 1.4 shows how the access priority is defined.

Power saving.   The HiperLAN draft standard addresses power saving in two ways. The first method is supported by an innovative two-speed transmission method. Packets have a short *low bit rate* (LBR) header, at 1.4706 Mb/s, which contains enough information to inform a node whether it needs to listen to the rest of the packet or not. Thus, even if the node is listening it can keep the error-correction, equalization, and other circuits powered off unless the LBR header informs it otherwise. Secondly, a node is not constantly receiving packets and hence can save power by receiving packets only at predetermined moments. This method of power saving is supported via the so-called *p*-saver method. In essence, a node can announce that it only listens periodically, with a short duty cycle for remaining powered on. This allows the node to power down most of its circuits at all other times. Other nodes wishing to transmit to it, namely *p*-supporters, only send packets to the *p*-saver at time instants when the *p*-saver is powered on.

TABLE 1.4   HiperLAN Access Priority Classes

NRMT (ms)	High user defined priority	Low user defined priority
<10	0 (highest)	1
10<NRMT<20	1	2
20<NRMT<40	2	3
40<NRMT<80	3	4 (lowest)
>80	4 (lowest)	4 (lowest)

HiperLAN/2 as ad hoc network. As mentioned in Sec. 1.7.2 a HiperLAN ad hoc configuration is enabled using the direct mode (DM). Mutual data exchange is supported using the direct link (DiL).

The TDMA/TDD scheme in HiperLAN requires a central controller in the network scheduling all traffic. In HiperLAN infrastructured configurations, it is the AP that orchestrates the transmission/reception duties of mobiles. In HiperLAN infrastructureless configurations, the CC is chosen randomly from the existing (CC-capable) MTs.[7] Just like an access point in the centralized mode, the CC in ad hoc mode is responsible for scheduling all traffic, maintaining the network, and granting resources.

The *PHY layer* does not change significantly in the ad hoc environment. To reflect the fact that there could now be spread delay difference between two stations, a guard time between two consecutive DiL phases is inserted by the CC. One preamble is also added before each DiL train.

In the *MAC layer*, access is requested by the MTs and granted by the CC. The addressing scheme is the same as in the direct mode; after the CC selection process, the CC assigns MACIDs to itself and other associating stations.

In the *RLC sublayer*, there are some additions introduced for the establishment of an ad hoc network. These are dynamic CC selection—to choose an arbitrary CC which is in charge of the resource control in the subnet, further frequency scan and probe mechanisms to find MTs in all available frequency channels, and functions to adjust the MTs to achieve the best connection quality. The PHY mode (e.g., connection speed) and the power level are negotiated directly between two peers. In addition, a link quality map is maintained by the CC.

In the *EC sublayer*, a fourth mode is supported, which is *forward error correction* (FEC). The other modes (acknowledged, repetition, unacknowledged) remain the same.

1.8 Bluetooth: A Technology for Wireless Body Area and Personal Area Networks

The Bluetooth technology is a de facto standard for low-cost, short-range radio links between portable devices such as mobile PCs and mobile phones [BISDIK01, MILLER00]. Bluetooth technology facilitates protected ad hoc connections for wireless communication of voice and data in stationary and mobile environments.

The Bluetooth system is a digital wireless data transmission standard operating in the 2.4 GHz unlicensed ISM band, which is generally

---

[7]There can be CC capable and noncapable MT.

available in most parts of the world. Historically, the name Bluetooth comes from the Danish King Harald Blatand (Blatand is Danish for Bluetooth). King Bluetooth is credited with uniting the Scandinavian people during the tenth century. Likewise, the Bluetooth wireless technology aims at uniting personal computing devices.

The Bluetooth specifications are released from a special interest group, called the Bluetooth *Special Interest Group* (SIG). The Bluetooth SIG was formed in September 1998 and counts a number of vendor heavyweights among its founding members. These include—in alphabetical order—3Com, Agere, Ericsson Mobile Communications AB, IBM, Intel, Microsoft, Motorola, Nokia, and Toshiba. Today, the core group has expanded significantly and the number of Bluetooth SIG members has reached to the thousands.

The Bluetooth SIG published version 1.0 of the specification in July 1999. Bluetooth, however, started five years earlier, in 1994, when Ericsson Mobile Communications began a study to find out how wireless technology can be used effectively as a cable replacement to link cellular phones with handsets and accessories [PAHLAV02]. The study focused on radio links.

Of the many requirements for Bluetooth, support for voice and data were key issues. Out of this study came the Bluetooth specification for Bluetooth wireless technology. The specification includes air interface protocols to allow several Bluetooth applications to intercommunicate simultaneously and to overcome interference both from other Bluetooth devices and external sources such as domestic and commercial microwave ovens.

1.8.1 Bluetooth network configuration

From a logical standpoint, Bluetooth belongs to the family of contention-free token-based multiaccess networks [MILLER00, BLUETH-SPEC]. Bluetooth specifies a 10-m (about 30 ft) radio range and supports up to seven devices. Bluetooth units can operate in one of two network configurations—as a master or as a slave. The master is responsible for assigning channel access priorities to slaves as well as for setting the frequency-hopping sequence that the slaves shall tune into. The master permits slaves to transmit by allocating slots for voice or data traffic. Data traffic is thus controlled by the master too.

Every Bluetooth device contains a unique Bluetooth device address and a Bluetooth clock. The *Bluetooth device address* (BD ADDR) is a 48-bit IEEE address. Hop sequences in the Bluetooth system are a function of a device's BD ADDR and a synchronized 28-bit clock. Slave units use the master's clock and address to synchronize to the master in time and frequency by applying the master's hopping sequence [BRAY00].

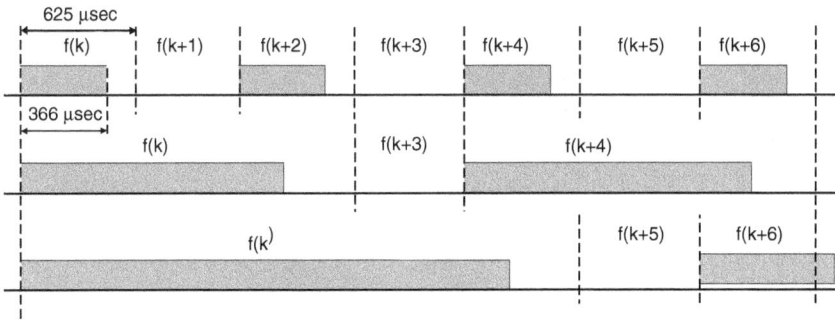

Figure 1.25   One-slot and multislot packet transmission.

Point-to-point *time division duplex* (TDD) communication is used between the master and the slave. The channel is thus divided into time slots, each of 625 μs in length. Time slots are numbered according to the Bluetooth clock of the master. The master has to begin its transmissions in even-numbered time slots, as odd-numbered time slots are reserved for the beginning of slaves' transmissions. The first row of Fig. 1.25 shows a snapshot of the master transmissions.

Additionally, the transmission of a packet nominally covers a single slot, but it may also last for three or five consecutive time slots, as illustrated in the second and third rows, respectively, of Fig. 1.25. For multislot packets, the RF hop frequency to be used for the entire packet is the RF hop frequency assigned to the time slot in which the transmission began.

Bluetooth uses a combination of circuit and packet switching. Two types of physical links are defined [BISDIK01]:

- *Synchronous connection-oriented (SCO).*   It is a point-to-point symmetric circuit-switched connection between a master and one slave. This type is mainly used for delivering delay-sensitive traffic (such as audio streaming) and is established through the reservation of duplex slots at regular intervals.

- *Asynchronous connectionless (ACL).*   It is a packet-switched connection between the master and all its slaves. It can support the reliable delivery of data by exploiting a fast *automatic repeat request* (ARQ) scheme. An ACL channel supports point-to-multipoint transmissions from the master to the slaves.

A maximum of one ACL link can be active between a master and a slave. A master can have up to three simultaneous SCO full-duplex voice links with one or several slaves, or a link that simultaneously supports asynchronous data and synchronous voice. Each voice link supports asynchronous data rate of 64 kb/s in each direction. The asynchronous link can support a maximal asymmetric data rate of 723.2 kb/s (and still up to 57.6 kb/s in the return direction), or a symmetric data rate of 433.9 kb/s. A slave can support SCO links with different masters at the same time, but only one SCO link with each.

1.8.2 Bluetooth protocol stack

A key function of the Bluetooth specification is to foster device interoperability and interconnectivity between different applications that may run over different protocol stacks and from a variety of vendor products. To allow the manufacturers interoperability, a number of service profiles define how Bluetooth-enabled devices should use the Bluetooth links and physical layer. All profiles define end-to-end services—cordless telephony, intercom, serial port, headset, dial-up, fax, LAN access (with PPP), generic object exchange, object push, file transfer, synchronization. In this regard, Bluetooth does not only define a radio system, but also introduces a layered protocol stack that enables applications to discover other Bluetooth devices in an area, discover what services they offer, and contain the capability to use those services.

The complete Bluetooth protocol stack (depicted in Fig. 1.26) comprises five main protocols—Bluetooth radio, baseband, *link manager protocol* (LMP), *logical link control and adaptation protocol* (L2CAP), and *service discovery protocol* (SDP).

The *radio layer* is responsible for the electrical interface to the communications media, coding/decoding, and modulation/demodulation of

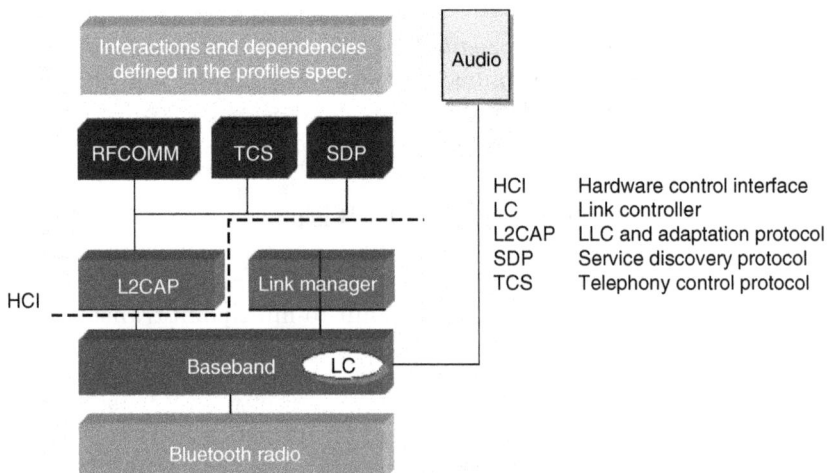

Figure 1.26 Bluetooth reference model.

data for transmission. The total ISM band provides 83.5 MHz of spectrum, offering 79 Bluetooth channels spaced 1 MHz apart, each providing around 500 kb/s of data bandwidth and signaling data at 1 Megasymbol per second (Ms/s) [TOH02]. The radio layer utilizes the FHSS as transmission technique. The hopping sequence is a pseudo-random sequence 79-hop in length, and is unique for each Bluetooth subnet (called piconet—see Sec. 1.8.3 for details on Bluetooth network topologies). The nominal rate of hopping between two consecutive RFs is 1600 hop/s. For a given Bluetooth device, this translates to a 500 kb/s interconnection (maximum) at any particular time. Within an area of 300 m^2 a total of 24 Bluetooth channels (i.e., 24 Bluetooth applications, through 24 different devices) could operate simultaneously (not 79 as this would make the frequency-hopping impossible to manage and would in any case negate its effects), resulting in a total of 12 Mb/s. As this would all be packet data (except for dedicated speech channels, which are 32Kb/s ADPCM), this provides a very efficient and flexible resource even when the Bluetooth device density is high. The FHSS system is chosen to combat the interference of nearby systems operating in the same range of frequency (for example, IEEE 802.11 W-LANs) and to make the link robust [GALI00, HAART99] in the presence of surrounding noise.

The *baseband layer* and *link controller* overlap to cover lowlevel data link control functionalities. Together they control the physical links via the radio layer by assembling packets, controlling frequency-hopping, and performing error checking and correction. The baseband controller also carries out the scan—page and inquiry procedures to discover—and synchronizes with devices in transmission range. The names baseband and link controller are often used interchangeably.

The LMP is the entity that is responsible for connection setup and maintenance. The link manager is responsible for connecting slaves to a piconet and creating their active member addresses [TOH02]. Additionally, the link manager serves to establish ACL data and SCO voice links. Other functions carried out by the link manager include QoS requests, power management, authentication, and encryption.

The L2CAP performs the services of a typical logical link control sublayer; these include the management and data flow control. The services provide both connection-orientated and connectionless links over the ACL link. Further, QoS, protocol multiplexing, segmentation and reassambly, and group management are supported. Protocol multiplexing is provided to distinguish between different protocols that run over the Bluetooth air interface. Segmentation and reassembly functions cope with packets exceeding the baseband packet size. Group management permits assigning and maintaining group properties to Bluetooth devices in order to support higher-level multicast operations.

The service discovery protocol permits the identification of services provided by and available through a Bluetooth device. These services, for instance, could be printing or LAN access. It provides knowledge about the availability, service type, and service class without the need of a central authority.

1.8.3 Bluetooth Topologies

Bluetooth wireless networks are classified into two network topologies named *piconets* and *scatternets*. Two or more devices (slave devices) sharing the same frequency-hopping sequence (channel) (i.e., synchronized to the same master) form a piconet. A unit can belong to more than one piconet, but can function as the master to only one. A piconet has a gross bit rate of 1 Mb/s, which is the channel capacity before considering the overhead introduced by the Bluetooth protocols and polling scheme. A piconet comprises one master station and up to seven *active* slaves that participate in data exchange. Piconets can either take the form of a point-to-point design, where only one slave and a master exist in a network, or of a point-to-multipoint design, where one master is connected to more than one slave in a network. Figure 1.27 shows a

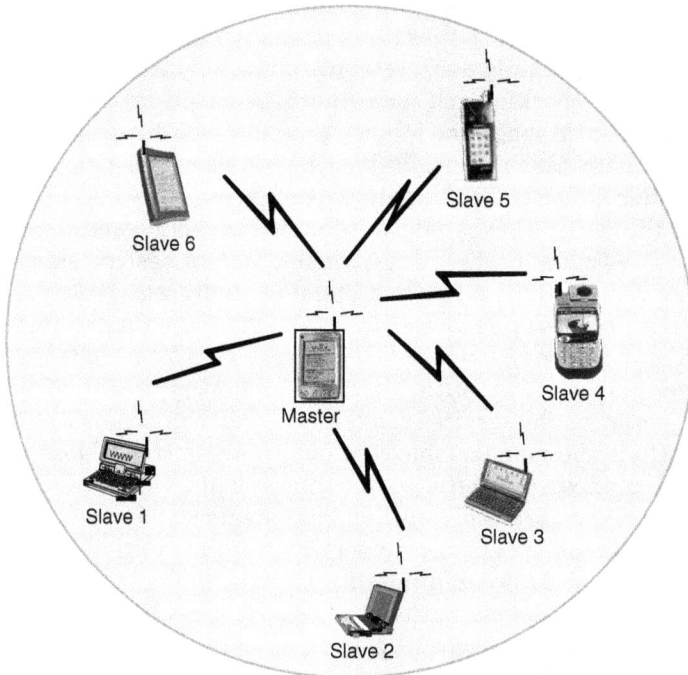

Figure 1.27 Bluetooth piconet architecture.

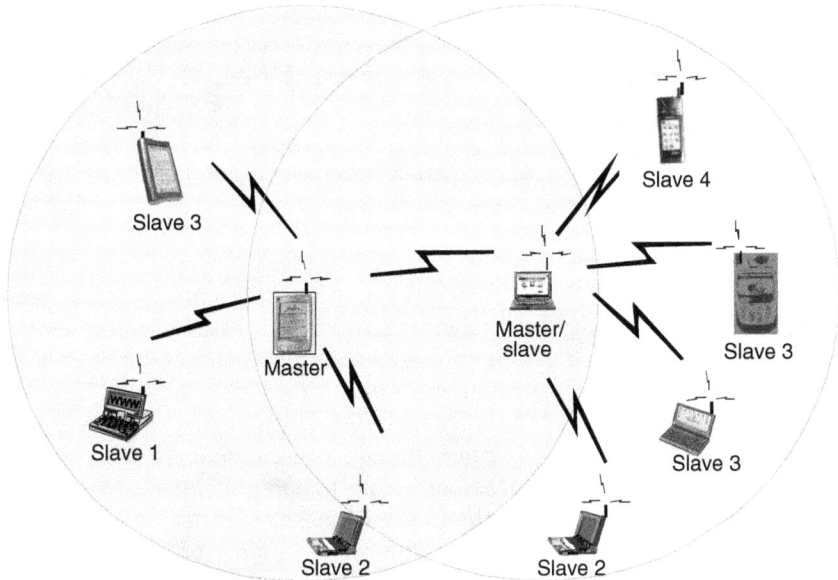

Figure 1.28 Bluetooth scatternet architecture.

point-to-multipoint architecture where the master becomes the head of the piconet and also serves as the central controller. Figure 1.28 shows an example of two partially overlapping piconets.

A direct link can exist only between a master and a slave but not between slaves. Thus, communication between slaves must be routed through the master.

Independent piconets that have overlapping coverage areas may form a *scatternet* (Fig. 1.28). A scatternet exists when a unit is active in more than one piconet at the same time. A slave may communicate with the different piconets it belongs to, only in a time-multiplexing mode. This means that for any time instant, a station can only transmit on the single piconet to which (at that time) its clock is synchronized. To transmit on another piconet it has to change the synchronization parameters. As Fig. 1.28 illustrates, when such an overlap occurs, a master of one piconet has to serve as a slave of the other piconet. No device can serve as a master of two piconets. When a slave from one piconet wishes to communicate with another slave from another piconet, both masters from each piconet are involved in the relay of packets across the piconets [BRAY00]. As additional piconets overlap, it is possible for one master to serve as a slave of two piconets. In such a scenario, this master/slave acts as a network bridge and router across piconets.

Multihop communications can thus be achieved through the scatter-net concept, where several masters from different piconets can establish links with each other. In this context, the master becomes the bottleneck [TOH02]. Therefore such a multihop scenario poses performance degradation issues due to the presence of time switching among piconets, local congestion situations as well as potential signal interference from adjacent piconets.

Bluetooth piconet formation.   Before starting a data transmission, a Bluetooth unit needs to discover if any other Bluetooth unit is in its operating space. To do this, the unit enters the *inquiry state*. In this state, it continuously sends an inquiry message, which is a packet that contains an access code. The inquiring unit can adopt a *general inquiry access code* (GIAC), which enables any Bluetooth device to answer the inquiry message, or a *dedicated inquiry access code* (DIAC), which enables only Bluetooth devices belonging to certain classes to answer the inquiry message. During the inquiry message transmission phase, the inquiring unit uses a frequency-hopping sequence of 32 frequencies derived from the access code. These 32 frequencies are split into two trains, each containing 16 frequencies. A single train is repeated at least 256 times before a new train is used. Several (up to three) train switches must take place to guarantee a sufficient number of responses. As a result of this inquiring policy, the inquiry state lasts for at most 10.24 s. A unit can respond to an inquiry message only if it is listening to the channel to find an inquiry message and its receiver is tuned to the same frequency used by the inquiring unit. To increase the probability of this event to occur, a unit scans the inquiry access code (on a given frequency) for a time long enough to completely scan for 16 inquiry frequencies. Obviously, a unit is not obliged to respond to an inquiring message. If it does, however, it has to send a special control packet, the FHS packet, which contains its Bluetooth device address and its native clock.

At the end of the inquiry phase, a Bluetooth unit has discovered the Bluetooth device address of the units around it and has collected information about their clocks. If it wants to activate a new connection, it has to distribute its own Bluetooth device address and clock. This is the aim of paging routines. The unit that starts the paging is (automatically) elected the master of the new connection, and the paged unit is the slave. The paging unit sends a page message, which is a packet that only contains the *device access code* (DAC). The DAC is derived directly from the Bluetooth device address of the paged unit. The paged unit is, therefore, the only one that can recognize the page message. Units can periodically listen to the channel to find a page message by tuning their receivers to the frequencies of the paging hopping sequence.

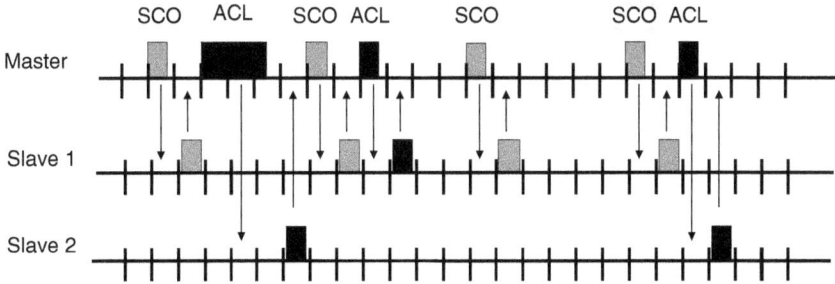

Figure 1.29   Transmission in a piconet.

At the end of the paging procedure, the slave has an exact knowledge of the master clock and of the channel access code. Hence, the master and that slave can enter the connection state.

When a connection is established, the active slaves maintain the synchronization with the master by listening to the channel at every master-to-slave slot. Obviously, if an active slave is not addressed, after it has read the type of packet, it can return to sleep for a time equal to the number of slots the master has taken for its transmission.

Polling is asynchronous for ACL links and synchronous for SCO links. To illustrate a typical transmission pattern within a piconet, consider the example of Fig. 1.29. There is one master and two slaves. Slave 1 has both a SCO and an ACL link with the master, while slave 2 has an ACL link only. In this example, the SCO link is periodically polled by the master every six slots while ACL links are polled asynchronously. Furthermore, the size of the packets on an ACL link is constrained by the presence of SCO links. For example, in Fig. 1.29 the master sends a multislot packet to slave 2, which replies with a single-slot packet. This is because the successive slots are reserved for the SCO link.

Bluetooth scatternet.   The Bluetooth specification defines a method for the interconnection of piconets—the *scatternet*. A scatternet can be dynamically constructed in an ad hoc fashion when some nodes belong, at the same time, to more than one piconet (interpiconet units). For example, the two piconets in Fig. 1.28 share a slave, and hence they can form a scatternet. The traffic between the two piconets is delivered through the common slave. Scatternets can be useful in several scenarios. For example, we can have a piconet that contains a laptop and a cellular phone. The cellular phone provides access to the Internet. A second piconet contains the laptop itself and several PDAs. In this case, a scatternet can be formed with the laptop as the interpiconet unit. By exploiting the scatternet, the PDAs can exploit the cellular phone services to access the Internet.

TABLE 1.5   Worldwide Service Revenue from Bluetooth

U.S. and western European carrier service revenue from Bluetooth (US$ '000s)

	2001	2002	2003	2004	2005	2006	Cumulative
PDAs	2,300	53,650	214,250	463,150	1,078,450	1,585,400	3,397,200
Notebooks	2,100	79,300	462,600	1,265,000	2,544,350	3,592,200	7,946,200
Headsets	700	13,000	60,750	145,850	294,950	540,100	1,055,350
Automotive	0	2,250	13,050	36,500	77,950	141,100	270,850
Total	5,100	148,200	750,650	1,911,150	3,995,700	5,858,800	12,669,600

SOURSE: The Zelos Group

The scatternet formation algorithms and the algorithm for scheduling the traffic among the various piconets remain hot research issues, see Refs.[LAW03] and [ZUSSM02], as these are not fully addressed in the current Bluetooth specification.

1.8.4   Bluetooth market developments

If two reports quoted extensively by the SIG are to be believed, then Bluetooth will be far more than a niche gadget used by technophiles. According to a U.S.-based research group Allied Business Intelligence, worldwide shipments of Bluetooth chipsets will increase from around 34 million in 2002 to 1.1 billion by 2007 with the majority of the early growth accounted for by mobile handsets.

The second report highlighted by SIG is from the Zelos Group, another U.S.-based research firm, which argues that the use of Bluetooth will significantly drive up cumulative mobile operator revenue in the 2002 to 2006 period (Table 1.5). In 2006 alone, Zelos projects Bluetooth service revenue to be US$2.6 billion in the United States and US$3.3 billion in Europe.

Taking handset growth first, the rapid decline in Bluetooth chipset costs is clearly a major factor. Not so long ago, chipsets were being sold for over US$30, which was well above the US$5 mark that was generally considered to be the point where the economics of Bluetooth started becoming attractive.

1.8.5   Bluetooth applications

Notable Bluetooth applications include the following:

- *The wireless headset.* One of the first prototype devices using Bluetooth technology is the wireless headset. Aimed first at the mobile phone, Bluetooth provides a wireless *handsfree* connection to the handset, which can be left in the briefcase (at least for incoming calls).

- *The Bluetooth equipped PC/PDA.* When Bluetooth becomes a standard interface that is installed in all portable computing devices, it will offer not only a quick and seamless access to LANs, but a new functionality too. To name but a few—the possibility of printing to any printer within range, communicating with a Bluetooth-equipped mobile handset to accept large file downloads from elsewhere on the planet, and to alert the user of appointments held in the PC calendar.

- *Wireless networked peripherals—printers, scanners, digital cameras, WebCams and the like.* Printers are a necessary adjunct to any computer but they will be more useful when they are untethered and accessible whenever they are within the range of a device needing them.

  Scanners and digital cameras (especially the latter) will benefit greatly from wireless downloads, because of the large amount of memory required to hold images. These images are always removed from the device's memory eventually, so the sooner they can be downloaded and passed to their correct depository, the more efficiently the digital camera can work.

  Other devices, traditionally wired to a computer to handle the data they provide, can now be positioned anywhere without trailing wires. By virtue of intelligent software behind these applications, the connection can be seamless, requiring a minimum amount of control by the user.

- *The Bluetooth "pad."* This is a concept device providing the users with a simple-to-use graphical interface to their new world of communication. A thin client (low inherent functionality), offering good "future proofing" and connected primarily to a 3G handset or 3G-equipped PC, for communication with the wider world. When within the range of other devices such as TVs and VCRs, the Bluetooth pad will control them directly; when out of range the Pad will control them indirectly via any other communication route it can find, with no user intervention required. In this context, when someone arrives late at the cinema and wants to see which films are available, he or she could do so by standing in the Bluetooth *hot spot* area and checking on the overhead display whether the film he or she wants to see has any seats left. If so, his or her phone's WAP browser displays a booking page from where he or she can pay for the seat over the GPRS network.

- *Bluetooth "white goods."* Not traditionally products that need to communicate with other equipment or humans, white goods can provide information making people's lives simpler. A refrigerator can track its own contents and alert the consumers when it is low on their favorite products. A washing machine can alert a busy person that it has finished its wash cycle. The interface for these communications would be the pad and the communication path would be Bluetooth.

- *Internet delivery.* The Internet is already playing a big role in communications, not only as a delivery mechanism in its own right but also as a source of data. Although Bluetooth does not impact the Internet directly, through devices such as the Pad and 3G personal terminals it will be possible to access the Internet more easily and make it our servant rather than our master.

1.9   Short-Range Ad Hoc Configurations

Depending on their coverage area, wireless ad hoc Bluetooth-like networks are classified into two main classes—*wireless body* and *wireless personal area networks* (Fig. 1.30).

In June 1997, a W-PAN project effort was initiated by the IEEE as part of the IEEE 802.11 project group. Subsequently, the first W-PAN specifications were published in January 1998. In 1998, Bluetooth responded to an IEEE invitation for participation in W-PAN standardization development. In March 1999, the IEEE 802.15 project committee was approved to handle W-PAN standardization. Bluetooth has since been selected as the base specification for IEEE 802.15 W-PANs [PAHLAV02].

1.9.1   Body area network

The miniaturization of microelectronics has enabled small and powerful handheld devices. Besides the PalmPilot and the family of Windows CE terminals, a new type of edge devices has gained momentum—wearables. These are small devices with a minimum power consumption that can be easily carried by human beings. Examples include head-mounted displays, microphones, and earphones. The idea behind wearables is to design gadgets out of simple electronic components to perform some functions formerly done only by computers.

A body area network (BAN) is a collection of wearable communication devices, which all together provide an integrated set of personalized services to the user. The communicating range of a BAN is analogous to the human body range, i.e., 1 to 2 m. As wiring a body is generally cumbersome, wireless technologies constitute the best solution for interconnecting wearable devices.

The main functional requirements of a BAN are [VAN01, DAM01]:

1. The ability to interconnect heterogeneous devices (e.g., a mobile phone with a microphone, display, and the like.)

2. Autoconfiguration capability—adding or removing a device from a BAN should be transparent to the user.

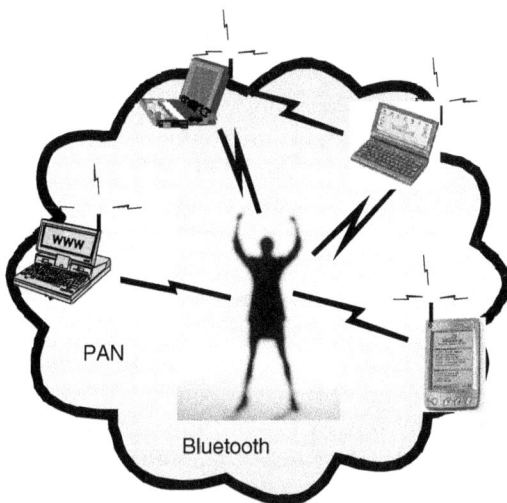

(a)

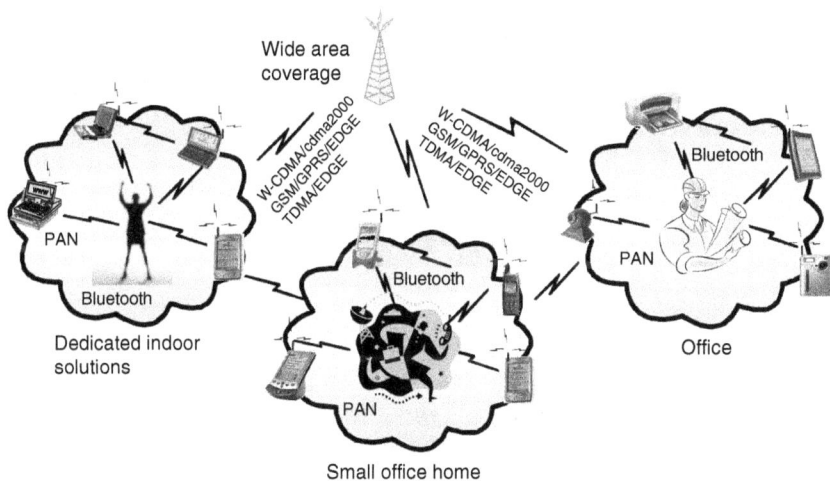

(b)

Figure 1.30 Relationship between (*a*) a body; and (*b*) a personal area network. (A scenario with three Bluetooth PANs is illustrated. The PANs are interconnected via wireless devices with Bluetooth links. The three PANs are also connected to an IP backbone network via GPRS/UMTS access.)

3. The ability to interconnect with the other BANs (e.g., to exchange data with other people) or PANs (e.g., to access the Internet).

Early examples of a body area network consist of wearable electronics (phone, MP3 player, headset, microphone, and controller) directly connected within a jacket. Different levels of integration among devices are achieved. For instance, the music to be muted when the phone rings and control of both phone and MP3 player is possible with a separate, easily accessible controller. Connection between the devices is established using wires integrated into the garment.

The BAN prototype developed by T.G. Zimmerman [ZIMMER96] provides data communications with rates up to 400 kb/s by exploiting the body as the channel. Zimmerman, in specific, showed that data can be transferred through the skin by exploiting a very small current (one billionth of an ampere). Data transfer between two persons (i.e., BAN interconnection) could be achieved through a simple handshake.

Besides, one sector where BANs may have widespread applicability is the healthcare sector where BANs are worn by healthcare professionals as well as patients [LIN99]. For a patient, BAN devices are worn on the body, perhaps embedded in clothing, jewelry, glasses, or even attached directly to the body (e.g., a cardiac monitor) or implanted. The BAN is intended to be an open, extensible platform which can be personalized not only to a class of patients (e.g., diabetic patients) but also to the particular set of (chronic and acute) problems of the individual person/patient. Early examples include the Citizen Health System (CHS - *lomiweb.med.auth.gr/chs/*) and TOPCARE (*www.topcare-network.com*) IST European Research projects.

1.9.2   Wireless personal area network

The use of numerous portable devices, such as laptops, mobile phones, PDAs, game terminals, MP3 players, and DVD players, is becoming more and more widespread in people's professional and private lives. For the most part, these devices are used separately, that is, their applications do not interact. Imagine, however, if they could interact directly—participants at a meeting could share documents or presentations; business cards would automatically find their way into the address register on a laptop and the number register on a mobile phone. Such small networks are often referred to as personal area networks (PANs). In the home environment the PAN devices may interact with appliances and home electronics for both information and control purposes.

In contrast to a BAN that addresses the interconnection of one-person wearable devices [see Fig. 1.30(a)], a Wireless PAN (W-PAN) is a personalized ad hoc wireless network in the environment around the person.

A PAN communicating range is typically up to 10 m. It thus enables the interconnection of BANs close to each other [see Fig. 1.30(*b*)] as well as the interconnection of a BAN with the environment around it.

The technologies for W-PANs offer a wide space for innovative solutions and applications that could create radical changes in everyday life. We can foresee that a W-PAN interface will be embedded not only in devices such as cellular phones, mobile computers, and PDAs, but also in every digital device. In this view, PANs can make possible the design of innovative pervasive applications. For example, let us imagine you carrying a PDA with a PAN interface that, upon your arrival at a location (such as home, office, and airport), automatically synchronizes with all the electronic devices within 10 m range of it. Furthermore, when you arrive at home, your PDA can automatically unlock the door, turn on the house lights while you are getting in, and adjust the heat or air conditioning to your preset preferences. Similarly, when you arrive at the airport you can avoid the line at the check-in desk by using your handheld device to present an electronic ticket and automatically select your seat.

In the near future, the role and capabilities of short-range data transaction are expected to grow, serving as a complement to traditional large-scale communication. Most man-machine communication as well as oral communication between human beings occurs at distances of less than 10 m and often, as a result of this communication, the two communicating parties have a need to exchange data. As an enabling factor, license-exempted frequency bands invite the use of developing radio technologies (such as Bluetooth) that admit effortless and inexpensive deployment of wireless communication.

In addition, access to the Internet via a (public) wireless LAN access point and/or via a 3G UMTS mobile phone would enable the devices in the PAN to be constantly online [see Fig. 1.30(*b*)]. For instance, commuters may have a public W-LAN access point, in a train, to their notebook computers and when exiting the train their notebook computers could remain online via the UMTS phone, while incoming e-mails could now be diverted to their PDAs through the PAN. Finally, as the user enters the office, the access could again, automatically, go through the notebook computer via a wireless access to the corporate campus network.

2

Dynamic Routing in Mobile Ad Hoc Networks

2.1 Routing in Communication Networks

A network, in its general context, consists of nodes and the links connecting them. Nodes can be functionally classified into two types—endpoints (also referred to as hosts or terminals) that act as sources and sinks of traffic, and switches (or routers) that forward traffic between a source and a sink node [STEEN95]. We will find it convenient here to refer to any component of the network (hosts, switches, and routers) that handles data packets simply as *nodes*, and to the communication channels (wireless, wireline) that connect adjacent nodes as *links*. Similarly, the terms *message*, *packet*, and *datagram* are used interchangeably to denote a single data or protocol packet, and the terms *flow* and *call* are used to denote a single data session.

Depending on whether or not the endpoints and network nodes are mobile, the network is aptly said to have a mobile infrastructure (also called mobile network) or a fixed infrastructure, respectively. In wireline networks the constituent links are fixed (etherlike) whereas their counterparts in wireless networks are wireless (e.g., radio or infrared).

To facilitate communication within a network, a routing protocol is needed to produce reliable and efficient routes between a pair of nodes so that messages may be delivered between them. Routing is defined as a set of several component functions including the following: constructing and selecting routes; monitoring network topology and services; distributing this information for use in route construction; locating session endpoints; and forwarding traffic along the selected routes.

Figure 2.1 shows a simple network with two hosts (H1 and H2) and four routers (R1, R2, R3, and R4). The role of the routing (network)

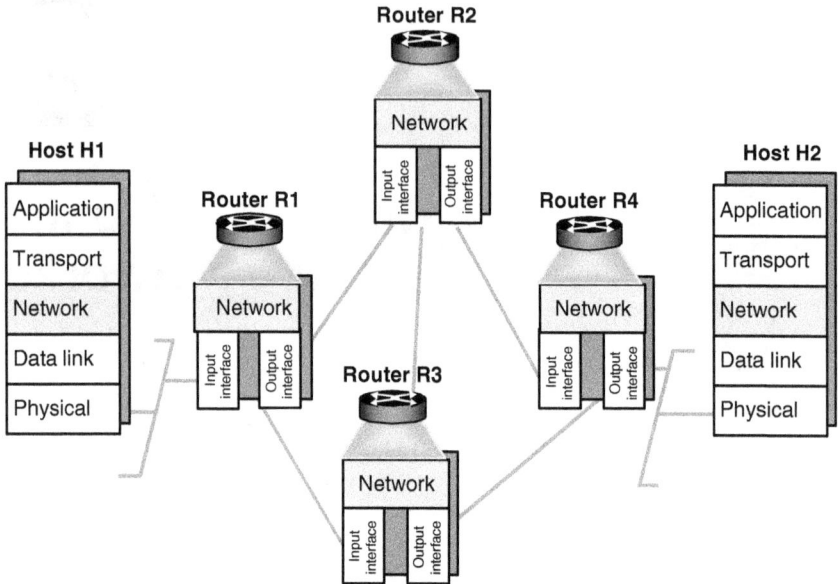

Figure 2.1 The network layer.

layer is to transfer the session packets from the sending host to the receiving endpoint. For example, if H1 is sending to H2, the network layer in host H1 transfers these packets to its nearby router, R1, which similarly transfers the packets to one of its immediate routers. The primary role of the routers is to "switch" packets from input links to output links. Eventually, the packet will be delivered to its intended destination host. At the receiving host (H2), the network layer receives the packet from its nearby router (in this case, R4) and delivers the packet up to the transport layer.

The role of the routing layer is thus deceptively simple—to transport packets between two hosts. To do so, two substantial network layer functions are identified:

- *Path determination.*   The network layer must determine the route or path that packets will travel. The algorithms that calculate these paths are referred to as *routing algorithms.* As an example, a routing algorithm would determine, for example, whether packets from H1 to H2 (see Fig. 2.1) flow along the path R1→R2→R4 or path R1→R3→R4 (or any other path between H1 and H2).

- *Switching.*   When a packet arrives at the input of a router, the router must "switch" it to the appropriate output link. For example, a packet arriving from host H1 to router R1 must be forwarded toward H2 either along the link from R2 to R4 or along the link from R3 to R4. Appendix A highlights the basic concepts of switching inside a router.

Before delving into the details of the path calculation, let us first take a broader view of the network model and consider what different types of *services* may be offered by the network layer.

2.2 Network Model

Perhaps the most important abstraction provided by the network layer to the upper layers is whether or not the network layer uses *virtual circuits* (VCs) or datagram packets. A virtual-circuit packet network behaves much like a telephone network, which uses real circuits as opposed to virtual circuits. There are three identifiable phases in a virtual-circuit packet network:

1. *VC setup.* During the setup phase the sender specifies the receiver address and waits for the network to set up the VC. The network layer determines the path between the sender and the receiver, i.e., the series of links and routers through which packets will travel. This typically involves updating information related to routing and data connections in each router along the path. The network layer may also reserve resources (e.g., bandwidth) along the path of the VC. A VC setup is illustrated in Fig. 2.2.

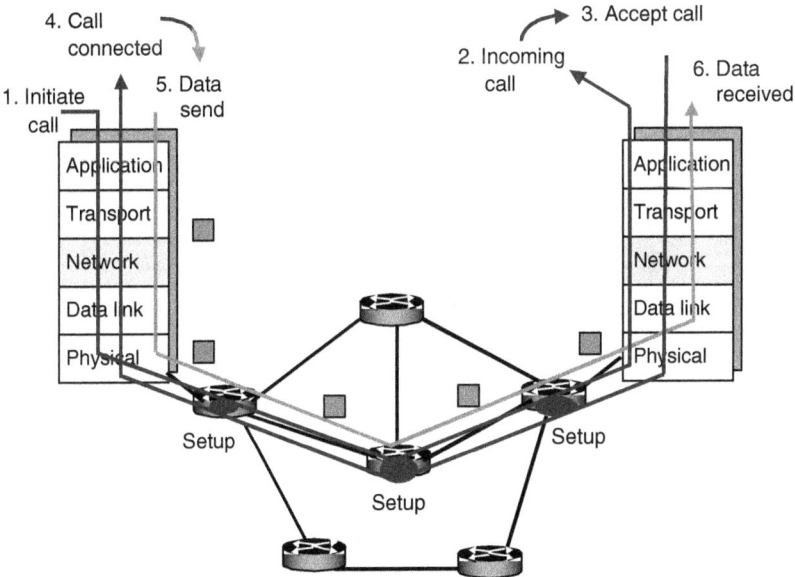

Figure 2.2 Virtual circuit service model.

2. *Data transfer.*   Once the VC is established, data can begin to flow along the VC.

3. *Virtual circuit teardown.*   This is initiated when the sender (or the receiver) informs the network layer of its desire to terminate the VC. The network layer will then typically inform the end system on the other side of the network about the call termination and update the tables in each switch along the path to indicate that the VC no longer exists.

ATM, Frame Relay, and X.25 are three networking technologies that use virtual circuits.

With a *datagram network layer*, an end system stamps its session packets with the address of the destination end system and then sends the packets into the network. Packet switches (routers) do not maintain any state information about VCs because communication is established without any VC setup (Fig. 2.3). Instead, packet routers route a packet toward its destination by examining the packet's destination address, indexing a routing table with the destination address, and forwarding the packet in the direction of the destination. Because routing tables can be modified at any time, a series of packets of the same session between a given host-destination pair may follow different paths through the network (and may then arrive out of order).

The Internet uses a datagram network layer.

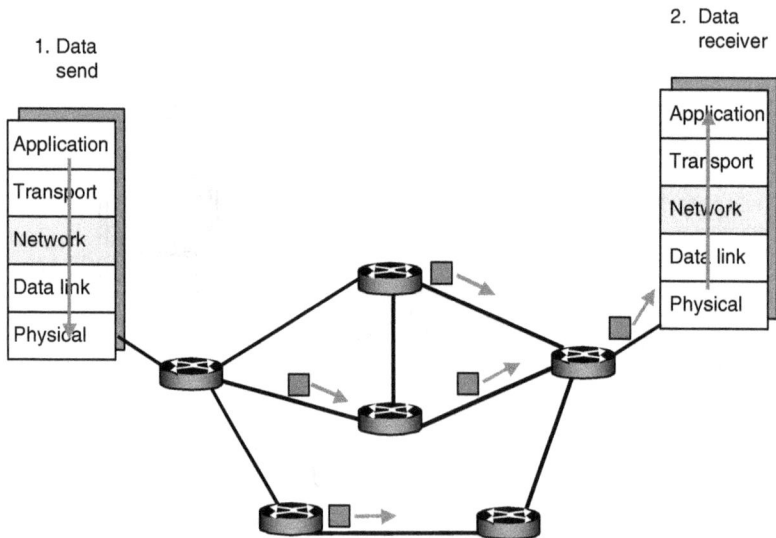

Figure 2.3 Datagram service model.

Alternative terminologies for VC service and datagram service are *network-layer connection-oriented service* and *network-layer connection-less service*, respectively. Indeed, the VC service is a sort of connection-oriented service, as it involves setting up and tearing down a connection and maintaining connection state information in the packet switches. The datagram service is a sort of connectionless service in that it does not employ any kind of connection or related state information.

2.3   A Review of Traditional Routing Protocol Families for Wired Networks

In order to transfer packets from a sending host to the destination host, the network layer must determine the *path* or *route* that the packets are to follow. Whether the network layer provides a datagram service (in which case different packets between a given host-destination pair may take different routes) or a virtual circuit service (in which case all packets between a given source and destination will take the same path), the network layer must, nonetheless, determine the path for a packet. This is the job of the network layer *routing protocol* [STEEN95].

At the heart of any routing protocol is the routing algorithm that determines the path for a packet. The purpose of a routing algorithm is simple—given a set of routers, with links connecting them, it finds a "good" path between a pair of nodes.

Routing protocols base their calculations on the available routing information for the network connectivity. Reachability (routing) information for remote stations is maintained in routing caches (also called *routing tables*). To guarantee that routing tables are up-to-date and reflect the actual network topology, nodes should continuously broadcast route updates and recalculate the paths.

Depending on whether routes are calculated in a distributed or centralized manner, routing protocols can be broadly classified in two general categories—*decentralized routing algorithm* and *global routing algorithm*.

2.3.1   Decentralized routing algorithm

A decentralized routing algorithm computes the least-cost path between a source and destination in an iterative, distributed manner. No node has complete information about the costs of all network links. Instead, each node begins with only the knowledge of the costs of its own directly attached links and then through an iterative process of calculation and exchange of information with its neighboring nodes gradually calculates the least-cost path to a destination, or set of destinations. A decentralized routing algorithm is commonly known as a *distance-vector* (DV) *algorithm*. It is called a distance-vector algorithm because a node never actually knows a complete path from source to destination. Instead, it only knows the direction (which neighbor) to which it should forward a packet in order to reach a given destination along the least-cost path and the cost of that path from itself to the destination.

Distance-vector routing algorithms.   Every node i maintains a set of distances for each destination j and for each neighboring node k. This information is stored in a data structure called the *distance table*. Let the distance to j as seen from i through its neighbor k be denoted as H_j^k. A node running a distance-vector protocol does not need to maintain routing information for the entire network topology.

Consider a node i that is interested in routing to destination j via its directly attached neighbor k. Node i's distance table entry H_j^k is the sum of the cost of the direct one hop link between i and k $c(i,k)$ plus the neighbor k's currently known minimum cost path from itself (k) to j. That is:

$$D^i(j, k) = c(i, k) + \min_n \{D^k(j, n)\} \tag{2.1}$$

The $\min_n$ term in Eq. 2.1 is taken over all k's directly attached neighbors.

In order to maintain up-to-date information, each host periodically broadcasts to each of its neighbors its current estimate of the shortest path to every other host in the network (its distance-vector). Route updates are sent periodically or when a topological change is detected.

The process of receiving updated costs from neighbors, recomputation of distance table entries, and updating neighbors on changed costs of the least-cost path to a destination continues until no update messages are sent. At this point, since no update messages are sent, no further distance table calculations occur and the algorithm resumes.

Let's consider an example that will help clarify the meaning of entries in the distance table as well as the operation of the distance-vector algorithm. Consider the network topology and the distance table shown for node E in Fig. 2.4. This is the distance table in node E once the distance-vector algorithm has converged. Let's first look at the row for destination A. Clearly the cost to get to A from E via the direct connection to A has a cost of 1. Hence $D^E(A, A) = 1$.

Let's now consider the value of $D^E(A, D)$—the cost to get from E to A, via its neighbor D. In this case, the distance table entry is the cost to get from E to D (a cost of 2) plus whatever the minimum cost it is to get from D to A. Note that the minimum cost from D to A is 3—a path that passes right back through E! Nonetheless, we record the fact that the minimum cost from E to A given that the first step is via D has a

E's Distance Table (D^E ( ))

j \ K	A	B	D
A	①	14	1
B	7	8	⑦
C	6	9	⑥
D	4	11	④

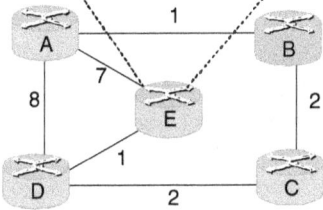

Figure 2.4   A distance table example.

cost of 5. We're left, though, with an uneasy feeling that the fact the path from E via D loops back through E may be the source of problems down the road.

Similarly, we find that the distance table entry via neighbor B is $D^E(A, B) = 14$. Note that the cost is *not* 15!

A circled entry in the distance table gives the cost of the least-cost path to the corresponding destination (row). The column with the circled entry identifies the next node along the least-cost path to the destination. Thus, a node's routing table (which indicates which outgoing link should be used to forward packets to a given destination) is easily constructed from the node's distance table.

Path calculation in distance-vector algorithms is based on a distributed version of the classical Bellman-Ford algorithm (DBF) [BERTS92, RFC 2453, JUBIN87, McQUILL77]. It is used in many routing algorithms in practice, including—Internet BGP, ISO IDRP, Novell IPX, and the original ARPAnet.

2.3.2 Global routing algorithm

A global routing algorithm computes the least-cost path between a source and destination using complete, global knowledge about the network. That is, the algorithm takes the connectivity between all nodes and all link costs as input. An algorithm that is aware of the state (cost) of each link in the network is commonly known as link-state algorithm [TANENB96, McQUILL80].

In practice, this is accomplished by having each node broadcast the identities and costs of its attached links to *all* other routers in the network. This link-state broadcast [PERLM99] can be accomplished without the nodes having to initially know the identities of all other nodes in the network. A node needs to only know the identities and costs of its directly-attached neighbors; it will then learn about the topology of the rest of the network by receiving link-state broadcast from other nodes. Such networkwide propagation of updates is commonly referred to as "flooding" [PETERS00] (in the sense that the network is flooded with protocol messages such as updates and protocol signaling).

The result of the link-state broadcasts is that all nodes eventually manage to establish an identical and complete view of the network. Each node can then run a least-cost path computation algorithm and compute the same set of least-cost paths as every other node.

Path calculation in link-state algorithms is based on Dijkstra's algorithm, named after its inventor. It computes the least-cost path from one node to all other nodes in the network. For a network with k destinations, Dijkstra's algorithm has the property that after the kth iteration, the least-cost paths to all destination nodes are computed; and among the least-cost paths, these k paths will have the k smallest costs.

2.3.3 A comparison of link-state and distance-vector routing algorithms

Let us conclude our discussions on link-state and distance-vector algorithms with a quick comparison of some of their attributes.

Message complexity.   The DV algorithm requires a periodic exchange of messages between directly connected neighbors. Whenever a link cost changes, the new link cost must be propagated to *all* nodes. It is shown that for a network of n nodes and E links, link-state routing generates $O(nE)$ message overhead [TANENB96, KUROSE02].

The following example highlights the link-state versus distance-vector overhead:

In its basic version, each link-state update includes the node ID of the original sender, a sequence number, and a list of neighbor nodes. The sequence number in the link-state updates needs to distinguish between current and old updates.

Let us assume a 10-bit node ID and a 16-bit sequence number. The routing information is thus:

Distance-vector

$$(10 + 1)N = 11\ N \text{ bits}$$

Link-state

$$[(10 + 16) + 10\ K]\ N = 26\ N + 10\ KN \text{ bits } (k = \text{nodal connectivity})$$

Link-state routing thus takes $(26 + 10\ K)/11$ times more data per period per node than distance-vector routing. For low connectivity ($K = 3$), this is a factor of 6.9; for high connectivity ($K = 10$), this is a factor of 11.45. It is evident then that link-state routing generates significantly more traffic than distance-vector algorithms.

A modification that can significantly reduce the overhead of link-state routing is *incremental updating*, where nodes send only data that have changed. Sequence numbers are incremented only when connectivity changes; neighbors are listed in an update only if their connectivity has changed. Updates include the most recent sequence number of each node in the network and any changes in connectivity.

Speed of convergence. Link-state is an $O(n^2)$ algorithm requiring $O(nE)$ messages [KETCHUM95], and potentially suffers from oscillations. Distance-vector protocols converge slowly (count-to-infinity problem) as they react badly to topological changes. In addition, they may introduce both short-lived and long-lived routing loops when the path length increases. Modifications to the distance-vector algorithm to ameliorate or eliminate these problems are reported in Refs. [CDPD, JAFFE82, ZAUMEN92].

Robustness. What happens if a router fails, misbehaves, or is sabotaged? In the case of link-state, a router could broadcast an incorrect cost for one of its attached links. A node could also corrupt or drop any link-state broadcast packets it receives as part of link-state broadcast. But a link-state node is only computing its own routing tables; other nodes are performing similar calculations for themselves. This means that route calculations are somewhat separated under link-state, providing a degree of robustness. In the case of distance-vector, a node can advertise incorrect least path costs to any/all destinations. At each iteration, a node's calculation is passed on to its neighbor and then indirectly to its neighbor's neighbor on the next iteration. In this way an incorrect node calculation can be diffused through the entire network. (Indeed, in 1997, a malfunctioning router in a small *Internet service provider* (ISP) provided national backbone routers with erroneous routing tables. This caused other routers to flood the malfunctioning router with traffic and caused large portions of the Internet to become disconnected for up to several hours [NEUMAN97].)

2.4 Routing in Mobile Wireless Networks

In mobile wireless networks, where communication terminals are mobile and the transmission medium is wireless, routing is one of the main problems. As we noted in Chap. 1, mobiles in a wireless ad hoc network share the same frequency channel, like IEEE 802.11 [IEEE-STDRD] and ETSI HIPERLAN Type 1 [HLAN-SPEC]. The limitations on power consumption, imposed by portable wireless radios, coupled with the fact that the communication infrastructure does not rely on the assistance of centralized stations, imply that terminals must communicate with each other either directly or indirectly using multihop routing techniques. The movement of nodes results in a distributed multihop wireless network with a time-varying topology.

Before delving into the details of the properties underlying dynamic routing in wireless networks, our primary issue is to find out whether a conventional routing protocol, like link-state or distance-vector, could apply in a wireless multihop environment. To respond, we first need to list several outstanding structural differences that exist between wired and wireless mobile networks and make routing very different in the two environments [JOHN94, PERK00, TOH02, PRAKAS98]:

1. In a mobile wireless network, the rate of topological changes is relatively very high as compared to wired networks. As is the case in wireline networks, the procedures for route selection and traffic forwarding in wireless mobile networks require accurate information about the current state of the network (e.g., node interconnectivity and link quality) and the session (e.g., traffic rate, endpoint locations) in order to direct traffic along paths that are consistent with the service requirements of the session and the service restrictions of the network.

However, changes in network or traffic sessions are likely to occur more frequently in mobile wireless networks than in stationary wireline networks. The degree of dynamism in route selection depends on several factors, including the type and frequency of changes in network and session state; the limitations on response delay imposed in assembling, propagating, and acting upon this state information; the amount of network resources available for these functions; and the expected performance degradation resulting from a mismatch between selected routes and the actual network and session state.

The routing mechanism must be able to quickly detect and respond to such state changes in order to minimize service degradation of existing traffic sessions whereas, at the same time, the algorithm must do so using a minimal amount of network resources so as to be able to maximize the overall network performance [RAMAN96].

The effectiveness of a routing protocol increases as network topology information becomes more detailed and up-to-date. To maintain up-to-date routing tables, a conventional routing protocol should be forced to continuously send and receive topology updates. In wireless ad hoc networks, however, the topology may change quite often, requiring *frequent exchanges of control information* (e.g., routes, route updates, or routing tables) among the network nodes. In an event-triggered link-state protocol any topological change would trigger a flooding, resulting in a flooding rate equal to the topological change rate. In this scenario, a blind route update mechanism could unnecessarily waste network resources since updates are sent even when no data transmission occurs in the network. In addition, as the number of network nodes can be large, the potential number of destinations is also large, thus requiring *a high volume of control information* exchange among the network nodes. As a consequence, the amount of update traffic can be even higher, the distribution of which can eventually saturate the network.

Notably, radio spectrum is a scarce resource, which means that packet-radio networks typically have limited bandwidth available. Because the wireless devices must share access to the radio channel, the bandwidth available to any node is even more limited. Relatively low bandwidth combined with the potential for routing algorithms to generate large numbers of packets means that efficiency is paramount in designing packet-radio routing algorithms. This observation is, however, in contradiction to the fact that all updates in the wireless communication environment travel over the air and are costly in resources. Even more disappointing is the fact that as the network size and the nodal mobility increase, the fraction of this total amount of control traffic used will decrease. This is so because the greater the mobility of the nodes, the shorter the residual lifetime of a link. Thus, the period for which the routing information remains valid decreases as well. Since the rate of link failure is directly related to node mobility, greater mobility increases both the volume of control traffic required to maintain routes and the congestion due to traffic backlogs. Thus, a crucial algorithm design objective for achieving routing responsiveness and efficiency is the minimization of reaction to mobility and of exchange of information.

On the other hand, when the rate of topological changes is extremely high, little can be done to ensure that routing algorithms converge fast enough to track topological changes [IWATA99, CORS95]. In this situation, flooding-based routing algorithms may be the only viable routing option. In this regard, we should also note that under extreme conditions, where the changes in network topology occur too frequently, finding a loop-free path may become impossible. We conclude then that the topology changes should occur sufficiently slowly in order to allow successful propagation of topology updates.

2. *Wireless links exhibit significant differences with wireline links.* The quality of a wireless link is admittedly less predictable than that of a wireline link and may fluctuate considerably depending on the network and environmental conditions. The factors that affect the quality of a wireless link include the distance between the link's endpoints, resulting in signal attenuation; the terrain between the link's endpoints, resulting in multipath signal propagation or even signal obstruction; externally generated noise, resulting in corrupted transmissions; and interference among multiple transmissions in a particular environment, resulting in lost transmissions. While many of these effects can be mitigated by lower-level solutions such as transmission power adjustments, error correction techniques (see App. B for details on error detection and correction), link-level retransmissions (see App. B for details on data link layer), and efficient channel access procedures, they cannot be eliminated entirely. Thus, it is the responsibility of the routing system to avoid routing traffic through low-quality links.

3. *Wireless links can be asymmetric and unidirectional [HAAS98a, CHAMBER02, PRAKAS98].* A link between two nodes i and j is called unidirectional when node i can properly receive traffic from a node j (in the sense, i can receive data from j above a certain *bit error rate* (BER) threshold and thus properly decode it), but j cannot receive traffic properly from i. As a consequence, a transmission that requires a handshake between i and j fails.

A link between two nodes i to j is called asymmetric when the transmission quality of the link (e.g., data rate) from i to j is different from that of the link from j to i.

4. *Broadcast transmissions are unreliable.* Since broadcast packets are not receiver directed, there is no way to reserve the wireless medium at the receivers before transmitting a broadcast packet (e.g., with the use of an RTS/CTS exchange—see Sec. 2.7.1). Consequently, broadcast packets are inherently less reliable than unicast packets.

This difference does not exist in wired networks, and presents a fundamental limitation to wireless networks that must be accounted for in the design of ad hoc network routing protocols. The authors in Ref. [BROCH98] demonstrate that over any single hop, 99.8 percent of *unicast data packets* are received successfully, while only 92.6 percent of *broadcast packets* are received. The difference between the two numbers is attributed to collisions.

5. Wireless devices present *technological limitations* on the use of resources—battery power, transmission bandwidth, and CPU time—compared to their wired counterparts.

6. *Security* is challenging in a mobile wireless environment [STAJNO99, PAPADI02, ZHOU99]. Due to the nature of radio transmissions, in the absence of any authentication mechanism, a malicious node can easily corrupt route tables, caches, and other information. For example, it can advertise false route information.

Applying these requirements in mobile networks is difficult because:

- Information about network flows is typically *not* available in datagram networks.
- Network topology can vary rapidly.
- Incremental delay and residual capacity change more quickly than the physical topology.
- Even if a radio generates timely routing information that reflects changes in delay and capacity, the delay in propagating that information throughout the network may be such that the information is stale by the time it reaches a distant node.
- The incremental delay and residual capacity of a radio link are affected by the traffic being carried on other radio links.

With these constraints in mind, routing protocols designed for wired networks cannot directly be applied to mobile wireless ad hoc networks. In fact, conventional routing protocols would perform very badly [KRISHN97, BARRET01, CHAMBER02], both from a practical standpoint of building such a network, and from a theoretical standpoint in terms of what seem to be promising routing algorithms, if used in a dynamic environment.

Link-state and distance-vector would probably work very well in an ad hoc network with low mobility, i.e., a network where the topology is not changing very often. The main problem with link-state and distance-vector is that they are designed for a static topology, which means that they would have problems converging to a steady state in an ad hoc network with a frequently changing topology. In addition, the problem that still remains with link-state and distance-vector is that they are highly dependent on periodic control messages. As the number of network nodes can be large, the potential number of destinations is also large. This requires large and frequent exchange of data among the network nodes. This is in contradiction to the fact that all updates in a wireless interconnected ad hoc network are transmitted over the air and thus are costly in resources such as bandwidth, battery power, and CPU.

Based on these considerations, the desirable qualitative properties of a mobile wireless routing protocol, or the basic requirements that specify the routing capability of a mobile wireless routing protocol, are to [KRISHN97]:

1. Cope with the dynamically changing network connectivity, owing to mobility

2. Ensure small convergence time, based on high collaboration among nodes

3. Be robust, given a common spectrum of MANET conditions such as high channel congestion and frequently changing topologies

4. Support communication with unidirectional links

5. Scale well in large mobile populations, in terms of storage, computational and transmission overhead

6. Be simple

Besides these, other more evident desirable qualitative properties include

- *Scalability.* In general, the scalability of a routing protocol is its ability to support the continuous increase in the network parameters (such as mobility rate, traffic rate, and network size) without degrading network performance [IWATA99, SANTINAV02]. We should distinguish, however, between *network scalability* and *routing protocol scalability*. In its general context, network scalability is what the network can support whereas routing protocol scalability is what the routing protocol can handle provided the network can.

Simply speaking, if the network can support thousands of nodes for a given traffic load, then for a routing protocol to be considered scalable, it should not break when run over that network of thousands of nodes with that traffic load. So, basically, routing protocol scalability means matching (or improving) the network scalability properties.

From Ref. [IWATA99], it is argued that routing protocol scalability is dependent on the scalability properties of the network the protocol is run over. That is, the network's own scalability properties provide the reference level as to what to expect from a routing protocol. Obviously, if the overhead induced by a routing protocol grows faster than the network rate but slower than the minimum traffic load, the routing protocol is not degrading network performance. The latter is in fact determined by the *minimum traffic load.*

Furthermore, scalability of a routing protocol does not merely depend on its performance (e.g., packet delivery ratio) versus network density, or traffic load, or control overhead, or some combination of performance measures. While we may assume that a protocol A that uses control signaling (control packets) more efficiently (number of packets delivered per control packet) than a protocol B, or else, both protocols deliver the same number of packets, but one must work harder to do so, can we say that protocol A is more scalable than protocol B, given that the performance of both the protocols is the same (at least from an external view)? Measuring the protocol's control overhead does not necessarily provide enough information to extrapolate the results to what will happen when network parameters (size, mobility, and the like) are increased. This is so because there are other factors, such as route suboptimality, which may become more relevant as traffic and network size increase. To this end, a routing protocol that produces less control overhead may be forming longer paths, which may not be an issue at your current traffic rate, but as the traffic rate increases the extra hops may be comparable (or greater than) to the control overhead.

- *Distributed operation.* The protocol should not be dependent on a centralized controlling node.

- *Demand-based operation.* To minimize the control overhead in the network and thus not waste network resources more than necessary, the protocol should be reactive. This means that the protocol should only react when needed and that the protocol should not periodically broadcast control information.

- *Multiple routes information.* To reduce the number of reactions to topological changes and congestion, multiple routes could be used. If one route becomes invalid, it is possible that another stored route could still be valid thus saving the routing protocol from searching for a new route.

- *Quality-of-service (QoS) support.* The general goal of routing in datagram networks is to get packets reliably from the source to the destination while maximizing the capacity of the network and minimizing the delivery delay. Optimal routing algorithms that maximize capacity or minimize delay typically need an estimate of the network flows, network topology, information about the residual capacity of links, and so on. Some sort of QoS support should then necessarily be incorporated into the routing protocol.

- *Power conservation and sleep period operation.* The nodes in a mobile network can be laptops and thin clients such as PDAs that are very limited in battery power and therefore use some sort of standby mode to save power. It is therefore important that the routing protocol has support for these sleep and temporarily inactive modes.

- *Loop freedom.* To avoid fraction of packets spinning around in the network.

- *Security.* Since the radio environment is especially vulnerable to impersonation attacks, then to ensure the wanted behavior from the routing protocol, some of preventive security measures are needed. Authentication and encryption is probably the way to go and the problem here lies within distributing keys among the nodes in the ad hoc network.

- *Network partition support.* Mobility of nodes together with wireless links of varying quality could lead to overly frequent routing changes, which could further cause some nodes to become completely disconnected from other nodes or the network.

In terms of network performance, wireless routing protocols need to fulfill a multitude of design and functional requirements (see Refs. [RAMAN96, BALAKR98, JINYA01, PAXSON97, RAPP96, SATYAN95]), including

1. High throughput (packet delivery ratio)

2. Low average latency (end-to-end delay, call setup times, and average queue size)

3. Heterogeneous traffic (e.g., data, voice, and video)

4. Preservation of packet order

5. Support for priority traffic

6. Adaptive to local and temporal variations in network characteristics

7. Low control overheads

For ad hoc networks lacking fixed infrastructure in the form of base stations, fulfilling the previously mentioned functional requirements becomes all the more difficult.

We should carefully note, however, that these specification requirements do not intend to specify or even bias the manner in which a service is delivered through a mobile wireless network. Well-conceived, useful protocols will find users, others will not. The minimum requirements as well as the primary attributes of the design of any dynamic routing protocol should be at the designer's discretion.

2.5 Routing and Mobility Management in Infrastructured Wireless Networks

Mobility management (also referred to as mobility tracking or location tracking) is the set of mechanisms by which location information is updated in response to terminal mobility. Location tracking typically consists of two operations:

1. *Updating (or registration).*   The process by which a mobile endpoint initiates a change in the location database (e.g., HLR/VLR and routing tables) according to its new location

2. *Finding (or paging).*   The process by which the network initiates a query for an endpoint's location to update the location databases

For wired environments, routing paths are usually fixed since end terminals are static; thus location tracking is not required. For infrastructured wireless (cellular) networks (see Chap. 4 for details on cellular wireless communications), endpoint mobility within a designated area is transparent to the network, and hence location tracking is only required when an endpoint moves from one domain to another (handover). For mobile wireless networks with no infrastructure and central authority (wireless ad hoc topology), mobility management is conducted at varying degrees of granularity.

2.5.1   Location tracking in personal communication systems

Two well-known personal communication systems (PCS) standards for location tracking are the North American standard IS-41 [ITU91, MOHAN94] and the European standard GSM [GSM92, MOHAN94]. Both employ a partitioning of the service area into location areas, and both are based on a two-level hierarchy. When a user subscribes to a PCS service provider, a user-specific information entry is created in its *home location register* (HLR)—see GSM in Sec. 4.3.2. When the user moves to a new location area, a temporary record is created in the *visitor location register* (VLR). HLRs and VLRs may be integral parts of the *mobile switching centers* (MSCs), or separate entities such that a single HLR or VLR serves multiple MSCs. An overview of the location-tracking mechanism in IS-41 follows; location tracking in GSM is considered in Sec. 4.3.2 and is thus not described here again.

In IS-41, once an endpoint enters a new (visiting) location area it sends a registration request to the MSC in that area. The MSC sends an authentication request message to its VLR which in turn forwards the request to the HLR associated with the visiting endpoint. The HLR's response is delivered to the MSC. If the endpoint is authenticated, the MSC sends a registration notification message to its VLR, which in turn forwards the message to the HLR. The HLR updates the location entry corresponding to the endpoint so that the entry points to its new-serving MSC/VLR. The HLR also sends to the VLR information related

Update strategies

Static

Dynamic

Location areas   Reporting cells

Extending
static

Endpoint
oriented

Dynamic LA   Dynamic RC

Time   Movement   Distance

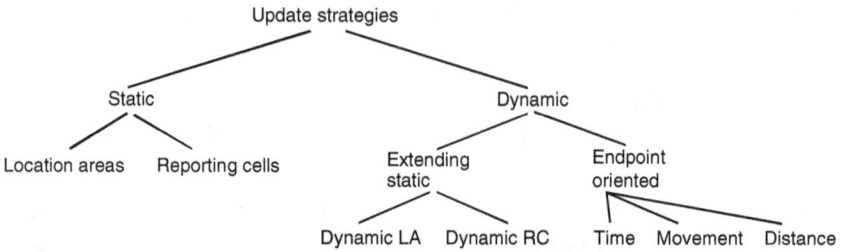

Figure 2.5   Classification of update strategies [RAMAN96].

to the endpoint's service profile. If the endpoint was registered previously in a different location area, the HLR sends a registration cancellation message to the previously visited VLR. Upon receiving this message, that VLR erases all entries for the endpoint and sends a cancellation message to the MSC which does likewise.

2.5.2   Updating the location database

The majority of the location-tracking techniques use a combination of updating and finding, in an effort to select the best trade-off between update overhead and delay in finding [MADHOW95]. To reduce the update overhead, updates are, usually, not sent every time an endpoint enters a new area. These are rather sent according to a predefined strategy such that the finding operation is restricted to a specific area. Figure 2.5 illustrates a classification of possible update strategies. These are briefly discussed as follows.

Static strategies.   In a static update strategy, there is a predetermined set of areas in which location updates may be generated. Whatever the nature of mobility of an endpoint, location updates may only be generated when, but not necessarily every time, the endpoint enters one of these cells. Two approaches to static updating are as follows.

1. *Location areas (also referred to as paging or registration areas) [KETCHUM95].*   In this approach, the service area is partitioned into groups of cells with each group as a location area. An endpoint's position is updated if and only if -it changes location areas. When an endpoint needs to be located, paging is done over the most recent location area visited by the endpoint. Location tracking in many second-generation cellular systems, including GSM [GSM92] and IS-41 [ITU91], is based on location areas [MOHAN94].

2. *Reporting cells (or reporting centers) [NOY93].*   In this approach, a subset of the cells is designated as the only one from which an endpoint's location may be updated. When an endpoint needs to be located, a search is conducted in the vicinity of the reporting cell from which the most recent update was generated. In Ref. [NOY93], the problem of which cells should be designated as reporting cells so as to optimize the cost function is addressed for various cell topologies.

The principal drawback of static update strategies is that they do not accurately account for user mobility and frequency of incoming calls. For example, even though a mobile endpoint may remain within a small area, it may cause frequent location updates if that area happens to include a reporting cell.

Dynamic strategies.   In a dynamic update strategy, an endpoint determines when an update should be generated, based on its movement. Thus, an update may be generated in any cell. A natural approach to dynamic strategies is to extend the static strategies to incorporate call and mobility patterns. The dynamic location area strategy proposed in Ref. [XIE93] determines the size of an endpoint's location area according to the endpoint's incoming call arrival rate and mobility. Analytical results [XIE93] indicate that this strategy is an improvement over static strategies when call arrival rates are user- or time-dependent.

The dynamic reporting centers' strategy proposed in Ref. [BIRK95] uses easily obtainable information to customize the choice of the next set of reporting cells at the time of each location update. In particular, the strategy uses information recorded at the time of the endpoint's last location update, including the direction of motion, to construct an asymmetric distance-based cell boundary and to optimize the cell search order.

In Ref. [NOY95], three dynamic strategies are described in which an endpoint generates a location update: (1) every T seconds (time-based); (2) after every M cell crossings (movement-based); or (3) whenever the distance covered (in terms of number of cells) exceeds D (distance-based).

Distance-based strategies are inherently the most difficult to implement since the mobile endpoints need information about the topology of the cellular network. As shown in Ref. [NOY95], however, distance-based updating outperforms both time-based and movement-based updating, for memoryless patterns on a ring topology. In Ref. [MADHOW95], a set of dynamic programming equations is derived and used to determine an optimal updating policy for each endpoint, and this optimal policy is in fact distance-based.

In addition, there exist a number of dynamic strategies that minimize location-tracking costs under specified delay constraints (i.e., the time required to locate an endpoint). As an example, in Ref. [ROSE95], a paging procedure is described that minimizes the mean number of locations polled with a constraint on polling delay, given a probability distribution for endpoint locations. A distance-based update scheme and a complementary paging scheme that guarantee a predefined maximum delay on locating an endpoint are both described in Ref. [TIA93]. This scheme uses an iterative algorithm to determine the optimal update distance D that results in minimum cost within the delay bound.

2.5.3   Location tracking in the Internet

The Internet protocol (IP) [RFC 791] itself does not support node mobility. To address this need, the Internet Engineering Task Force (IETF) has been standardizing a protocol, called *mobile IP* [MOBILE-IP], which provides support for mobile hosts in the Internet. A comprehensive overview of mobile IP is provided in Ref. [JOHNS95].

In mobile IP, mobile nodes are allocated permanent IP addresses on a "home" network. If a node moves to a "foreign" network, it shall obtain a new temporary forwarding address (called the *care-of address*). The direct way of obtaining the care-of address is through a *foreign agent* in the visited network (whose existence is ascertained through an agent discovery protocol). Nodes register with the foreign agent, and the IP address of the foreign agent is used as the care-of address. Another way of obtaining the care-of address is through an address discovery protocol, such as the *dynamic host configuration protocol* (DHCP) [RFC 1541]. The use of DHCP in supporting portable and mobile computing is discussed in detail in Ref. [PERK95].

A node wishing to send a message to a mobile node sends the message to the permanent (home) address of the node. If the mobile is currently roaming away from its home network, the message is encapsulated and tunneled to its new location.

2.6   Routing and Mobility Management in Mobile Wireless Networks

In mobile networks with *stationary infrastructure*, as in mobile IP [JOHNS95] and GSM cellular network [GSM92], the main component of location tracking for mobile endpoints is the handover (or handoff) of active calls. In mobile networks with *mobile infrastructure*, such as mobile radio networks, communication terminals are free to move, causing frequently changed routing paths. Mobiles must thus keep track of each others' locations and interconnectivity as they move.

In its general context, mobility management in MANETs involves three mechanisms—route discovery, route selection, and route maintenance.

In brief, when a mobile node has data traffic to send to some destination, it first consults its route cache to determine whether it already has a route to the destination. If so, the radio can use this route to send the traffic. On the other hand, if an unexpired route to the destination does not exist, the node initiates a *route discovery* procedure. This procedure is completed when one or more routes are found or all possible route permutations are examined.

Route selection in packet radio networks is mainly centered on distributed adaptive procedures, which take into account local or global information about the network state in selecting the next hop to a destination. When only local network state information is available, a node selects the next hop that is the "best" choice within its neighborhood. When global network state information is available, a node selects the next hop which lies on the "best" route to the destination.

Route maintenance is responsible for reacting to topological changes in the network so that in the event of a link failure the affected data sources are informed. Alternatively, the link in error may be repaired locally at the point of failure with no further notification to affected data source node(s). In the latter approach, the affected session is rerouted through a new path in a way transparent to the data source.

2.6.1 Route discovery

As we mentioned previously, in order to route incoming calls to a mobile endpoint, the system must first discover a route to the destination terminal. *Route discovery is then denoted by the system process that is responsible for searching a route.* A straightforward implementation of route discovery is to use a query dissemination mechanism similar to *flooding* [SZE99], where a networkwide query packet is propagated (broadcast) throughout the entire network area. In this context, the data source node broadcasts a query (*route request*) packet, which is flooded through the network. A receiving node should help propagate the request if (1) the request has not been forwarded previously, and (2) the node is not the destination of the searching procedure.

The transit nodes, upon receiving the route request, may use this packet to extract reachability information for the source node. To accomplish this, they use the mobile from which the query was obtained as the next hop to reach the source node. (This technique of learning routes during the propagation of route requests is derived from LAN bridge routing, which is route discovery via backward learning.)

When the route request is eventually received by the destination, a *route reply* message is sent back to the source, indicating the route to the destination. The propagation of route request messages effectively stops at this point. The route reply then travels in the reverse direction of the discovered route. This is accomplished by using a route request table at each node that records information for the newly received route requests. When a node receives a route reply, the matched route request is retrieved from the route request table and the route reply is forwarded to the node from which the initial route request message was received.

The route reply may also contain route-cost information, so that recipients can select routes based on specific costs. Similar to the backward learning exercise during the propagation of route request packets, each recipient of the reply may update its route to the destination using as next hop the node from which the reply is obtained. A node may also maintain multiple routes to other nodes, but route acceptance shall be done in such a way as to guarantee freedom from loops.

Optimizations to flooding-based route searching. A plethora of studies on route discovery schemes in MANETs exist in the literature [ROYER99]. A common objective that motivates the design of these route-finding strategies is that the algorithm should compute stable routes with minimal communications overhead and computational complexity. As we noted previously, the wireless transmission capacity is a scarce resource; hence, the procedures for searching for as well as distributing network routing information are designed with one general goal in mind—to produce routes with the minimum amount of network resources.

While these are the key design objectives of route searching schemes in wireless multihop networks, in the process, however, they often impact other aspects of performance in ways that are not always desired. In fact, a number of route discovery methodologies select a point that is a three-way trade-off between overhead, route searching latency, and simplicity.

As we discussed in the previous section, because of the ever-changing topology of MANETs, flooding [IPFLOOD-ID] is a fundamental communication primitive, essential to wireless ad hoc routing algorithms (e.g., [PERK02, JOHNS96]) for route discovery [INRIA02]. Flooding is well suited for MANETs as no prior assumptions of the network topology are required to provide routing between any pair of nodes. In mobile networks where the routing infrastructure is highly dynamic, this is a particularly attractive property as it can enable, with limited complexity, both the dissemination of control information as well as the delivery of data packets.

However simple in its concept and straightforward in implementation, flooding is far from optimal, as it generates a high number of redundant messages, thus wasting valuable limited resources such as bandwidth and energy supplies. As the entire network is flooded every time a node needs to find a route, the cost in terms of network resources can in fact be very high; and the problem becomes more pronounced as the number of network users increases. Therefore, given the respective cost for terminal searching, *an efficient route discovery mechanism that reduces the excessive overhead induced by flooding is highly desirable.*

In this regard, there exist two common mechanisms for suppressing the propagation of route-search signaling overhead:

- *Query quenching.* To reduce the network overhead induced from a networkwide path search, intermediate nodes, on receiving a route query, may themselves reply to the query by sending a route reply message back to the source on behalf of the destination, given that they maintain valid routing information for the destination in search.

    Early quenching of route search stops, in effect, the spreading of the query flooding at some intermediate node. To a large extent, query quenching may reduce the route discovery overhead as well as the inherent route acquisition latency. However, on the other hand, since the route queries are not broadcast end-to-end, the routes constructed by the mechanism may not always be the optimum routes (i.e., with fewest hops, most long-lived, least congested, and like attributes). The source node may actually be prevented from discovering a better route even if one exists. Besides, when the freshness of routes is more crucial or when some up-to-date end-to-end information must be collected to assist in proper route selection, such as end-to-end bandwidth availability and an individual node's energy reserve, this is not a desired option.

- An alternative to suppressing the propagation of route queries is the *expanding ring search.* In an expanding ring search, several route discovery attempts of limited scope are made before a networkwide flooding is triggered. At each attempt, the searching scope is increased by some factor. This process continues until the searching scope reaches a maximum threshold, after which the query is flooded throughout the network, or the node in search is successfully located. One potential drawback of the expanding ring search mechanism is the increase in route discovery latency when the initial attempt to discover a route fails and a new route discovery cycle is initiated.

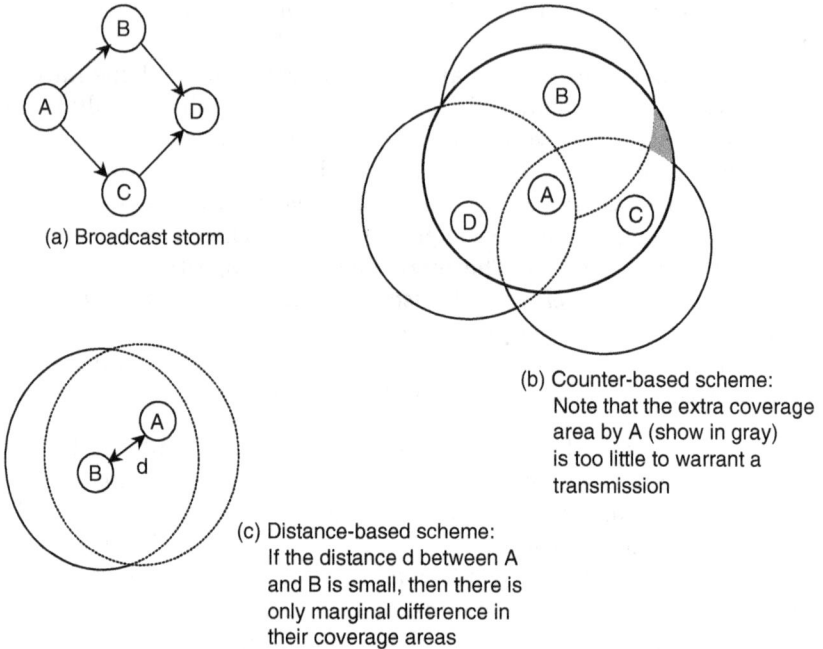

(a) Broadcast storm

(b) Counter-based scheme:
Note that the extra coverage
area by A (show in gray)
is too little to warrant a
transmission

(c) Distance-based scheme:
If the distance d between A
and B is small, then there is
only marginal difference in
their coverage areas

Figure 2.6   (a) Broadcast storm; (b) counter-based scheme; (c) distance-based scheme.

Apart from the excessive overhead induced from a networkwide path search, flooding a message in the entire network area results in many *unnecessary message retransmissions* as well as processing in areas where the messages are completely unproductive. The mechanism described for the suppression of duplicate route queries, where nodes avoid rebroadcasting copies of the query they have already seen, does not fully address the problem of redundant transmissions. Indeed, even if nodes rebroadcast only those queries that they receive for the first time, some of the rebroadcasts may be redundant. As shown in Fig. 2.6(a), suppose node A broadcasts a new query to nodes B and C, which in turn rebroadcast to node D. Hence, node D receives two copies of the same query, one of which is redundant. Moreover, if nodes B and C are close to each other and both transmit at the same time, channel contention could occur.

Furthermore, as the number of communicating nodes increases, the level of congestion, contention, and collisions increases accordingly as a result of the excessive use of the control access layer. If the underlying MAC does not provide collision detection capability (the CSMA/CA [LAM80] protocol, for example, cannot listen while sending), packet collisions could be damaging.

To this end, several schemes are proposed [SZE99] to alleviate the resulting redundancy, contention, and collision problems related to broadcast transmissions.

1. *Probabilistic-based.*   Each node rebroadcasts a message that is received for the first time with some fixed probability p.

2. *Counter-based.*   Each node rebroadcasts a message only if the same message has not been heard for more than c times. The rationale behind this approach is that if the message has been rebroadcast several times by the node's immediate neighbors, then the extra coverage contribution from its own rebroadcast is probably too low to be worth transmitting. This is illustrated in Fig. 2.6(*b*).

3. *Distance-based.*   Each node rebroadcasts a message only if the physical distance between itself and the node from which the message is received is not less than d [Fig. 2.6(*c*)]. The signal strength of the received message is used for distance estimations.

4. *Location-based.*   A message is rebroadcast only if the extra area, which is expected to be covered from this broadcast, is greater than some parameter, say a. Nodes use location information (from GPS) to determine the area of this extra coverage.

5. *Cluster-based.*   Only cluster-heads and gateway nodes are able to rebroadcast the message. Nodes may use any of the other schemes to determine whether or not to rebroadcast the message.

These schemes are mostly effective in densely populated networks, where nodes are communicating in close proximity to each other. An inherent problem with these techniques, however, is that all the threshold values are fixed, which can result in some messages not being broadcast to the destination under certain conditions, such as when the network is sparse. Further improvements to adapt the threshold values of these schemes to changing node density are proposed in Ref. [TSENG01].

Additionally, flooding protocols may *construct routes that use unidirectional links*. This is attributed to the fact that floodings are made with link broadcast transmissions, which in general use no acknowledgements (see Sec. 2.4). During the query-reply process, when the reply travels from the destination back to the source, the packet would eventually get lost on unidirectional links, and the source has then to initiate another flooding. This produces an overhead proportional to the percentage of unidirectional links.

Furthermore, when flooding is used as a mechanism for acquiring shortest-path routes, it is shown that flooding often *produces suboptimal routes* (see the analytical study in Ref. [INRIA02] for details).

Such nonoptimal routes generate a *nonnegligible overhead* that is proportional to the data load of the network. With these in mind, we intuitively conclude that the total overhead incurred by a routing protocol consists of two elements—overhead in the form of control traffic generated by the protocol, and overhead from data traffic, forwarded through routes of nonoptimal length.

To summarize, flooding is a simple mechanism, but however overly congestive, wastes nodal resources, constructs unidirectional links, and produces suboptimal routes. The excessive bandwidth overhead produced by query flooding during route discovery has made such an approach impractical and has thus prompted protocol designers to seek new ways to limit the extent and effects of query flooding.

Route selection and forwarding.   For a routing protocol to select a route from among a number of available routes, one or more criteria may be considered. The chosen route is often characterized as the *optimal route*, meaning that the preferred route can deliver a given flow with optimal performance given these specific criteria.

In wireline networks, the path length, measured in hops, is the metric often in use [TANENB96]. A hop denotes a *direct* physical or logical link between two end points (routers, switches, end-terminals, and the like). Conventional routing protocols produce minimum-hop, or else shortest-path routes with no explicit attention paid to link quality, link lifetime, or link reliability. Compared to a longer route, a minimum-hop path is always preferred as it involves less buffering, less propagation time, fewer resources, and potentially lower data-delivery delay.

However, in wireless mobile networks with unreliable and low-quality links, it is doubtful whether hop count can be considered a good indication for route selection. Authors in Ref. [COUTO02] verify this conjecture; they observed that the design of a routing protocol for mobile wireless networks whose semantics are carried over from those of minimum-hop routing, fails to compute effective and efficient routes in a wireless mobile environment. This is mainly attributed to the fact that in a mobile wireless network laid out with no goals other than convenience and basic connectivity, a node can expect to be in radio contact with other nodes for a wide range of distances and signal strengths. In this context, simple shortest-path routing is not appropriate since it does not distinguish between good links and bad links. The implication of this is that in many cases a path with many forwarding hops may have better links and thus be of higher quality than a path with fewer, but worse-in-quality, links.

The following example illustrates the effectiveness of this observation. Consider the following simple network configuration (Fig. 2.7):

Nodes B and C can both hear A; C hears A with an excellent signal-to-noise ratio and B hears A with a poor signal-to-noise ratio, such that

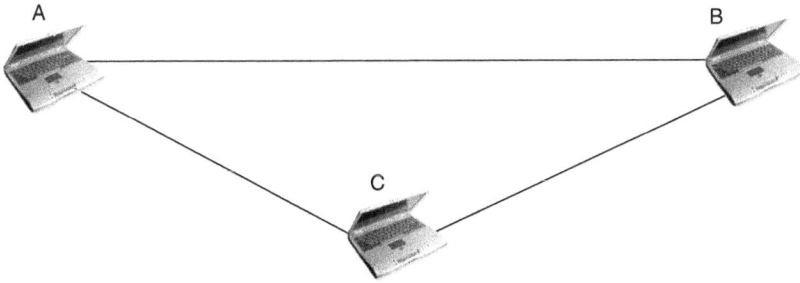

Figure 2.7   Topology update due to a link failure.

a small percentage of packets (less than some threshold) can actually traverse A→B. High loss rates obviously result in poor throughput. In this scenario it can be argued that A→C→B is much more preferable than A→B as seen from the application. Hence, adding one more hop in the example creates a significantly improved route.

Moreover, if network links are of varying quality, the unreliable delivery of protocol updates could lead to overly frequent routing changes and even cause periods of complete disconnection. In the same line of thought, preferring paths with few hops may force a routing protocol to choose long-distance links that may be operating at the edge of their reception ranges. These links will in general be more susceptible to noise and interference and thus be less reliable and more lossy than shorter links.

Besides hop count, several different selection criteria (metrics) exist for selecting routes in packet radio networks, including capacity, per-hop or end-to-end delay, link or path error ratios, route reliability, link stability, and various measures of interference. Given this spectrum of route selection criteria, hop count can be considered instead as being a tie-breaker among routes of equal link qualities. In this regard, wireless routing protocols often allow the selection of paths that are shortest *with respect to a metric*.

Overall, it seems that there is no single metric that improves the network performance in terms of raw network capacity, end-to-end throughput, end-to-end delay, power consumption, reliability, and so on. A key point, however, is that a wireless mobile routing protocol performs better with some metrics; the exact order of improvement of each single metric is dependent on several factors, including user and communication scenarios and application and services requirements.

2.6.2 Route maintenance

As mobiles roam about, network connectivity continuously changes. Mobiles shall adapt their routing accordingly in order to minimize the disruption of the ongoing traffic sessions. *Route maintenance* is the mechanism that detects whether the network topology has changed such that a data path is no longer viable and a route reconstruction is required.

The reachability relation between two nodes may change for various reasons. For example, a node may wander too far out of range, its battery may be depleted, the wireless channel could get lossy (e.g., due to obstructions), or it may suffer a software or hardware failure. For a node to detect broken links a mechanism similar to the beaconing protocols for detecting adjacent neighbors is often used. In this case, during the propagation of data traffic, each forwarding node is responsible for confirming the receipt of each packet to the previous-hop node by a link-layer acknowledgment. If a forwarding node retransmits a packet the maximum number of times and no receipt confirmation is received from its next-hop node, it indicates that the link is broken.

When a node detects a broken route, it sends a *route error* message back to the data source node identifying the link over which the packet could not be forwarded. The sender then removes the broken link from its cache. If the source node has another route to the destination in its route cache, it switches the flow over the new route immediately; otherwise, it may invoke a *route discovery* to find a new route.

Alternatively, on detecting a failure in the primary route, mobiles may actively seek out an alternate route so as to shift quickly to it, with no further notification to the data source. This would minimize data delivery delays as well as potential data disruptions.

2.6.3   Routing protocol categories

Routing has traditionally used the knowledge of instantaneous connectivity of the network with emphasis on the state of network links. This is the so-called *topology-based* approach [MAUVE01]. A different approach, called *location-based (or position-based)* routing, uses information related to the physical position of nodes to help the task of routing [BASAGN98, BOSE99, MAUVE01, CAPKUN01, XUE01, COUTO98, HAAS99, KARP01, KARP00, KO98, KO99, JANNOT00].

An alternative to these approaches, called *power/energy-aware routing*, uses information related to the remaining battery lifetime of mobiles with the aim to produce paths that comprise nodes with a high value of the remaining lifetime as well as to help them adjust their transmission power so as to keep the energy required to complete the routing task to a minimum.

This sequel highlights the specifics of each category, exposing their advantages as well as potential limitations. To develop an intuitive feel of the concepts behind each category, we tabulate a few typical routing techniques per category.

Topology-based routing protocols.   In the topology-based approach, the associated routing protocols can be classified into the following three general categories, based on the timing when the routes are discovered and updated—*proactive* (also called *table-driven), reactive* (also called *on-demand),* and *hybrid.*

Proactive (table-driven) routing.   The proactive approach is similar to the connectionless approach of traditional datagram networks (see Sec. 2.2). In proactive schemes, nodes—based on a periodic update process [ROYER99, McDONA99]—attempt to compute a priori and provide consistent, up-to-date routing information to every other node in the network, regardless of whether the routes are being used for carrying packets.

A routing protocol is then proactive in the sense that nodes calculate all possible paths to all destinations independently of their effective use such that when a packet needs to be forwarded, the route is already known and can be immediately used.

As is the case for wired networks, each node maintains a routing table. Routing tables typically contain a list of addresses for all possible destinations, next-hop nodes, and the number of hops to reach each destination node. The routing table is constructed using either link-state or distance-vector algorithms.

Properties.   The main advantage of proactive routing protocols is that, when an application needs to initiate a data call, routing information is immediately available thus eliminating route acquisition delays. In fact, this can be useful in various cases, as in interactive applications.

Although this approach can ensure high-quality routes in a static topology, as that of a wireline network, it does not scale well to large, highly dynamic networks. In fact, proactive protocols require each node to maintain a large table to store routing information for the entire network. The constant propagation of routing incurs substantial signaling traffic and power consumption. Given that both bandwidth and battery power are scarce resources in mobile devices, pure proactive schemes may not be the appropriate solution for a mobile wireless environment with a large number of nodes [ROYER99, PEARL99]. A more disappointing observation is that the overhead expended to establish and/or maintain a route between a source-destination pair is wasted if the data source never requires a data path.

Also, since proactive schemes rely on periodic broadcasts, they need some time to converge before a route can be used. This convergence time is probably negligible in a static wired network, where the topology is not changing so frequently. In a mobile wireless network, on the other hand, where the topology is expected to be very dynamic, this convergence time will probably mean a lot of dropped packets before a valid route is detected.

In the following section, we review some proactive approaches for routing in MANETs.

The distance source distance-vector (DSDV) routing protocol [PERK94]

Description. In DSDV, every node in the network maintains a routing table in which all the possible destinations within the network as well as the number of hops to reach each destination are recorded. Each route entry is marked with a sequence number; these are assigned by the destination node and are called *destination sequence numbers.* Sequence numbers enable nodes to distinguish stale routes from new ones, thereby avoiding the formation of routing loops. An odd number indicates a distance equal to infinity and is used for the destinations that become unreachable, while even numbers are used by the destination to stamp route updates.

Nodes periodically transmit routing table updates throughout the network in order to maintain table consistency. Route updates contain the address of some node (destination), the number of hops to reach the destination, the destination sequence number, as well as a sequence number that uniquely identifies the update. On receiving a route update for a destination, say D, a node updates the corresponding entry for D in its routing table with the new received information, if and only if the route update has a more recent sequence number. In the event that two updates have the same sequence number, the route with the smaller metric (number of hops) is used in order to optimize the path (shortest path). A route entry is deleted from the table if no updates are received for a given number of update intervals.

At some node A, the sequence number is increased for destination D when A detects that the route to D has broken. If so, A will advertise the route to D with an infinite hop count and a sequence number that is larger than the present one.

DSDV basically is a distance-vector protocol with small adjustments to make it better suited for ad hoc networks. These adjustments primarily address triggered updates, which are sent when the topology changes in the time between the scheduled updates. DSDV employs two strategies for determining when to send triggered updates. In the first strategy, DSDV-SQ, a node sends a triggered update each time it receives a new sequence number for some destination node. The second scheme requires that triggered updates are sent only when a new metric is received for a destination. In the latter case, however, link breakages are not detected as quickly as in DSDV-SQ.

To reduce the potentially large amount of network traffic caused by the periodic update mechanism, route updates can employ two possible types of packets—full and incremental dump. The *full dump* carries all available routing information. During periods of occasional movement, these packets are transmitted infrequently. The *incremental dump* packets carry only the information that has changed since the last full dump.

To summarize, the key elements of DSDV are:

1. An aging mechanism based on monotonically increasing sequence numbers, which indicates the freshness of the route and which is used to avoid routing loops and the count-to-infinity problem

2. The use of full route updates, sent periodically every update interval, or incremental route updates sent on topological changes

3. The delay of route updates for routes that are likely to be unstable, i.e., those for which a new update is on the way toward a node

Properties.   The main source of inefficiency in DSDV is the requirement for periodic update transmissions, regardless of the number of changes, as well as active data sources in the network topology.

The wireless routing protocol (WRP) [MURHY96]

Description.   WRP belongs to the general class of path-finding algorithms (PFA) [HUMBL91, MURHY96, CHENG89], that calculate shortest-path routes using the length and the second-to-last hop (predecessor) of the path to each destination.

In WRP, each node in the network is responsible for maintaining four tables: (a) distance table, (b) routing table, (c) link-cost table, and (d) *message retransmission list* (MRL) table. The distance table of node i is a matrix containing, for each destination j and each neighbor of i (say k), the distance to j and the predecessor node reported by k.

The routing table of a node i is enriched with the predecessor and the successor of the path to the destination node, as well as a tag that specifies whether the entry corresponds to a simple path, a loop, or a destination that has not been tagged. With this information at hand, WRP is able to avoid count-to-infinity effects and to highly reduce routing loops; the latter is envisioned by forcing each node to perform consistency checks.

Mobiles inform each other of link changes through the use of update messages. An update message is sent only between neighboring nodes and contains a list of updates (the destination, the distance to the destination, and the predecessor of the destination), as well as a list of responses indicating which mobiles should acknowledge (ACK) the update. Mobiles send update messages after processing updates from neighbors or detecting a change in a link to a neighbor.

In the event of the loss of a link between two nodes, the nodes send update messages to their neighbors. The neighbors then update their distance table entries and check for new possible paths through other nodes. Any new paths are relayed back to the original nodes so that they can update their tables accordingly. Nodes therefore learn about the existence of their neighbors from the receipt of acknowledgments and other messages.

In case no route update is received within a suitable time interval, WRP relies on the transmission of HELLO messages to detect the connectivity with neighbors. When a mobile receives a HELLO message from a new node, that new node is added to the mobile's routing table. The receiver of the HELLO message then sends the new node a copy of its routing table information.

Finally, the MRL is used to store the identification of the nodes that have not acknowledged a route update. Each entry of the MRL contains the sequence number of the update message, a retransmission counter, an acknowledgment-required flag vector with one entry per neighbor, and a list of updates sent in the update message. The MRL records which updates in an update message need to be retransmitted and which neighbors should acknowledge the retransmission [MURHY96].

Properties.   Even though it belongs to the class of pathfinding algorithms, WRP has an advantage over the classical pathfinding algorithms as it avoids the formation of temporary routing loops. This is achieved through the verification of predecessor information.

The main drawback of WRP seems to be the requirement for nodes to maintain four routing tables, as well as the use of updates and HELLO messages. The former can lead to substantial memory requirements, especially when the number of nodes in the network is large. The latter contributes to a significant amount of network bandwidth consumption.

The optimized link-state routing (OLSR) protocol

Description.   OLSR is an optimization over the classical link-state protocol tailored for operation in mobile wireless networks [OSLR-ID, QAYYUM01]. The key idea behind OLSR is to reduce duplicate broadcast packets in the same region (see section "Proactive (table-driven) routing"). This is achieved with the use of the so-called *multipoint relay* (MPR) nodes.

Each node i selects a minimal (or near minimal) set of multipoint relay nodes, denoted as MPR(i), from among its one-hop neighbors. The nodes in MPR(i) have the following property—every node in the symmetric two-hop neighborhood of i must have a symmetric link toward MPR(i). In other words, the union of the one-hop neighbor set of MPR(i) contains

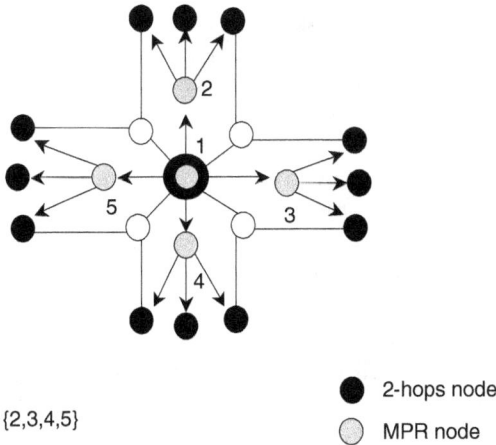

MPR(*i*) = {2,3,4,5}

● 2-hops node

○ MPR node

Figure 2.8   An example of flooding using MPR nodes.

the whole 2-hop neighbor set (see Fig. 2.8). The multipoint-relay principle is thus ruled as follows: a node selects, among its neighbors, a set of nodes called *multipoint-relays* such that any node in the 2-hop neighborhood is reachable through at least one multipoint relay.

The goal behind the multipoint-relay principle is to achieve efficient flooding. This is done as follows: When a node *i* wants to flood a message, it sends the message only to the nodes in MPR(*i*), which in turn send the message to their MPR nodes and so on. A node retransmits a message if it has not received the message before, and the node is selected as multipoint relay by the node from which the message is received.

Furthermore, the *multipoint relay selector* set of a node comprises the set of neighbors that have selected it as MPR. Each node periodically floods its MPR selector set with a special type of control message called a *topology control* (TC) message. Using TC messages, a node announces its reachability relation to the nodes of its MPR selector set. Also, if a node has an empty MPR selector set, it may not generate any TC message. However, when its MPR selector set becomes empty, it should still send empty TC messages for a period of time in order to invalidate the previous TC messages. Information gained from TC messages is used to build the network topology and the routing tables.

To increase the reaction to topology changes when a change in the MPR selector is detected, the time interval between two consecutive TC message transmissions shall be decreased to a minimum.

Reactive (on-demand) routing.   The philosophy behind on-demand routing protocols is to evaluate the network on as-needed basis and create routes only when there is a need for carrying data traffic. If no data traffic is generated, then the routing activity should be totally absent. Based on the assumption that not all the routes are used at the same time, the need for a route thus triggers a route search in a reactive routing strategy.

Reactive protocols are characterized by the elimination of the conventional routing tables at nodes, and consequently the need of routing table updates to track changes in the network topology. As a result, an on-demand process for discovering routes is a prerequisite; a path discovery is triggered asynchronously when there is a need for data packet and no path to the intended node is known.

The discovery procedure is often based on a query-reply cycle—the data source node floods the network with a query packet to discover a route to the data destination. Assuming no network disconnections, the query will eventually reach the destination. Upon receiving the query, the destination sends a reply back to the source. Since multiple copies of the same query may arrive at the destination via alternate paths, the destination may send more than one reply back to the source, each producing a different route. The source then selects the optimum route, using its own route selection criteria (see Sec. 2.3). The process is completed once a route is found or all possible route permutations have been examined. Once established, the data source uses the route to send its data packets to the destination.

During the data forwarding phase, routing information is maintained by a maintenance procedure until either the destination becomes inaccessible along every path from the source or the route is no longer desired [ROYER99]. At any time during the data exchange, should the route maintenance indicate that the route is broken, a new route discovery cycle is triggered to set up a new route.

When the communication ends, the source does not attempt to further maintain the connectivity to the destination. Hence, network resources are not expended to maintain routes that are not actively used for data traffic.

Properties. On-demand routing strategies create and maintain routes between a pair of source-destination only when necessary. Therefore, in contrast to the proactive approach, in reactive protocols the control overhead as well as the routing table storage is drastically reduced when traffic is sparse [BROCH98, DAS98, DAS00, DAS00a, JACQUET00, JOHANS99, MALTZ99, JOHNS96]. On-demand routing does, thus, scale well to large populations as it does not maintain a permanent routing entry to each destination.

However, similar to connection-oriented communications, a route is not available when needed but on completion of the route discovery phase. This may introduce an initial route setup latency. This latency

TABLE 2.1   Overall Comparisons of On-Demand Versus Table-Driven Based Routing Protocols

Parameters	On-demand	Table-driven
Availability of routing information	Available when needed	Always available regardless of need
Routing philosophy	Flat	Mostly flat except for CSGR
Periodic route mobility	Not required	Yes
Coping with mobility	Using localized route discovery and in ABR and SSR	Inform other nodes to achieve consistent routing table
Signaling traffic generated	Grows with increasing mobility of active routes (as in ABR)	Greater than that of on-demand routing
Quality of service support	Few can support QoS	Mainly shortest path as QoS metric

may in fact be detrimental to certain applications such as interactive applications (e.g., distributed database queries). Moreover, the quality of the data path (e.g., bandwidth, delay, and the like) is not known prior to call setup. It can be discovered only while setting up the path, and must be monitored by all intermediate nodes during the session, thus paying the related latency and overhead penalty. Such a priori knowledge is, however, desirable in multimedia applications for call acceptance control decisions as well as bandwidth negotiations [IWATA99].

Because of the long route setup delays as well as the lack of path quality information prior to call setup, pure reactive routing protocols may not be applicable to real-time communications [GIORD00].

Table 2.1 lists some of the basic differences between the two classes of algorithms.

In the following section, we review some reactive protocols for routing in MANETs.

The dynamic source routing (DSR) protocol [BROCH98, JOHNS96]

Description. DSR is based on the concept of source routing [PETERS00]. In source routing each packet carries in its header the complete ordered list of nodes the packet should pass through the network. This is done by having each node maintain a cache with source routes to destinations.

Route discovery–route selection. Route discovery is based on flooding the network with a route request (RREQ) packet. An RREQ message includes the following fields: the sender address, the target address, a unique number to identify the request, and a route record listing the addresses of each intermediate node through which the RREQ is forwarded [BROCH98]. On receiving an RREQ control packet, an intermediate node can:

1. Reply to the query source node (S) with a route reply (RREP), if a valid path to the destination is stored in its cache. To return the route reply, the responding node must have a route to the initiator in its route cache. If so, it may use that route. However this assumes symmetrical links. If symmetric links are supported, the node may reverse the route that is carried in the RREQ in route record. If symmetric links are not supported, the node shall initiate its own route discovery to S, and piggyback the route reply on a new route request.

2. Discard the packet, if the same RREQ is already received. To limit the number of route requests, a node only forwards the route request if the same request has not been forwarded again. To examine this case, a node simply needs to check whether its address already appears in the route record carried in the RREQ.

3. Append its own ID into the route record and relay the packet to its neighbors. Figure 2.9 illustrates the formation of the route record as the route request propagates through the network.

On receiving the RREQ packet, the destination replies to the originator with an RREP packet. An RREP packet is routed back in a similar way as in case (1). Once a route is found, it is stored in the cache with a time stamp and the route maintenance phase begins.

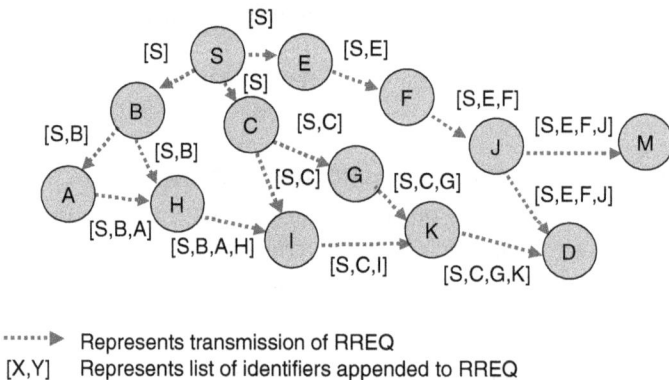

┈┈┈▶ Represents transmission of RREQ
[X,Y]   Represents list of identifiers appended to RREQ

Figure 2.9 An example of a routing propagation request in DSR.

Route maintenance. When an intermediate node detects that the link to its next-hop node toward the destination is broken, it immediately removes this link from its route cache and returns a route error control message to the data source node (S). The source S then triggers a new route discovery. When a node receives a route error packet, the hop (link) in error is removed from its route cache, and all routes containing this hop are truncated at that point.

Properties. DSR uses the key advantage of source routing to learn routes by scanning the routing information in the header of received packets. For example, a route from A to D through B and C means that A not only learns the route to D but also to B and C. Source routing also helps in that nodes B and C learn the routes to A, C, and D, and to A, B, and D, respectively. This form of active learning can be very efficient and reduce network overhead. This is so because intermediate nodes do not need to maintain up-to-date routing information to route packets as the packets themselves already contain all the routing information.

Additionally, DSR allows nodes to keep multiple routes to a destination in their cache. Hence, when a link on a route is broken, the source node can check its cache for another valid route. If such a route is found, route reconstruction does not need to be reinvoked thus saving network bandwidth as well as reducing route acquisition latency.

DSR also supports unidirectional links with the use of piggybacking source routes on new request. This can be critical, particularly when a unidirectional link is the only link available to reach a destination, and thus failure to use it may result in repeated unsuccessful route discoveries.

On the other hand, the need to carry full routing information (entire route) in each packet causes a significant increase in network control overhead. This overhead grows in size as the data path increases in length.

DSR-specific optimization techniques. Several improvements to the base mechanism of DSR protocol are proposed in the literature:

1. *Nonpropagating route requests.* The goal of this optimization technique is to limit the number of route discoveries. Specifically, DSR protocol limits the number of route discoveries to two attempts. In the initial attempt, the source broadcasts a nonpropagating RREQ with the maximum propagation limit (hop limit) set to zero, prohibiting its neighbors from rebroadcasting it. At the cost of a single broadcast packet, this mechanism allows a node to query the route caches of all its neighbors. In case the first attempt is not successful in finding a route to destination node, the source node broadcasts a "propagating"

RREQ with no hop limit, which essentially floods the network. It should be relatively straightforward to extend this dual-phase search to an expanding ring search by allowing the hop limit to increase in incremental steps.

2. *Rate limiting of repeated route discovery.*   To avoid the possible generation of an overwhelming amount of control traffic due to repeated unsuccessful route discoveries to some (temporarily) unreachable destinations, DSR uses an exponential back-off scheme to limit the rate at which the source node may repeat a route discovery to the same destination.

3. *Promiscuous mode operation.*   A node can operate its radio interface in promiscuous mode by disabling the interface address filtering and thus allowing the network protocol to receive all packets that the interface overhears. These packets are scanned for useful source routes or route error messages, and then discarded. Hence, less discovery attempts are conducted since, through this passive updating mechanism, nodes update their routing information more frequently.

   On the other hand, the promiscuous learning of routes may incur high processing overhead as well as require large memory storage. In addition, running the interfaces in promiscuous mode is a serious security issue, since the address filtering of the interface is turned off and all packets are scanned for information. A potential intruder could listen to all packets and scan them for useful information such as passwords and credit card numbers. Applications have to provide security by encrypting their data packets before transmission. The routing protocols are prime targets for impersonation attacks and therefore must also be encrypted.

4. *Preventing route reply storm with random delays.*   Due to the nature of broadcast transmission, many nodes around the broadcasting node may receive the RREQ and send RREPs simultaneously. This may result in what is dubbed an RREP "storm," in which a great number of nodes attempt to send RREPs from their caches simultaneously, causing local congestion and excessive packet collisions. Having some nodes delay sending their RREPs may mitigate this problem. The delay time is specified to be $d = Hx(h - 1 + r)$, where h is the number of hops of the returned route, r is a random number between 0 and 1, and H is a small constant delay to be introduced per hop.

5. *Salvaging.*   Instead of sending a route error message back to the source, an intermediate node can use a path to the destination from its own cache when the link to the next hop is broken.

Ad hoc on-demand distance-vector routing protocol (AODV) [PERK99, PERK00, PERK02]

Description.   An AODV protocol builds on the DSDV algorithm described in section "DSDV routing protocol." AODV is essentially a reactive protocol, while DSDV is proactive. Whereas in DSDV nodes maintain a complete list of routes [PERK99], in AODV nodes create routes on an on-demand basis. AODV thus does not require nodes to maintain routes to destinations that are not actively used in communications. As in DSDV, the use of destination sequence numbers guarantees that a route is "fresh."

The AODV routing protocol is based on the concept of *next-hop* routing model. Each host should keep a routing table that indicates the next host to be used as the immediate relay to reach a destination. Each node, on receiving a data packet, only needs to check its routing table and if a valid route for the destination of the data exists, it forwards the packet to the next node as this is indicated in its routing cache entry associated with the packet's destination IP address.

Route discovery–route selection.   During the route discovery phase, the data source node floods the network with an RREQ packet. After broadcasting an RREQ, the node waits for a route reply. If the reply is not received within a certain time, the node may rebroadcast the RREQ or assume that there is no route to the destination. To ensure that all routes are loop-free and up-to-date, AODV uses destination sequence numbers. Similar to DSDV, each node maintains its own sequence number, as well as a broadcast ID. The broadcast ID is incremented for every RREQ the node initiates, which together with the node's IP address, uniquely identifies an RREQ. Along with its own sequence number and the broadcast ID, the source node includes in the RREQ the most recent sequence number it maintains for the destination.

When a node receives an RREQ message, it checks whether it has already seen the RREQ message by noting the source address/broadcast ID pair. If so, it discards the message. Otherwise, it sets up a reverse path pointing toward the data source. The reverse path setup is achieved as follows: intermediate nodes record in their route tables the address of the neighbor from which the first copy of the broadcast packet is received, thereby establishing a reverse path to the query initiator. Once the RREQ reaches the destination node, a route reply is sent back to the source along the reverse path.

In addition, intermediate nodes can reply to the RREQ only if they have a route to the destination, whose corresponding destination sequence number is greater than or equal to that carried in the RREQ. Figure 2.10 shows an example of a route discovery using the AODV routing protocol.

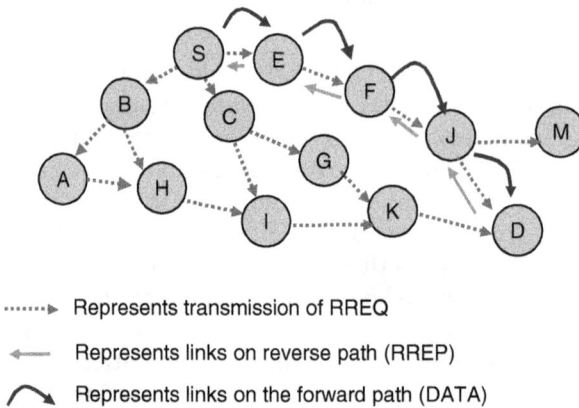

······▶   Represents transmission of RREQ

◀━━━   Represents links on reverse path (RREP)

⌒⬎   Represents links on the forward path (DATA)

Figure 2.10   An example of propagation of RREQ in AODV and reverse path setup.

As the RREP is routed back along the reverse path, nodes along this path set up forward route entries in their route tables to destinations that point to the node from which the RREP was received. These forward route entries indicate the active forward (data) route. The routing state created in each node along the path from the source to the destination is the hop-by-hop state; that is, each node remembers only the next hop and not the entire route, as is the case in source routing. Data source node (S) can begin data packet transmissions as soon as the first RREP is received.

Associated with each route entry is a route timer. A forward path is deleted if not used within a given route expiration time interval. Each time the route is used, the expiration time interval is reset.

Route maintenance.   In order to maintain routes and neighborhood information (local connectivity), AODV requires that each node periodically transmit a locally (one-hop) broadcast message, called *HELLO message.* The default rate of HELLO transmissions is set to one per second. Each time a node receives a HELLO message from a given neighbor, it updates the information associated with the neighbor in its route table. The failure to receive a number of consecutive HELLO messages from a neighbor is used as an indication that the node has moved away; the link to this neighbor is then marked as broken. Alternatively, the AODV specification proposes that a node may use physical layer or link layer methods to detect link breakages with its neighbors [PERK02].

When a link is marked as broken, any active upstream node that uses that link is notified via an unsolicited route reply. For this purpose, AODV uses an active neighbor list to keep track of the neighbors that are using a particular route. All active upstream neighbors invalidate the routes using the broken link from their routing caches. These nodes in turn propagate the link failure notification to their upstream neighbors, and so on until the affected data source nodes are reached. If a route is still desired, the data source node(s) may then choose to reinitiate a route discovery for that destination.

Properties.   The advantage of AODV, compared to classical distance-vector and link-state routing protocols, is the significant reduction of routing messages. The protocol achieves this by using a reactive approach to discovering and maintaining routes.

However, AODV uses HELLO messages at the IP-level to constantly update nodes' knowledge about their local (one-hop) topology. HELLO messages add a significant overhead to the protocol.

In addition, AODV does not support unidirectional links. A node, on receiving an RREQ, sets up a reverse route to the source by using the node that forwarded the RREQ as next-hop. This means that the RREP, in most cases, is unicasted back the same way as the RREQ used. Unidirectional link support would make it possible to use all links and not only the bidirectional links.

AODV-specific optimization techniques: expanding ring search. Expanding ring search is a technique that allows for suppressing the propagation of RREQs before a networkwide flooding is triggered. In an expanding ring search, several route discovery attempts may be conducted. The data source initially conducts a route discovery that searches only a region of limited hops from itself. If a route to the destination is not found after some time-out, the source attempts another route discovery, but this time with a greater searching scope than the preceding attempt. This is achieved by increasing the *time-to-live* (TTL) value of the RREQ. This process continues until the TTL reaches a maximum threshold, after which the RREQ is flooded throughout the network.

However, this technique is effective only if the destination can be found within the initial attempts. If not, the route discovery overhead can be even higher than with flooding on the first attempt, in addition to increasing route discovery latency due to multiple timeouts.

Associativity-based routing (ABR) protocol [TOH96]

Description.   In ABR, the key selection criterion is the longevity of routes, instead of the route length; this metric is known as the *degree of association stability*. Association stability is related to the connection stability of one node with respect to another, over time and space. A high degree of association stability may indicate a low state of node mobility, while a low degree may indicate a high state of node mobility.

A fundamental objective of ABR is to derive longer-lived routes. A longer-lived route is preferred to a shorter-lived one, even if the length of the latter is less than that of the former. Using this approach, the protocol seeks to calculate more stable routes, resulting in fewer reconstructions.

The technique proposed to capture associativity is as follows: each node periodically broadcasts a local beacon to signify its existence. Nodes count the beacons received from their neighbors to update their *associativity ticks*. For each beacon received, the associativity tick of the current node with respect to the beaconing node is incremented. Associativity ticks are reset if the beacon signal is not received for a certain period of time.

Route discovery–route selection. A data source node acquires a new route by broadcasting a route request control packet. As the packet is propagated, intermediate nodes append their own ID and route quality values. The latter are the associativity ticks.

The destination node waits for a suitable period of time, after receiving the first route request packet, so as to receive other request packets forwarded along other paths. The destination is then able to select the best route by examining the associativity ticks along each of the paths. Route selection is then primarily based on the aggregated associativity ticks of nodes along the path. A route in ABR is then *selected by the destination node.*

Once a route is selected, the destination sends a route reply control packet through the selected route. As the route reply is propagated backward to the data source, intermediate nodes involved in its retransmission are able to set up a route entry for the destination, thus activating a forward path to the destination in a hop-by-hop fashion.

Route maintenance. Route maintenance is invoked when the association stability relationship is violated. If so, a route error message is sent to the source and a new discovery phase is triggered. An alternative to reporting the failure is to apply a local discovery close to the vicinity of the failure and try to repair the corrupted portion of the data subpaths with no notification to the data sources.

When a route is no longer desired, the data source node initiates a *route delete* (RD) broadcast so that all nodes along the route update their routing tables. As an alternative to such a costly deletion procedure, ABR also proposes a less expensive one, called the *soft state approach*, in which the route table entries are deleted upon time-out.

Properties. The major benefit of ABR relies on the "longevity" concept of routes. Stable routes with a long life result in fewer route reconstructions.

ABR, however, relies on periodic beaconing. The beaconing interval should be short enough so as to accurately reflect the spatial, temporal, and connectivity state of mobiles. This beaconing requirement results in additional network overhead and power consumption [TOH96].

Signal stability routing (SSR) protocol [CORS95, DUBE97]

Description.   The SSR protocol is a logical descendant of ABR. SSR is broadly divided into two cooperative protocols—the *dynamic routing protocol* (DRP) and the *static routing protocol* (SRP).

DRP is responsible for maintaining a *signal stability table* (SST) and a *routing table* (RT). The SST records the signal strength of neighboring nodes, which is obtained through periodic link-layer beacons from neighboring nodes. All transmissions are received by and processed in the DRP. After updating all appropriate table entries, DRP passes a received packet to the SRP.

If the node is the intended destination of the packet, SRP passes the packet up to the protocol stack for further processing, or otherwise a routing table looking up is performed to forward the packet to its next hop.

Route discovery–route selection.   A node initiates a route discovery by broadcasting a route request packet throughout the network. A node that receives an RREQ to relay, forwards the RREQ only if it is received over a strong channel and has not been previously processed; otherwise, the RREQ is dropped. This practice assures that the paths of strongest signal stability are always chosen. As in ABR, while the paths selected by this algorithm are not necessarily shortest in hop count, they do tend to be more stable and longer-lived, resulting in fewer route reconstructions. If the route discovery initiator receives no route reply within a specific time period, it initiates a new route discovery. This time, however, the query originator indicates in the new RREQ (PREF field) that weak channels are also acceptable so that the RREQ can propagate through links of lower quality as these may be the only ones over which the packet can be propagated.

The destination chooses the first arriving RREQ to send back an RREP. The DRP then reverses the selected route and sends an RREP message back to the initiator. The DRP of the nodes along the path updates their routing tables accordingly.

Route maintenance.   When a failed link is detected, the intermediate nodes send a route error message back to the source indicating the link in error. The source then initiates a new route discovery process to find a new path to the destination. The source, by virtue of source routing, also sends an erase message to notify all nodes that share the broken link.

Relative-distance microdiscovery ad hoc routing protocol [AGGE02, RDMAR-ID]

Description.   To ensure network resource efficiency, changes in the network topology should have local effect only. In this regard, creation of a new link at one end of the network is an important local event but, most probably, not a significant piece of information at the other end of the network. In many reactive routing protocols, the route discovery process is based on distributing topological changes on a global scale. Although networkwide propagation of topological changes is the simplest approach to updating nodes' view of network topology, this is far from efficient in a wireless mobile environment (see Sec. 2.4).

In this line of thought, the relative distance microdiscovery ad hoc routing (RDMAR) protocol proposes a distributed mechanism for discovering and repairing routes within a local region of the network. To accomplish this, a simple distributed route searching algorithm, called *relative distance microdiscovery* (RDM), is used. RDM uses a stochastic model for estimating the relative distance between two nodes as the basis for routing decisions. Relative distance (RD) between two nodes is their hop-normalized distance. Knowledge of this RD is leveraged by the RDMAR protocol to improve the efficiency of a reactive route discovery/repair mechanism.

As demonstrated through simulations [AGGE02], the performance of the RDM model is very close to that of an optimal route searching policy while under a variety of conditions (see Ref. [AGGE02]) the query localization protocol is able to reduce the routing overhead significantly, often in the neighborhood of 48 to 50 percent of a flooding-based scheme.

Route discovery–route selection: Illustration of the relative distance microdiscovery (RDM) procedure. Let us say that a node *src* initiates the route discovery phase to another node *dst* (see Fig. 2.11). Here, the source of the route discovery (*src*) has two options—either to flood the network with a query in which case the query packets are broadcast into the entire network area or, instead, to limit the discovery in a smaller region of the network if some kind of location prediction model for *dst* can be established. The former case is straightforward. In the latter case, *src* refers to its routing table in order to retrieve information on its previous relative distance (RD field) with *dst* and the time elapsed (TLU field) since *src* last received routing information for *dst*. Let us denote this time by t_{motion}. Based on this information and assuming a moderate velocity, Avg_Velocity, and a moderate transmission range, Avg_Comm_Range, node *src* is able to estimate its new RD to *dst* (New_RD$_{\text{SRC,DST}}$ in Fig. 2.11). Thereafter, node *src* can confine the propagation of route queries by using the hop-normalized value of the newly estimated RD in the TTL field of the header of the route request packet. This is the *relative-distance microdiscovery* (RDM) procedure.

One can easily observe that the New_RD$_{\text{SRC,DST}}$ is not a straightforward numerical operation by simply adding or subtracting the new distance offset to the previous relative distance (RD$_{\text{SRC,DST}}$). The calculation of

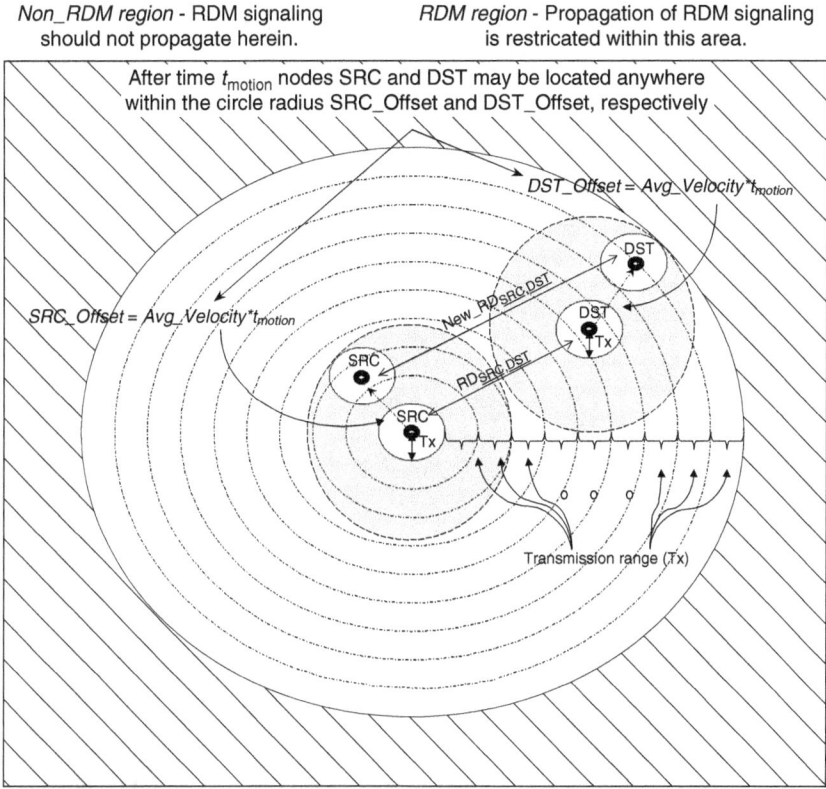

Figure 2.11   Illustration of the relative distance micro-discovery (RDM) procedure.

New_RD$_{SRC,DST}$ rather depends on the moving directions of the two terminals during t_{motion}. To estimate the new RD between *src* and *dst*, a stochastic model is used to derive expressions for the estimation of the relative position of nodes as a function of time. The stochastic method for the modeling and analysis of the RDM algorithm is presented in App. D. In this model an iterative algorithm, called the *relative distance estimation* (RDE) algorithm, is introduced and is used to determine the optimal relative distance between two nodes.

The logical relationship between the RDE algorithm, the routing mechanism, and other network-layer entities is depicted in Fig. 2.12. The RDMAR algorithm resides logically between the routing-layer and upper layers of the OSI system or could operate together with the *Internet MANET encapsulation protocol* (IMEP) [IMEP-ID], for example, for the encapsulation of the RDMAR-specific control messages.

Clearly, the ability of the protocol to terminate a query thread relies heavily on the efficiency of the RDE algorithm in estimating correct RDs.

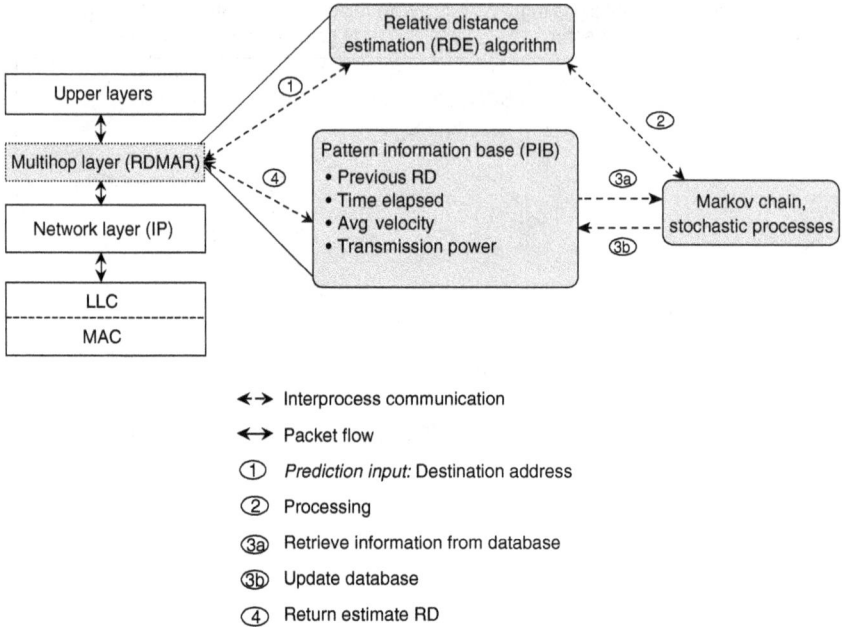

Figure 2.12 Logical relationships among RDMAR network-layer entities.

The effects of the parameter (RD) are tightly coupled to the hit (or, else, the success) ratio of the route discovery algorithm, making it difficult to select optimal values. Large values for RD seem desirable as they increase the path availability, and hence the hit ratio, but they also increase the routing overhead and computational requirements of route maintenance and discovery. Small values of RD, on the other hand, imply a considerable reduction in the routing overhead for the discovery of new paths with an inherent small hit ratio. Consequently, the accuracy of RD calculations trades off protocol efficiency with route optimality.

A concise description of the methodology of the RDE algorithm as well as a detailed performance evaluation of the RDMAR protocol is reported in Ref. [AGGE02]. For the sake of completeness, we illustrate some results of the RDE performance.

RDE performance.   The minimum searching cost is achieved by adjusting the TTL, or else the RD, value of a route request to the shortest value that is sufficient to locate a destination terminal. The optimal RD value RD_{ideal} is defined as the smallest possible RD for the successful route discovery. Let us address the route-searching scheme that is capable of estimating the minimum RD as the ideal route discovery scheme (RD_{ideal}).

In the event of RDM failure, two heuristics can be followed for discovering a route—either use flooding or use RDM again. In the latter case, the new RD is set equal to the previous one increased by some

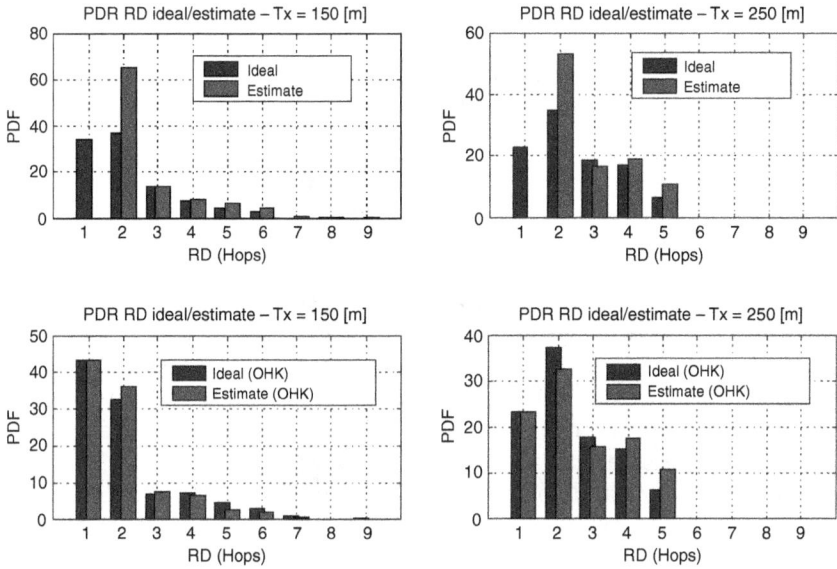

Figure 2.13   Performance of RDE algorithm for low and high Tx values.
Top - Basic RDMAR, Bottom – RDMAR with one-hop knowledge (OHK)
Nodes = 30, Tx = Variable, Mobility pattern = Random walk

factor k (RDM_INC). A route discovery miss may eventually result in much higher network traffic than when using flooding alone. This is evident since the routing overhead on the unsuccessful RDM attempt adds up to the overhead from subsequent searching attempts.

The performance of the RDE algorithm is examined in Fig. 2.13. As figures illustrate, for lower transmission values (T_x) RD estimations are distributed through the entire RD spectrum, whereas for higher T_x values, RD estimations result in smaller values. This is evident as higher T_x values result in lower searching scopes, called *virtual wireless rings* in RDMAR terminology (VWRs—see App. D for details) and thus the network is partitioned into a smaller number of VWRs.

An interesting observation on the results illustrated in Fig. 2.13 (upper plots) is that RDE fails to make accurate estimations when the RD_{ideal} of two terminals is one-hop distance (i.e., nodes are neighbors). To further investigate this effect, neighbor detection capability (one-hop knowledge—OHK) is added at the data link layer/MAC layer and a set of experiments is run under the same assumptions to examine the level of improvement on RDE estimations. The results are illustrated in Fig. 2.13 (lower plots). The significant improvements in RDE estimations are clear as the distribution of the estimated RD values is very close to the optimal, RD_{ideal}.

Packet forwarding and route repair. On receiving a data packet an intermediate node *i* first processes the routing header and then attempts to forward the packet to the next hop. If forwarding is successful, node *i* performs two tasks— (1) a small network layer control message called Bidir_Link_Req is sent to the previous node, and (2) the IP address of the previous node is recorded in the "Dependent Neighbor" list indexed from the destination node of the packet. The recipient of a Bidir_ Link_Req message, in turn, replies with a Bidir_Link_Ack message. The significance of the Bidir_Link_Req/Ack exchange lies in the fact that a data forwarding node in RDMAR determines whether a bidirectional link exists with the node from which the packet is received, such that routing information will exist to send the future acknowledgment back to the source. Therefore, RDMAR includes a possibility for asymmetric operation during the data transfer. The importance of the Dependent Neighbor is illustrated as follows.

If, on the other hand, forwarding is not successful (i.e., the link to the next-hop node is broken), node *i* initiates the route repair phase. During the route repair phase, node *i* exploits the spatial relationships between itself and the source and destination nodes of the data packet. Depending on its relative position from the source and destination nodes of the packet, *i* may optionally initiate an RDM procedure (local repair) or notify instead the source of the active call. To put this into perspective, if, at the time when the failure occurs, *i* determines that its RD to the destination of the data packet is smaller than to the source of the packet, *i* proceeds and initiates the RDM procedure;[1] otherwise, *i* proceeds and notifies the data source about the failure to deliver the packet through this path. It is evident that *i* is able to perform the RD comparison only if valid routing information for both the source and destination nodes of the data stream exists in its routing table; otherwise, *i* proceeds and notifies the source about the failure to deliver the packet through this path. This phase is called the *failure notification* (FN) phase.

In response to an RDM_RR request, a route reply shall be received fast so that a quick rerouting is performed in a manner seamless to the data source. This also prevents the data source from retransmitting the packets that have not been acknowledged and whose acknowledgment timer has expired. If, however, a route reply is not received within some time-out, the FN phase or a new RDM_RR is initiated instead with a higher RD value.

Local repair is useful especially in large networks where routes could be long and thus more susceptible to link failures. By performing route

---

[1]To distinguish this RDM phase from the one triggered from a source node, let us designate this as RDM_RR.

rediscovery from the *point of failure* rather than from the data source, the amount of control traffic generated and the route repair latency could be significantly reduced, especially if the link failure occurs close to the destination. One downside of this technique is that the new route calculated from the RDM_RR may no longer be the shortest route available to the destination and thus may degrade the performance of ongoing data sessions in terms of packet transfer and delivery rate. In addition, if the initial local repair fails, the source has to perform a new route discovery. In this case, the route reconstruction delay and overhead add to the previous searching attempts, thus potentially rendering the overall searching effort even more expensive than a flooding-based approach.

Route maintenance.   If the FN phase is triggered, node i generates and unicasts an FN message to all nodes listed in the Dependent Neighbor list. The importance of this mechanism is that all data source nodes that currently use the failed route are notified about the failure. When the FN message is returned to the data sender, the node first examines whether new routing information to the destination exist. If so, the node immediately proceeds and sends the pending data packet(s) over the new route. However if no route to the destination is available, route discovery is triggered.

Temporally-ordered routing algorithm routing (TORA) protocol [PARK97]

Description.   TORA belongs to a general family of "link reversal" algorithms [CORS95, PARK97]. The key design concept of TORA is the localization of control messages to a small set of nodes near the occurrence of a topological change. To accomplish this, nodes need to maintain routing information about their adjacent (one-hop) nodes.

During the route creation and maintenance phases, nodes use a "height" metric to establish a *directed acyclic graph* (DAG) rooted at the destination. The DAG is obtained by assigning a logical direction to the links, on the basis of a height or reference level assigned to nodes. If (i, j) is a direct link of the DAG, i is called the upstream node and j the downstream node. Thereafter, links are assigned a direction (upstream or downstream) based on the relative height metric of neighboring nodes, as shown in Fig. 2.14. The DAG has the following property: there is only one sink node (the destination), while all other nodes have at least one outgoing link.

TORA provides only the routing mechanism and is layered on top of the IMEP, which allows for other functions such as link status sensing, control message aggregation and encapsulation, network-layer address resolution, MANET router identification, interface identification, and addressing.

To reduce overhead, the IMEP attempts to aggregate TORA and IMEP control messages (which the IMEP refers to as objects) into a single

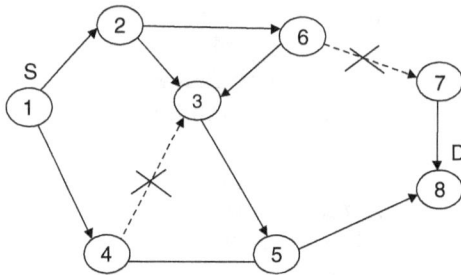

Figure 2.14 An example of topological changes that do not require maintenance.

packet before transmission (as an object block). Each block carries a sequence number and a response list of the other nodes from which an acknowledgment is expected; only these nodes acknowledge the block on receiving it. In addition, the IMEP retransmits each block and continues to do so, if needed, for some maximum total period, after which time, the link to each unacknowledged node is declared "down." For link status sensing, nodes periodically transmit a BEACON packet, which is answered by every node that hears it.

Route discovery–route selection.   TORA associates a height with each node in the network. All messages in the network flow downstream, from a node with higher height to a node with lower height. Routes are discovered using *query* (QRY) and *update* (UPD) packets. When a node with no downstream links needs a route to a destination, it broadcasts a QRY packet. This QRY packet propagates through the network until it reaches a node that has a route, or the destination itself. Such a node then broadcasts a UPD packet that contains the node height. Every node receiving this UPD packet sets its own height to a height greater than that specified in the UPD message. The node then broadcasts its own UPD packet. This results in a number of directed links from the originator of the QRY packet to the destination. This process can result in multiple routes.

Route maintenance.   If the DAG route to some destination (D) is broken, route maintenance is necessary to reestablish a new DAG rooted at D. Route maintenance is based on a finite sequence of link reversal operations. A key feature of TORA is that many topological changes may trigger no reaction at all. In fact, if one of the outgoing links of a node break, but the node has at least one other downstream node, the destination can still be reachable through another path, and thus no repairing activities are required (see Fig. 2.15). On the other hand, when a node detects that it has no downstream nodes, it adjusts its height so that it is a local maximum with respect to its neighbors' (new reference level). The new reference level is propagated into the network (UPDATE message), causing partial link reversal for those nodes,

(a)

(b)

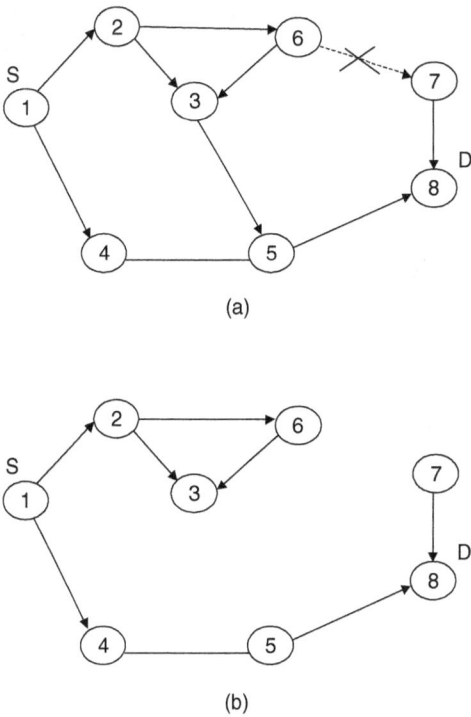

Figure 2.15 (a) An example of topological change that requires link reversal operations; (b) DAG obtained after the link reversal operations.

which, as a result of the new reference level, have lost all routes to the destination. If the node has no neighbors of finite height with respect to this destination, the node instead attempts to discover a new route as described earlier. At the end of the repairing activities, localized near the node, the DAG is reestablished (see Fig. 2.15).

Properties.   One of the advantages of TORA is its support for multiple routes. Route reconstruction is not necessary until all known routes to a destination are considered invalid. Fewer route rebuilding exercises in turn result in significant network bandwidth savings.

On the other hand, TORA's operation relies heavily on synchronized clocks. This requirement inherently limits its applicability. If a node does not have a GPS positioning system or some other external time source, it cannot use the algorithm. Additionally, if the external time source fails, the algorithm will cease to operate. In addition, there is a potential for oscillations to occur, especially when multiple sets of coordinating nodes are concurrently detecting partitions, erasing routes, and building new routes based on each other. Because TORA uses internodal coordination, its instability problem is similar to the count-to-infinity problem in distance-vector routing protocols, except that such oscillations are temporary and route convergence will ultimately occur. This can lead to potentially lengthy delays while waiting for the new routes to be determined.

Hybrid routing [ZRP-ID, RAMAN98, KRISHNA97, LIN97, GERLA95]. Mobility of nodes in infrastructureless wireless networks raises organizational problems quite different from and rather more challenging than those for infrastructured wireless networks. Mobile wireless networks differ in the frequency and degree at which the topology changes. A protocol that works well in one MANET may not work well in another with a different density, size, and the like. The diverse applications of ad hoc networks, however, pose a challenge for a single protocol that operates efficiently across a wide range of operational conditions and network configurations. Purely proactive or purely reactive protocols perform well in a limited region of this range. For example, reactive routing protocols are well suited for networks where the call-to-mobility ratio is relatively low. Proactive routing protocols, on the other hand, are well suited for networks where this ratio is relatively high. The performance of either class of protocols degrades when the protocols are applied to regions of ad hoc network space between the two extremes.

Regardless of the type of routing protocol preferred, there will be a set of circumstances under which it will not perform well. Consequently, despite being designed for the same type of underlying network, it does not seem that a routing strategy based exclusively on proactive or reactive routing can achieve the objectives required for ad hoc routing [ROYER99]. A desirable design objective for an architectural framework capable of supporting routing in large ad hoc networks subject to high rates of node mobility shall balance the trade-off between reactive and proactive routing while minimizing the shortcomings of each [McDONA99]. So, what is ideally needed is a single routing protocol that has the intelligence to adjust its behavior *dynamically* based on the rate of changes (mobility) and the activity (rate of data) so as to match the specific mobility/activity ratio.

Researchers advocate that the issue of efficient operation over a wide range of conditions can be addressed by a *hybrid* routing approach, where the proactive and reactive behavior is mixed in the amounts that best match these operational conditions. Given multiple protocols, each suited for a different region of the ad hoc network design space, it does make sense to capitalize on each protocol's strengths by combining them into a single framework (that is, hybridization). In the most basic hybrid framework, one of the protocols would be selected based on its suitability for the specific network's characteristics. Although not an elegant solution, such a framework would perform as well as the best-suited protocol for any scenario and outperform either protocol over the entire ad hoc network design space. However, by not using both protocols together, this approach fails to capitalize on the potential synergy that would make the framework perform as well or better than either protocol for any given scenario.

A more promising approach for protocol hybridization is to have the base protocols operate simultaneously, but with different "scopes." For the case of a two-protocol framework, protocol A could operate locally, while the operation of protocol B would be global. The key to this framework is that the local information acquired by protocol A is used by protocol B to operate in a more efficient manner. This framework can be tuned to network behavior simply by adjusting the size of protocol A's scope. In one extreme configuration, the scope of protocol A is reduced to nothing, leaving protocol B to run by itself. As the scope of protocol A is increased, the information provided to protocol B increases as well, thereby decreasing protocol B's overhead. At the other extreme, protocol A is made global, eliminating the load of protocol B altogether. So, at either extreme, the framework defaults to the operation of an individual protocol. In the wide range of intermediate configurations, the framework performs better than either of the protocols on their own.

Notably, most basic protocols exhibit, to some extent, some degree of multiscope behavior. Many proactive routing protocols monitor the status of neighbor connectivity through broadcast beacons, which occur at a faster rate than the global link-state (or distance-vector) advertisements.

Framework tuning.   As mentioned earlier, hybridization, multiscope operation, and local tuning form the basis for scalable, adaptable routing. Specifically, the idea behind hybrid and multiscope routing is to provide a framework that can be configured to match the network's operational conditions, an integral component of the framework being a tuning mechanism. The three basic ingredients for a tuning mechanism are: a means for measuring relevant network characteristics, a mapping of these measurements to a framework configuration, and a scheme to update the configurations of affected nodes.

The most basic approach to tuning is to determine the proper network characteristics and configuration offline, prior to the network deployment. Typically, the configuration parameters would be determined through network simulations and subsequent parameter optimizations. Nodes are loaded with the proper configuration and then activated. When it is not possible to preconfigure all nodes individually, a small number of nodes may be configured, and this configuration can be shared with other nodes as part of an automatic configuration procedure.

The main advantages of preconfiguration are that it requires limited network intelligence and low real-time processing overhead and ensures stable and consistent configuration. However, for many applications, preconfiguration is not an option. Preconfiguration requires a central configuration authority, which may not exist for distributed applications. In addition, the network characteristics may not be known a priori, or may vary over time, preventing the offline analysis and reducing the effectiveness of the static configuration.

Wireless ad hoc networks naturally lend themselves to dynamic reconfiguration. Through the course of normal operation, nodes directly measure (or infer) local network characteristics. Each node may then use its own local measurements for independent self-configuration. Alternatively, the measurements could be relayed to a central configuration node or shared with surrounding nodes for a distributed configuration approach. At first glance, centralized dynamic reconfiguration may appear to prevent inconsistent configuration, as is the case for centralized static configuration. However, the multihop nature of ad hoc networks makes it impossible to reliably perform tightly synchronized configuration updates for all nodes. This means that for some period of time the network could be in an inconsistent state. As this also affects distributed and independent reconfiguration, it is necessary that a dynamically tunable routing framework be able to deal with, and potentially exploit, nonuniform node configurations.

With support for nonuniform configuration, reconfiguration decisions and the associated measurement/control traffic can be kept local (or eliminated altogether in the case of independent reconfiguration), making the tuning mechanism scalable. Furthermore, the framework can be fine-tuned to adapt to regional changes or even to nodal behavior, rather than broadly tracking average network behavior. This can lead to significant performance improvements, especially in the case of networks where node behavior has regional dependencies.

Routing architectures.   The configuration of an ad hoc network can be either *hierarchical* or *flat*. In a *hierarchical* network, the network nodes are partitioned into groups, called *clusters*. Generally, there are three kinds of nodes in a cluster—the cluster-head node, the gateway node, and the cluster-member (internal) node. Cluster-head nodes basically emulate the functionalities of a base station as in cellular networks [KWON99]. All nodes in a cluster can communicate with their cluster-head and (possibly) with each other (of the same cluster) [IWATA99].

The cluster-head election (the cluster setup) phase can use various different heuristics, such as node addresses, node degrees (neighbor connectivity), transmission power and mobility, or more sophisticated node weights combining these attributes (see Refs. [CHATTER02, SINGH98]).

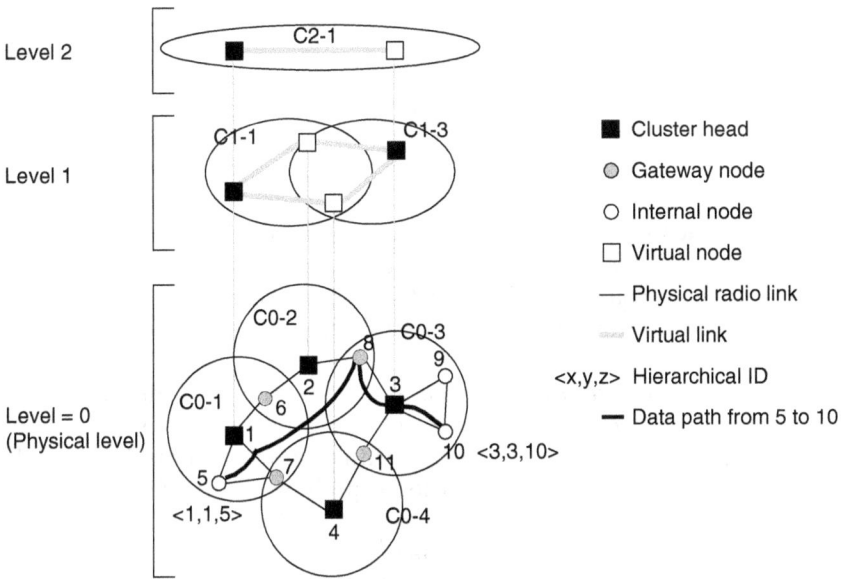

Figure 2.16   An example of physical/virtual clustering.

Gateway nodes are used to provide connectivity among clusters. To communicate within a cluster, a gateway must select the frequency or code used by that cluster. Gateway nodes can communicate with multiple cluster-heads and thus can be reached via multiple paths. Consequently, similar to a router in the wired Internet, which is equipped with multiple subnet addresses, a gateway may have multiple hierarchical addresses.

Depending on the number of hierarchies (levels), the depth of the network can vary from a single tier to multiple tiers. Figure 2.16 illustrates a two-tier example. At Level = 0, there appear 4 physical-level clusters: C0-1, C0-2, C0-3, and C0-4. Level 1 and Level 2 clusters are generated by recursively selecting cluster-heads [IWATA99].

For a node A in a cluster X, to establish communication with some node B at some different cluster, say Y, its traffic must first be routed to its cluster-head. From the cluster-head, traffic is then routed to a gateway node, to another cluster-head, and so on until the cluster-head of the destination node (cluster Y) is reached. Traffic is then delivered to the destination node.

In a *flat* ad hoc network, on the other hand, there is no grouping[2] and all nodes have equal responsibilities. Connections are established

_____

[2]Flat networks could however be perceived as a special case of "zero-tier" hierarchical networks.

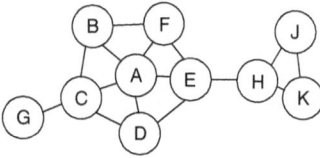

Figure 2.17   A flat ad hoc network.

between nodes that are in close enough proximity to allow sufficient radio propagation conditions to establish connectivity. Routing between any two nodes is constrained only by the connectivity conditions and, possibly, by security limitations. An example of a flat network is shown in Fig. 2.17.

Hierarchical versus flat architecture.   The major advantage of the *hierarchical* architecture is the ease of the mobility and resource management process. Mobility of nodes within an infrastructureless mobile wireless network raises organizational problems quite different from and rather more challenging than those for wireless communication (cellular) networks [GSM92]. As there is no centralized administrative control, rapid response to nodal movement requires *adaptive, autonomous,* and *distributed* organization mechanisms that involve minimal manual intervention.

The aggregation of nodes into clusters provides a convenient framework for the development of important features, including channel reuse among clusters (in terms of frequency, time, or spreading code) [GILH91], channel access, and channel and bandwidth allocation [GERLA95, GILH91]. In cluster-based networks, a cluster-head node acts as a local coordinator within its cluster—it keeps track of the member nodes of its cluster (network management), accounts for resources so that bandwidth reservations can be placed on them (in a deterministic or statistical sense), and assists in locating nodes outside its cluster during the course of a data transmission [GERLA95]. This approach is in line with that followed in cellular communication networks, where such resource accountability is facilitated by the fact that all stations learn of each other's requirements through a control station (e.g., base station in cellular systems). In addition, to reduce the overhead for routing and processing [IWATA99], complete routing information is maintained only for intracluster routing [LIN97], whereas for intercluster routing the topology details are hidden through hierarchical aggregation techniques. For networks composed of a large number of mobile devices, a hierarchical network configuration would be a practical and scalable solution [KLEINR97, XU98].

Furthermore, clustering can provide a nice methodology for accomplishing QoS guarantees in a mobile network. Ensuring QoS communications in a wireless mobile ad hoc network is apparently highly dependent on routing and resource management control—conforming to QoS measures, such as delay bounds, depends on the quality of the chosen route, whereas, in addition, complying with QoS guarantees imposes the use of a MAC method that guarantees the successful transmission of packets under high mobility and/or heavy load circumstances. This can hardly be made feasible without a central regulatory authority to carry out the significant functions of routing and channel (resource) management in wireless mobile ad hoc networks.

The major advantage of clustering is that some information about the state of the network is kept local [BROCH98, McDONA99]. Even in a very dynamic situation, not all changes need be propagated throughout the network. Specifically, by confining location update propagation to the lowest level (in the hierarchy) containing the moving endpoint, costs can be made proportional to the distance moved. As with most large organizational problems, specialized node roles and regional node addressing help hierarchical routing protocols to scale with network size, especially when there is a structure in the underlying network connectivity (for example, group mobility) that can be exploited [McDONA00].

Notably, mobility and traffic have a great impact on cluster size and design. In a large MANET, local conditions can vary significantly; so in some regions a larger cluster size might be advisable. There is no single or default parameter, or set of parameters, such as cluster size and cluster merge/split threshold that can be fitted for all real networks. Besides, it is unlikely that a single default parameter would be acceptable for all clustered MANETs. In this context, a single initial value would be a good starting point and during operation different heuristics are used to locally optimize cluster parameters.

There are also several features in cluster-based wireless networks that are potentially complex to implement and thus hinder scalability [IWATA99]. First, cluster IDs are dynamically assigned. This assignment must be unique—not an easy task in multihop mobile environment, where the hierarchical topology is continuously reconfigured. Second, each cluster can dynamically merge and split, based on the number of nodes in the cluster. This feature causes frequent changes of the cluster-head, thus degrading the network performance significantly. Since the diameter of a cluster is variable, it is also difficult to predict the time it takes to propagate clustering-control messages among nodes. As a consequence, it is difficult to bind the convergence time of the clustering algorithm. Third, the paging and query/response approach used to locate mobile nodes may lead to a nonnegligible amount of control message overhead. Fourth, if the cluster-head leaves its current cluster, this function migrates to another location manager. This requires a complex consistency management between the original and the new cluster.

Finally, the determination of the cluster-heads shall be done in such a way that the reconfigurations of the network topology are minimized. This is an important issue, since an essential criterion in cluster-based algorithms is *cluster stability*. Frequent cluster-head changes may adversely affect the performance of other functions such as scheduling and resource allocation, that rely on it.

Under the "extreme" scenario where mobility rates are high and mobility patterns random, such that all nodes in the field are moving very rapidly in different random directions, each cluster stays intact for only a very short period of time. In this scenario, it seems that clusters would need to be constantly created or modified, thus rendering a lot of cluster maintenance overhead. Therefore the bulk of overhead actually becomes cluster maintenance. See Fig. 2.18.

In *flat* architectures, each node maintains a routing table with entries for all the nodes in the network [IWATA99]. Flat networks require only one type of equipment as all nodes have to perform the same operation. That is, all nodes are treated as network members, similar to cluster-member nodes in hierarchical clustered architectures, with all having the same responsibilities. In addition, nodes in flat networks transmit at a significantly lower power than the transmission power of a cluster-head, which is reasonably higher in order to cover its cluster territory.

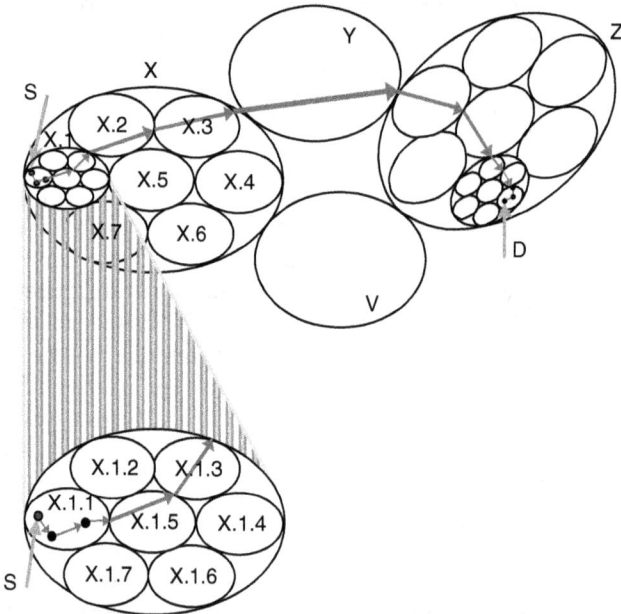

Figure 2.18 Hierarchical routing.

Operating a network at low power levels has several implications. First, the battery power of the nodes in ad hoc networks is preserved. Second, the wireless spectrum can be better reused, leading to more network capacity. Third, and possibly most important, a larger degree of low probability of interception and detection can be achieved, resulting in a more secure network operation.

On the other hand, a flat architecture is acceptable if the user population is small. As the number of mobile hosts increases, so does the overhead, thus creating scalability concerns when applied to large networks.

To conclude, there are surely circumstances in which a flat MANET architecture is preferable to a hierarchical one, but one always has the size factor backward. Hierarchal clustering tends to localize the impact of state changes, so it tends to be more useful in larger networks than in smaller ones. If each node is one or two hops away from every other node, clustering probably is not worth the overhead. However, if the network diameter is 10 or 20 hops, a methodology is needed to reduce the size of the routing problem, and clustering seems to be an effective way of achieving this.

The cluster-based routing protocol (CBRP) [CBRP-ID]. The idea behind CBRP is to divide the nodes into a number of overlapping or disjoint clusters, where the elected cluster-heads maintain the membership information for their cluster.

Nodes are aware of their bidirectional links to their neighbors as well as the unidirectional links between them and their neighbors. To handle this, nodes maintain a neighbor table (similar to Table 2.2).

Nodes periodically broadcast their neighbor table in a HELLO message, which is used to update the neighbor tables at each node. A HELLO message contains the node ID, the nodes role (cluster-head, cluster member, or undecided), and the neighbor table. If no HELLO message is received from a certain node, that neighbor entry will be removed from the table.

Cluster formation. The cluster formation algorithm in CBRP is as follows: the node with lowest node ID is elected as the cluster-head. Nodes use the information carried in the HELLO messages to decide

TABLE 2.2   Neighbor Table

Neighbor ID	Link status	Role
Neighbor 1	Bi/unidirectional link to me	Is 1 a cluster-head or member
Neighbor 2	Bi/unidirectional link to me	Is 2 a cluster-head or member
—	—	—
Neighbor n	Bi/unidirectional link to me	Is n a cluster-head or member

whether or not they are the cluster-heads. The cluster-head regards all nodes with which it has a bidirectional link as its member nodes. Similarly, a node regards itself as a member node of a particular cluster if it has a bidirectional link to the cluster-head.

Route discovery–route selection.   Routing in CBRP is based on source routing. To establish intercluster routes (routes between clusters) a flooding approach is used. In contrast to networkwide flooding, where all nodes in the network receive and relay the broadcasts, route queries in CBRP are propagated only to the cluster-heads and not to member nodes of clusters. This selective flooding scheme assures that the number of nodes disturbed is much fewer. The clustering solution in CBRP scales well to large ad hoc networks. One remaining question, however, is "How large each cluster should be?" This parameter is critical to the behavior of the protocol.

Zone routing protocol (ZRP) [HAAS97, ZRP-ID, PEARL99].   ZRP divides the network into several routing zones (Fig. 2.19). A zone $Z(k, n)$ for a node n with radius k, is defined as the set of nodes at a distance not greater than k hops.

$$Z(k, n) = \{i \mid H(n, i) \; \hat{U} \; k\}$$

where $H(i, j)$ is the distance in number of hops between node i and node j The node n is called the central node of the routing zone, while node b

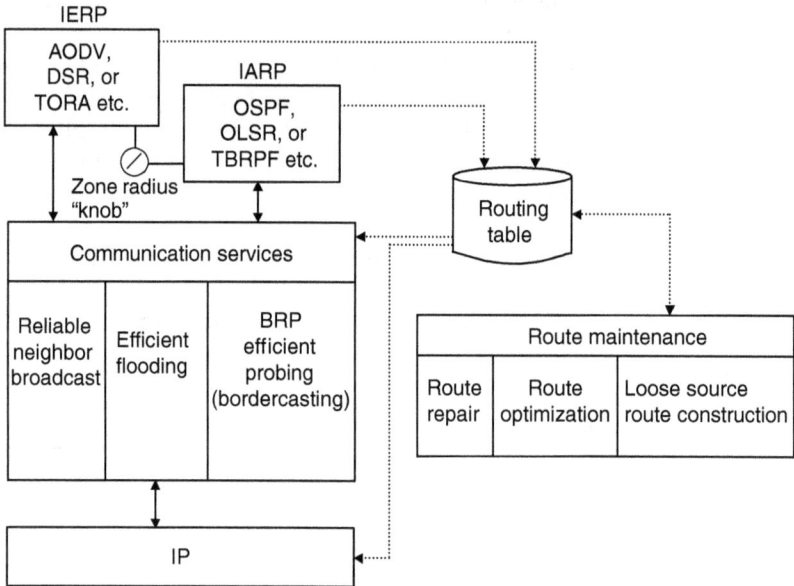

Figure 2.19   Architecture of the zone routing protocol.

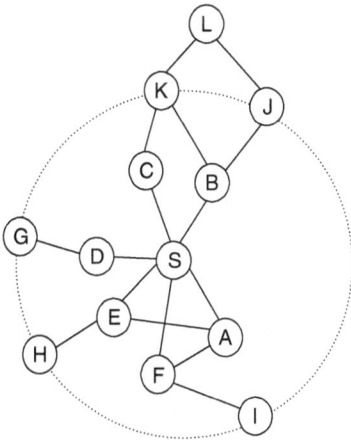

Figure 2.20 A routing zone of radius 2 hops.

such that $H(n, b) = k$ is called the peripheral node of n (see Fig. 2.20). In Fig. 2.20, nodes G-K are peripheral nodes of node S. The value for k is usually small compared to the network diameter and can be optimized under different mobility and traffic scenarios [PEARL99].

The protocol's architecture is organized into four main components— the *intrazone routing protocol* (IARP), the *interzone routing protocol* (IERP), the *bordercast resolution protocol* (BRP), and a layer-2 *neighbor discovery/maintenance protocol* (NDP) [IERP-ID, IARP-ID].

NDP implements neighbor discovery functions; it typically operates through the periodic broadcasting of HELLO messages. The reception (or quality of reception) of a HELLO message can be used to indicate the link status to the beaconing neighbor.

The IARP operates inside the routing zone and is aware of the minimum distance and routes to all the nodes within the zone. IARP can in fact include any number of proactive protocols, such as distance-vector or link-state routing. Different zones may operate with different intrazone protocols as long as the protocols are restricted to those zones. In addition, a change in topology means that update information propagates only within the affected routing zones instead of affecting the entire network.

For the nodes located at a distance $k' > k$ from the source, ZRP relies on IERP to calculate on-demand interzone paths (that is, routes between different routing zones). The IERP uses a form of selective flooding to exploit the underlying zone structure generated by the IARP. Specifically, flooding is based on sending query packets only to the peripheral nodes, using a special kind of multicast protocol called the bordercast resolution protocol. Bordercasting is the process of sending datagrams from one node to all its peripheral nodes. The BRP keeps

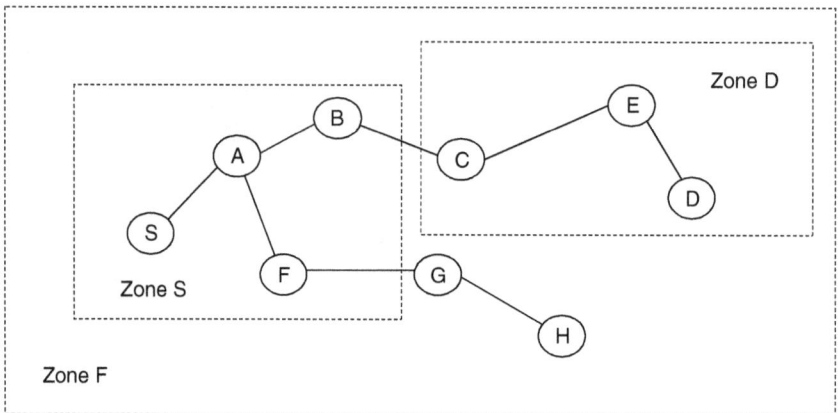

Figure 2.21 Network using ZRP. (The dashed squares show the routing zones for nodes S and D)

track of the peripheral nodes and resolves a bordercast address to the individual addresses of the peripheral nodes.

When a node receives a query packet, it can either reply to the source, if the node in search is a member of its routing zone, or bordercast the query packet to its peripheral nodes. This procedure is repeated until the requested node is found and a route reply is sent back to the source indicating the complete route. A route can be accumulated in the query and reply packets.

Consider the network in Fig. 2.21. Node S needs to send a data packet to node D. Since D is not in the routing zone of S, a route request is sent to its peripheral nodes B and F. Each peripheral node checks to see if D is in its routing zone. Since neither B nor F find the requested node in their routing zone, the request is forwarded to the respectively peripheral nodes. F sends the request to S, B, C, and H while B sends the request to S, F, E, and G. Now the requested node D is found within the routing zone of both C and E. Thus a reply is generated and sent back toward S.

To summarize, ZRP maintains proactive routing information for a local scope (the routing zone), while reactively acquiring routes to destinations beyond the routing zone. At a local level, a proactive routing protocol provides a detailed and fresh view of each node's surrounding local topology. The knowledge of local topology is used to support services such as proactive route maintenance, unidirectional link discovery, and guided message distribution. Bordercasting is used in place of traditional broadcasting to improve the efficiency of a global reactive routing protocol.

Packet forwarding along an interzone path adopts a modified source routing. A routing path contains only the peripheral nodes that have to be traversed.

Properties.   Since ZRP is a hybrid of the proactive and reactive schemes, this protocol uses advantages of both. Routes can be found very fast within the routing zone, while routes outside the zone can be found by efficiently querying selected nodes in the network. ZRP also limits the propagation of information about topological changes to the neighborhood of the change only, as opposed to fully proactive schemes that basically flood the entire network with route queries when a change in topology occurs.

The proactive intrazone routing protocol is not specified. Hence, the use of different intrazone routing protocols would mean that the nodes would have to support several different routing protocols. This is not a wise approach when dealing with thin clients though. It is better to use the same intrazone routing protocol in the entire network. In addition, since there is no coordination among nodes, zones may heavily overlap. As a consequence, a node can be a member as well as a peripheral node of many zones. Hence, the basic search mechanism can perform even worse than a standard flooding method.

Finally, the routing zone tuning is not done dynamically, but is instead set by the administration of the network or with a default value by the manufacturer. The performance of the ZRP depends quite a lot on this decision.

Position or location-aided routing protocols [IWATA99, MAUVE01, JAIN01, KARP00, LIN99, STOJME01].   Position-based routing protocols use the geographic position of nodes to make routing decisions. Location information can be obtained through GPS (global positioning system) or some other type of positioning mechanism[3] [CAPKUN01, KAPLAN96].

Position-based routing can be divided into two distinct tasks—the location service (discovering the position of the destination) and the actual routing of data packets (based on location information). The location service is required by the data sender to find the location of the destination. The resulting location is included in the header of the packet for routing decisions. Intermediate nodes may not need to consult the location service again to obtain a more accurate position of the destination, but simply route the packet using the location information carried in its header. However, if an intermediate node maintains a more accurate position for the destination, it may well choose to update the position information in the header of the packet prior to forwarding it.

Position or geographical routing [STOJME02] thus allows radios to operate nearly stateless; nodes neither have to store routing tables nor

---

[3]An overview of GPS technology can be found in App. C.

transmit messages to keep routing tables up-to-date. This is important if the topology of a network changes fast, as is the case with mobile wireless ad hoc networks. Instead, forwarding decisions are made locally, based only on the node's own position, the positions of its neighbors, and the position of the destination.

One potential weakness of this family of location-aware protocols is the dependence on GPS for obtaining one's location. This may be problematic in certain environments since direct line-of-sight access to GPS satellites may not always be possible due to blockage by objects such as buildings and foliage. The problem can be remedied, however, by using some non-GPS techniques as proposed in Ref. [CAPKUN01].

There exist three common routing strategies for position-based routing —*greedy forwarding, directed flooding,* and *hierarchical routing.* For the first two strategies, a node forwards a given packet to one (greedy forwarding) or more (directed flooding) one-hop neighbors that are located closer to the destination than the forwarding node itself. The third forwarding strategy aims at structuring mobile nodes into a hierarchy. Hierarchical mechanisms may use different types of routing protocols at different levels of the hierarchy (e.g., a non-position-based ad hoc routing protocol at one level and a position-based protocol at another).

In the following section, we review some routing protocols for MANETs that take advantage of the location information of the hosts. Examples of position-aided routing schemes are discussed in the next section.

Distance routing effect algorithm for mobility (DREAM) [BASAGN98]

Description.   DREAM is based on the observation that the greater the distance separating two nodes, the slower they appear to be moving with respect to each other. Source and intermediate nodes calculate the direction of destination D and, based on the mobility information about D, associate an angular range with the routing entry pointing to D. The direction toward the destination is determined by means of the so-called "expected region." As depicted in Fig. 2.23, the expected region is a circle around the position of D as is seen from a forwarding node N. Since this position information may be outdated, the radius r of the expected region is set to $(t_1 - t_0)V_{max}$ where t_1 is the current time, t_0 is the timestamp of the position information that N has for D, and V_{max} is the maximum nodal speed. Given the expected region, the "direction toward D" (see Fig. 2.22) is defined by the line between S and D and the angle φ.

A message is forwarded to all neighbors whose direction belongs to the selected range. The neighboring hops repeat this procedure using their view for D's position. The range is determined by the tangents from A to the circle centered at D and with radius equal to a maximal possible movement of D since the last location update.

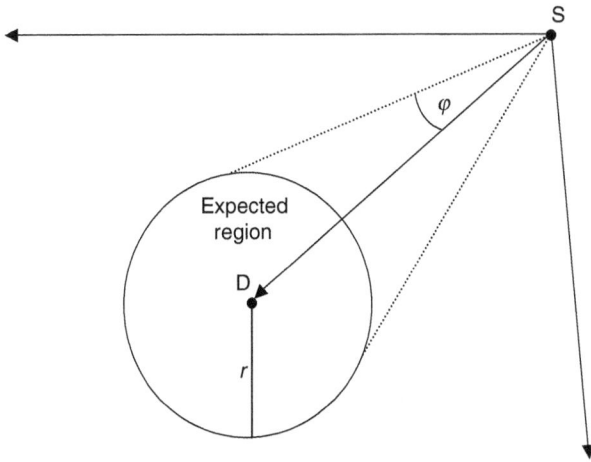

Figure 2.22 Directed flooding in DREAM.

Nodes in DREAM broadcast position update messages to update the position information by other nodes. A node can control the accuracy of its position information available to other nodes by (1) modifying the frequency with which it sends position updates and (2) indicating how far a position update packet is allowed to travel before being discarded. The temporal resolution of the updates is coupled with the mobility rate; the higher the speed of a node, the more frequent the updates it sends. Location updates with a high maximum hop count are sent less frequently than updates that only reach nearby nodes. Thus, a node provides accurate location information to its direct neighborhood and less accurate information (because of fewer updates) to nodes farther away. The reason for this update strategy is that the greater the distance separating two nodes, the slower they appear to be moving with respect to each other (termed the distance effect) [BASAGN98]. Temporal and spatial resolutions are used to control the accuracy of location information.

The distance effect is a reasonable paradigm when intermediate hops are allowed to update the position information carried in a packet. The closer the packet gets to its final destination, the more accurate the position information contained in the packet header.

Properties. DREAM is highly robust with respect to link/path failures. This qualifies the protocol for applications that require high reliability and fast message delivery for very infrequent data transmissions. Also, DREAM works well in combination with an all-for-all location service that provides more accurate information close to the destination. This reduces the size of the expected region and thus the area in which the packet is flooded.

On the other hand, DREAM uses directed flooding to limit the flooding to the direction of the destination. To a certain level, however, directed flooding restricts routing redundancy which may prevent the routing protocol from discovering the shortest paths.

Location-aided routing (LAR) protocol [KO98]

Description.   LAR assumes that the data source node knows the location and roaming speed of the destination node. Suppose that the data source (S in Fig. 2.23) knows the destination's location (D in Fig. 2.23), denoted as (Xd, Yd), and speed, denoted as v, at time t_0 and that the current time is t_1. *Expected zone* is defined as the circular area in which host D may be located at time t_1 with radius $R = v(t_1 - t_0)$. In addition, the *request zone* is derived from the expected zone and is the shaded rectangle as shown in Fig. 2.23 (surrounded by corners S, A, B, and C).

The LAR protocol basically uses restricted flooding to discover routes. Specifically, only the hosts in the request zone help forward route-searching packets. When S initiates a route discovery process, it should include the coordinates of the request zone in the packet. A receiving host simply needs to compare its own location to the request zone to decide whether or not to rebroadcast the route-searching packet.

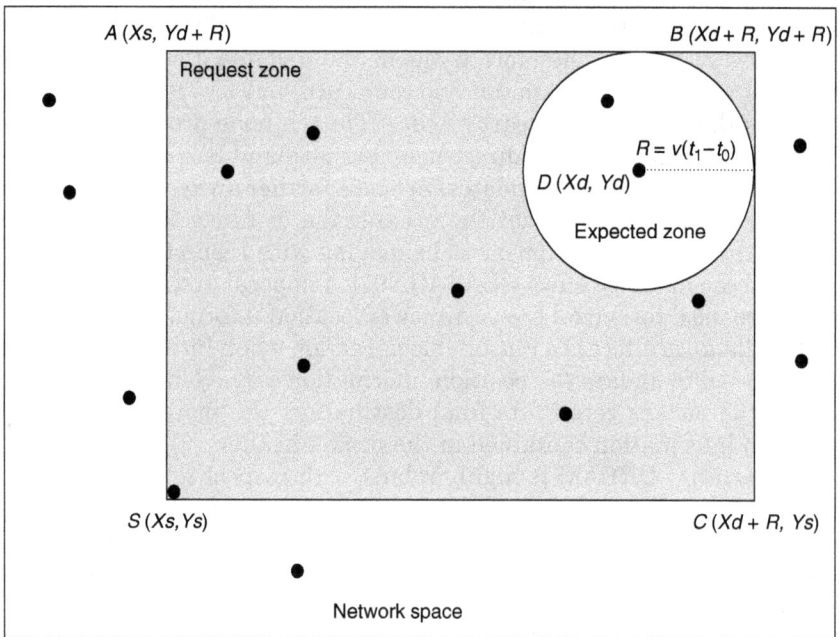

Figure 2.23   Request and expected zones in the LAR protocol.

When D receives the route request packet, it sends a route reply packet back to S. When S receives the reply, the route is established. If the route cannot be discovered within a specified time-out period, S can initiate a new route discovery with an expanded request zone. The expanded request zone should be larger than the previous request zone. In the extreme case, it can be set as the entire network. Since the expanded request zone is larger, the probability of discovering a route is increased with a gradually increasing cost.

A variant of the basic protocol assumes that the data source uses location information to compute the distance between the destination and itself, and then attaches this distance value to the query. When a node receives the query, it computes its own distance to the destination and then forwards the query only if it is closer to the destination than the node from which the message is received. Hence, the query will only get progressively closer to the destination after each relay.

Properties.   One potential weakness of the protocol is that prior knowledge of the destination's location may not always be available at the source. In this case, the protocol may require mobiles to communicate their locations more frequently, or alternatively enlist the aid of a distributed location service [JANNOT00], if necessary. Savings from the reduced flooding of queries may far outweigh the costs of retrieving location information.

Furthermore, underlying the location-aided routing (as well as DREAM) protocol is the notion that a route to the destination can be found by searching in the general direction of the destination. Terrain features such as buildings, hills, and foliage are, however, not considered. The presence of such objects can in fact obstruct or substantially weaken the transmission of radio signals, making communication across them difficult, if not impossible, even though the communicating nodes may be physically close. Lack of terrain awareness may render the use of these techniques less effective in real-life scenarios.

The following example illustrates that when obstructions are not considered, route discovery may be problematic. In Fig. 2.24, obstructions are illustrated by rectangular boxes in gray. Let us assume that these are impenetrable by radio waves. Suppose that node S initiates a route discovery to node D by broadcasting a query message. Node S floods this query to its request zone only to find that node D is unreachable because no queries rebroadcast by other nodes in the request zone have reached node D due to obstructions. In fact, a route does exist through node K. However, this route is not discovered since node K is lying beyond node S's request zone. If the obstructions are known a priori, then node S may (for example) increase the search space around edges of the obstruction at node D, so that node K can be encompassed within node S's request zone.

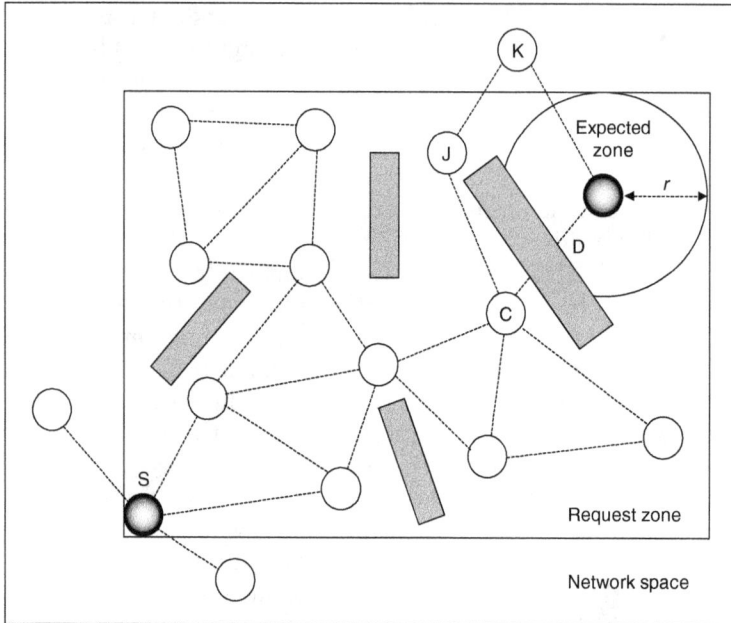

Figure 2.24 Route discovery with LAR in the presence of obstructions.

Grid location service (GLS) [JANNOT00, LIAO01]

Description.   In GLS the geographic area is hierarchically partitioned into squares called *grids*. Each mobile node periodically updates a small set of nodes called *location* servers with its current location. A node sends its position updates to its location servers with no prior knowledge of their actual identities. Instead, nodes use a predefined ordering of node identifiers and a predefined hierarchy.

To determine where to store position information, GLS establishes the notion of near-node IDs, defined to be the least ID greater than a node's own ID within a given n-order square. ID numbers wrap around after the highest possible ID. Considering the example illustrated in Fig. 2.25, when node 10 in the example intends to distribute its position information, it sends position updates to the node with the nearest ID in each of the three surrounding first-order squares. Thus, the position information is available at nodes 15, 18, 73, and all nodes that are in the same first-order square as 10 itself. In the surrounding three second-order squares, again the nodes with the nearest ID are chosen to host the node's position; in the example these are nodes 14, 25, and 29. This process is repeated until the area of the ad hoc network is covered.

Position information is forwarded to nodes with nearer IDs in a process closely resembling position queries. Since GLS requires that all nodes

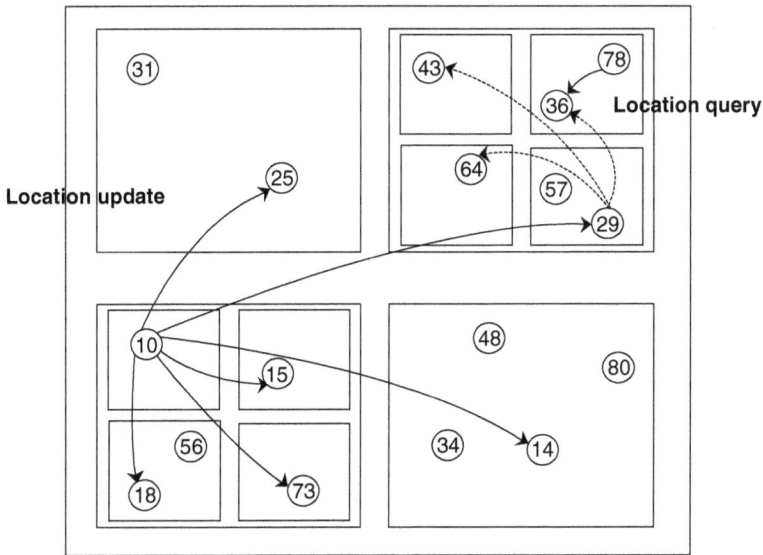

Figure 2.25 Grid location service.

store the information of some other nodes, it can be regarded as an all-for-some approach. Let us assume that node 78 needs to obtain the position of node 10. It therefore should locate a nearby node that knows about the position of node 10. An ideal candidate in the example would be node 29. It is therefore useful to have a look at the position servers for node 29. Its position is stored in the three surrounding first-order squares in nodes 36, 43, and 64. Note that each of these nodes and node 29 are automatically also the ones in their respective first-order squares with the nearest ID to 10. Thus, there exists a "trail" with descending node IDs to the correct position server from each of the squares of all orders. Position queries for a node can now be directed to the node with the nearest ID the querying node knows of. In our example this would be node 36. The node with the nearest ID does not necessarily know the sought-after node, but it will know a node with a nearer node ID (node 29, which already is the sought-after position server). The process continues until a node that has the position information available is found.

Power and energy-aware routing.   Mobile/portable devices are inevitably battery powered; battery lifetime thus becomes crucial for wireless communications and mobile computing. Battery technology has lagged compared to the advancements in communication and computing technology, in the past decade. Now that the capacity of batteries cannot be significantly

TABLE 2.3   Power Consumption Comparison among Different Wireless LAN Cards

Card	Tr mA	Rv mA	Idle mA	Slp mA	Power Sup.V
RangeLAN2-7410	265	130	n/a	2	5
WaveLAN(11 Mb/s)	284	190	156	10	4.74
Smart spread	150	80	n/a	5	5

SOURCES: Refs. [RANGELAN, FEENEY01, ADCON].

improved, efforts should be made to design energy-efficient software and hardware.

A portable device has, typically, several main hardware components, such as the display monitor, CPU, memory, and *wireless network interface card* (WNIC), that consume battery power. The WNIC component can in fact consume 10 to 50 percent of the overall system energy [KRAVETS98], which explains why the lifetime in the batteries of notebooks reduces significantly when inserted with WNICs. Thus, it is essential to support some low-power modes in WNICs.

Power consumption in a wireless network can be classified into three categories [LI01]: (1) the power utilized for the transmission of a message; (2) the power utilized for the reception of a message; and (3) the power utilized while the system is idle. Table 2.3 lists doze/receive/transmit power-consumption numbers for several wireless cards [RANGELAN, FEENEY01, ADCON].

Apart from the transmission/reception modes of operation, energy consumption management functions schedule wireless stations to sleep at certain time intervals. Base stations and wireless access points, being central controllers of receptions/transmissions within a service area, are in charge of buffering all incoming packets to sleeping nodes. As illustrated, even the sleep mode incurs a nonnegligible consumption [XU00, BENJIE01]—for RangeLAN2, the power consumption for doze mode is 5 mA.

In a wireless mobile ad hoc network, the lack of a centralized authority dictates that the wireless stations must always stay awake. Hence, when a message is broadcast through the system, all nodes receive and process the message, either directly or through relaying. Power is thus consumed unnecessarily, which is principally due to overhearing other nodes' transmissions.

Energy efficiency may be the most important design criterion for mobile networks—the operation time of a mobile radio is basically restricted by its battery capacity. Limited battery power restricts the communications (transmission) range as well as the duration of the active period for the nodes. Below some critical threshold for battery power, a node will not be able to function as a router, thus immediately affecting the network connectivity, possibly isolating one or more segments of the network. Fewer routers almost always mean fewer routes and therefore increased likelihood of degraded performance in the network. In fact, communication becomes meaningless if a node is not able to communicate owing to low battery power. Since exchange of messages necessarily means power consumption, many ad hoc networking mechanisms, especially routing and security protocols, explicitly include minimal battery power consumption as a design objective [WANG97, TOH01, SINGH98, RODOPLU99, CHANG00, BAHL01, LI01, XUE01].

Besides, in infrastructured wireless networks, where a wireless link is limited to one hop between an energy-rich base station and a mobile node, the goal of energy conservation can be largely achieved by relocating power-intensive network operations to the base station. However, the wireless link-only routing path in a MANET makes energy savings difficult to achieve. The corresponding reduction of the lifetime of nodes directly affects the network lifetime since mobile nodes themselves collectively form a network infrastructure for routing. *This is the primary design principle of energy-aware routing—to equally balance energy expenditure among mobile hosts to prolong network lifetime, while at the same time conserving overall power consumption as much as possible.*

To illustrate the effectiveness of energy-aware routing in wireless ad hoc networks, let us consider the routing scenario in Fig. 2.26. Based on the hop count metric, host A selects the shortest route A-F,.. E to reach E. Similarly, B chooses route B-F-D to reach D. Both communication pairs go through F, which may quickly drain off F's battery energy. This may make F die earlier, forcing G to become isolated from the network. Consequently, energy cost should be taken into account to lengthen the network lifetime.

Decreasing power in wireless networks, on the other hand, is useful for decreasing the system's interference and thus making more efficient use of the RF spectrum [ADLER00, LI01, KAWAD03]. In a system with base stations or access points, the advantages seem obvious. In wireless ad hoc networks, however, the advantage does not seem so obvious. Data paths in ad hoc networks are composed of mobile hosts where wireless links have to be fashioned out of the radio by nodes choosing the power levels at which they transmit. Transmit power control can reduce (or increase) the local scope (one-hop neighbors) of nodes, and as a result the channel contention at the medium access control (MAC) layer [KAWAD03]. Decreasing the transmitted power means increasing the number of hops between a pair of nodes involved in the average data path. This further implies that more choke points are potentially introduced in the network.

Besides, there is a mutual dependence of power control and routing in the network layer—power control impacts on the routes employed

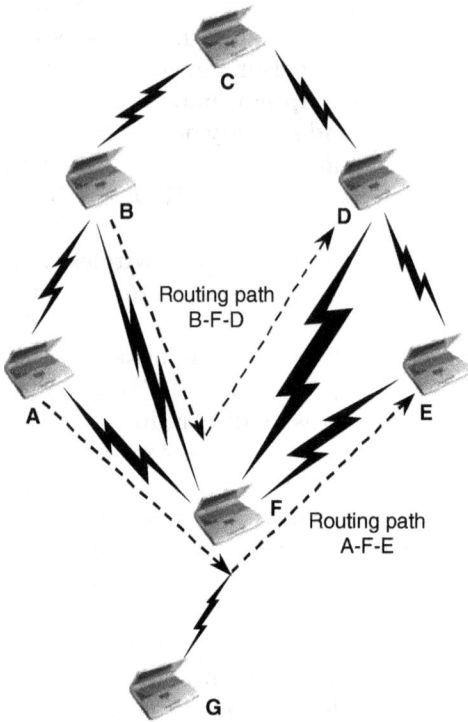

Figure 2.26 A scenario that may quickly drain off the energy of host F, an articulation point in the network.

since the power level dictates what links are available for routing and, vice versa, and the power control protocol needs connectivity information which is provided by the routing layer. This mutual dependency motivates the need for a joint solution for power control and routing, or else *power-aware routing.*

The difficulty in power control for wireless ad hoc networks is that it infringes on several layers in the layered hierarchy. Clearly, power control impacts the physical layer because of the need for maintaining link quality. However, power control also impacts the network layer, as shown in Fig. 2.27(*a*). If all nodes transmit at 1 mW, then the route from node N1 to node N5 is N1→N2→N3→N4→N5. However, if all nodes transmit at 30 mW, then one can choose the route N1→N3→N5. Transmit power thus also affects routing latency as longer (in hops) routes are constructed choosing low power levels.

Furthermore, power control also impacts on the transport layer. In Fig. 2.27(*b*), transmissions from node N1 to N2, with increased power, would increase the interference levels at the transmission of N4 to N3.

(a) Routing

(b) Transport

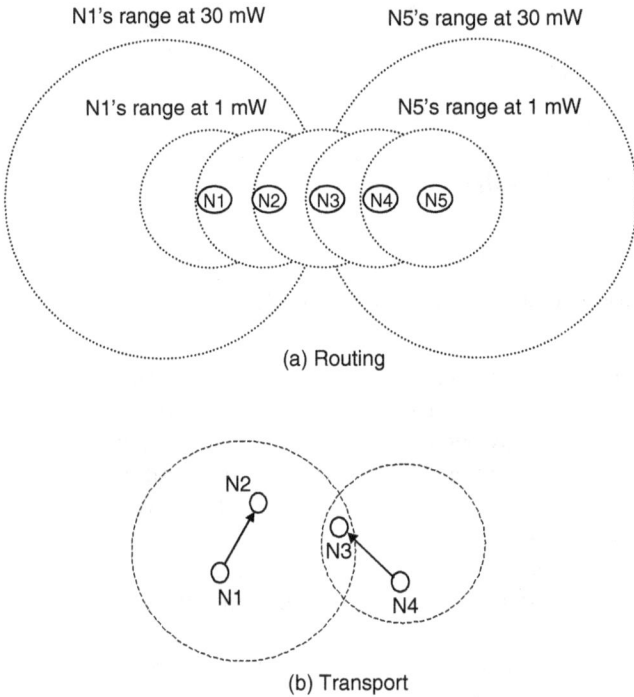

Figure 2.27   Power control implications on the physical, routing, and transport layers.

This would result in a loss of several packets on the communication link from N4 to N3.

In summary, choosing an excessively high power level causes excessive interference as seen in Fig. 2.28(a). This reduces the traffic carrying capacity of the network in addition to reducing battery life. On the other hand, in Fig. 2.28(b), having a very small power level results in fewer links and hence network partitioning effects may result. When the power level is just right, the network is still connected and there is no excessive interference as shown in Fig. 2.28(c).

It should be made clear though that reducing transmitter power does not imply a decrease in the receiver's sensitivity. By reducing output power, one will reduce the area over which the transmitter will interfere with other transmissions. However, reducing sensitivity will only add to the problem of data collisions. In addition, reduced sensitivity requires higher transmit power and offers less protection against Raleigh fading, multipath, and the like.

There are three major issues involved in power/energy-aware routing protocols. First, the goal is to find the path that either minimizes or

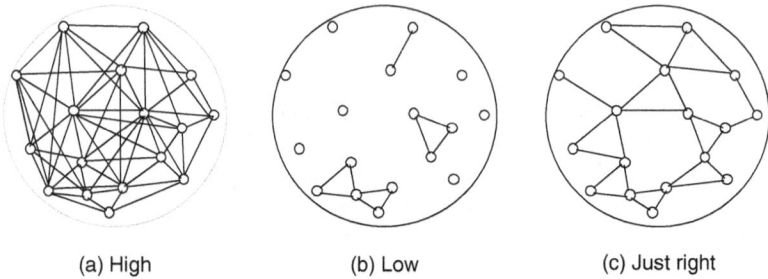

(a) High   (b) Low   (c) Just right

Figure 2.28   Choice of power level affects network connectivity and level of interference.

balances the power consumed. Balanced energy consumption does not necessarily lead to minimized energy consumption; it simply prevents a certain node from being overloaded, thus ensuring longer network lifetime. Second, energy awareness can be either implemented at purely the routing layer or at the routing layer along with other layers such as MAC or application layer. For example, information from the MAC layer, such as interference levels, is beneficial as MAC often supports power saving features that the routing protocol can exploit to provide better energy efficiency. Third, some routing protocols assume that the transmission power is controllable and nodes' location information is available (e.g., via GPS). With these observations in mind, the problem of finding a path with the least-consumed power becomes a conventional optimization problem on a graph where the weighted link cost corresponds to the transmission power required for transmitting a packet between the two ends of the link.

Based on these observations, we could broadly classify wireless networks into two categories—simple or constant-power radio networks, where mobile hosts use fixed transmission powers, and power-controlled or variable-power radio networks, where mobiles are able to vary their transmission power [ADLER00].

In the *variable-power case*, nodes can dynamically vary their transmitter power levels; the chosen transmission power depends on the distance between the transmitter and receiver nodes. Authors in Ref. [BANERJ02] observed that the choice between a path with many short-range hops and a path with fewer long-range hops is nontrivial, but involves a trade-off between the reduction in the transmission energy for a single packet and the potential increase in the frequency of retransmissions. Even though all links have identical error rates, it is not always true that splitting a large-distance (high-power) hop into multiple small-distance (low-power) hops results in overall energy savings. In fact, the analysis in Ref. [BANERJ02] shows that if the number of hops exceeds an optimal value, the rise in the overall error probability negates any apparent reduction in the transmission power.

The idea of allowing mobile hosts to vary their transmission power is already used in power-controlled CDMA systems, where the base station can direct mobiles to reduce their power so as to reduce the system's interference and thus allow more users on the system [JACOBS96, GIBSON96].

For the *fixed-power case*, the transmission power of a node is chosen independent of the distance or quality of the link [RFC 2453, RFC 2328]. If communication links are assumed to be error-free, conventional minimum-hop routing could fit in constant power/energy-efficient routing. If this is the case, the minimum-hop path may not be the most energy-efficient approach, since an alternative path with more hops may prove to be better, given a sufficiently low overall error rate. Similarly, for the variable-power case, a path with a greater number of smaller hops may not always be better; the savings achieved in the individual transmission energies (given by Eq. 3 in Ref. [BANERJ02]) may be nullified by a larger increase in link errors and consequent retransmissions.

The latter issue, that of retransmissions, is of paramount importance—wireless links typically employ link-layer frame recovery mechanisms (e.g., link-layer retransmissions, or forward error correcting codes) to recover from packet losses. Additionally, protocols such as TCP or SCTP employ additional source-initiated retransmission mechanisms to ensure a reliable transport layer. The energy efficiency associated with a candidate path should thus reflect not merely the energy spent in just transmitting a *single* packet across a link, but also the *total effective energy* spent in packet delivery, which includes the energy spent in potential retransmissions as well.[4] It is thus important to consider the links' error rate as a part of the route selection algorithm, since the choice of links with relatively high error rates can significantly increase the effective energy spent in reliably transmitting a single packet.

Given the respective intricacies of mobile multihop radio networks and the semantics of power-aware routing, the question that naturally arises then is—What is the smallest common power level at which the entire network becomes connected? Or else, how can the power control problem be implemented in a distributed asynchronous fashion by the nodes participating in the network to optimally resolve multihop radio connectivity?

Ideally, energy-aware routing protocols manage to operate all nodes at a common power level that is chosen to be the smallest power level at which the network is connected [KAWAD03]. A common power level, however, does not guarantee a common *signal-to-noise plus interference*

---

[4]This is especially relevant in multihop wireless networks, where variable channel conditions often cause packet error rates as high as 15 to 25 percent [BANERJ02].

ratio (SINR) at all receivers [KAWAD03, BANERJ02]. In cellular systems there exists a feasible choice of power levels that achieve equal SINR for all the transmitters to a single receiver (the base station). However, in ad hoc networks, there is no choice of power levels that achieve a common SINR for all transmitter-receiver pairs. It all depends on the density of the system's hosts and the patterns of communications. If there exist a very large number of hosts in a rather small geographical area, that would inherently generate increased contention for the RF channel. By reducing power, the potential for interference among hosts on the same channel is greatly reduced, but in significantly different geographical locations.

In Ref. [GUPTA00], the authors show that when nodes are homogeneously dispersed in space, which is the case when a large number of nodes are *uniformly distributed* (see Fig. 2.29(*a*)), then the choice of a common transmit power level has several appealing features and properties.

However, when nodes are nonhomogeneously dispersed, as in Fig. 2.29(*b*), power control becomes more complicated. For example, let us consider that the nodes in a MANET are organized in clusters, which are about 2 km apart. Within a cluster, each node is normally within a few meters of at least one other node. When the communications is within a cluster, it would be prudent to limit the transmit power to the level necessary to carry out communications within that cluster (see Fig. 2.30), but not high enough to cause interference between clusters. On the other hand, when communication is between different clusters, a higher power level would be needed.

Obviously, the lowest common-power level that assures network connectivity implies that the common-power level is dictated by outlying nodes; those which are far from others (that is, nodes X and Y in Fig. 2.30). All nodes within clusters A and B are mutually reachable at 1 mW except X and Y. Without the loss of generality, we can say that these nodes form a 1 mW cluster. Nodes X and Y can be mutually reached by using a higher power level of, say, 150 mW. A power-aware network layer protocol, which converges to the lowest power level such that the network is connected, will in this case converge to 150 mW. Thus every

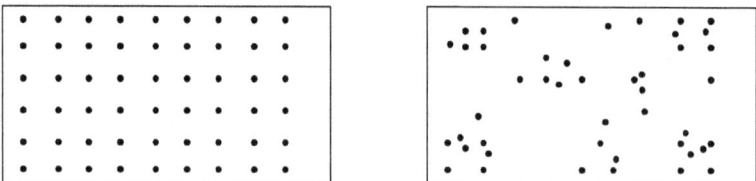

(a) Homogeneous spatial dispersion of nodes   (b) Nonhomogeneously dispersed

Figure 2.29   Homogeneous vs clustered networks.

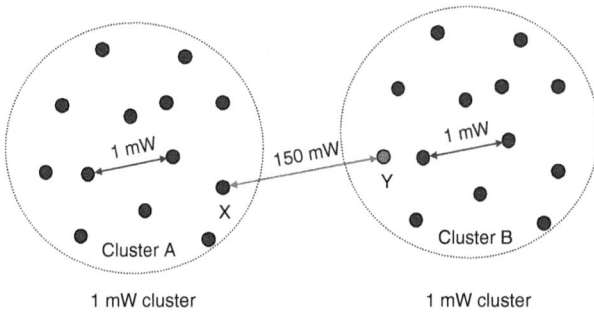

Figure 2.30 A common power level is not appropriate for nonhomogeneous networks.

node in the network will be forced to use 150 mW even though 1 mW is enough for most communications.

Power-aware routing.   Following are some power-conservative protocols for routing in MANETs.

Clusterpow control [KAWAD03].   The *clusterpow control protocol* is designed for power-aware routing in nonhomogeneous networks. A single route in clusterpow can consist of hops of different transmit power such that the clustered structure of the network is maintained. The algorithm aims simply at using the lowest transmit power level p, such that the destination is reachable by using a power level not greater than p. This algorithm is executed at the data source and at every intermediate node along the data route. The route resulting on running this algorithm in a typical clustered network is illustrated in Fig. 2.31.

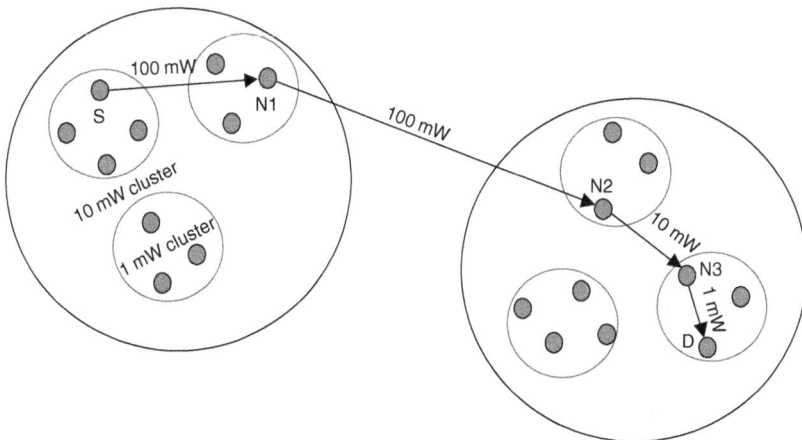

Figure 2.31   Routing by clusterpow in a typical nonhomogeneous network.

As illustrated, the network has three levels of clustering, corresponding to power levels of 1 mW, 10 mW and 150 mW; the whole network
being the 150 mW cluster. To get from source node S to destination D,
a power level of 150 mW is used until the packet gets to the 10 mW cluster to which the destination belongs. Then 10 mW is used until the 1 mW
cluster to which the destination belongs is reached, and finally a 1 mW
hop delivers the packet to destination D. Thus, transmit power control
leads to an automatic clustering of the network.

Clusterpow architecture. The architecture of clusterpow involves
running multiple routing daemons, each corresponding to a power level
P_i. These routing daemons build their own separate routing tables RT
P_i by communicating with their peers at other nodes. To achieve this,
HELLO packets are used and transmitted at power P_i. This idea of parallel modularity at the network layer is illustrated in Fig. 2.32.

Nodes in clusterpow protocol maintain two routing tables—the
kernel routing table and the user-space routing table. The kernel
routing table has a transmit power field for every entry that indicates the power to be used when routing packets for that destination.
The user-space routing table contains reachability information. The
next-hop node is determined by consulting the lowest-power routing
table in which the destination seems to be reachable. That is, for every
destination D, the entry in kernel routing table is copied from the
lowest-power routing table in which D is reachable, i.e., has a finite metric.

The user space routing tables at each power level and the kernel routing table for the node topology are illustrated in Fig. 2.32.

The routing procedure is as follows. At node S, the destination D
appears in the 150 mW routing table using N1 as the next hop. This
entry is thus copied in the kernel routing table and used for routing.
Similarly, for N1 the destination appears only in the 150 mW routing
table using N2 as the next hop. At N2, however, the lowest power level
at which D is reachable is 10 mW. So this is used for routing and the
packet is sent to N3, which has D in its 1 mW routing table. So the final
hop of the packet is at 1 mW.

Alternate path routing (APR) protocol [PEARL00]. APR indirectly
balances energy consumption by distributing network traffic among a
set of diverse paths for the same source-destination pair. This pair is
called an *alternate route set*. APR's performance greatly depends on the
quality of the alternate route set, which can be measured by the so-
called *route coupling*; that is, how many nodes and links two routes
have in common. Since the movement of a common node breaks the two
routes altogether, a good alternate route set consists of decoupled
routes. A decoupled alternate route set can be constructed as shown
in Fig. 2.33. When node S searches for a routing path to D, it may
obtain three alternate routes—[S→A→B→C→D], [S→A→E→C→D],

Node S

1 mW routing table		
Dest	NextHop	Metric
D		Inf

10 mW routing table		
Dest	NextHop	Metric
D		Inf

100 mW routing table		
Dest	NextHop	Metric
D	N1	3

Kernel IP routing table

Dest	NextHop	Metric	TxPower
D	N1	3	100 mW

Node 1

1 mW routing table		
Dest	NextHop	Metric
D		Inf

10 mW routing table		
Dest	NextHop	Metric
D		Inf

100 mW routing table		
Dest	NextHop	Metric
D	N2	3

Kernel IP routing table

Dest	NextHop	Metric	TxPower
D	N2	3	100 mW

Node 2

1 mW routing table		
Dest	NextHop	Metric
D		Inf

10 mW routing table		
Dest	NextHop	Metric
D	N3	2

100 mW routing table		
Dest	NextHop	Metric
D	D	1

Kernel IP routing table

Dest	NextHop	Metric	TxPower
D	N3	2	100 mW

Node 3

1 mW routing table		
Dest	NextHop	Metric
D	D	1

10 mW routing table		
Dest	NextHop	Metric
D	D	1

100 mW routing table		
Dest	NextHop	Metric
D	D	1

Kernel IP routing table

Dest	NextHop	Metric	TxPower
D	D	1	1 mW

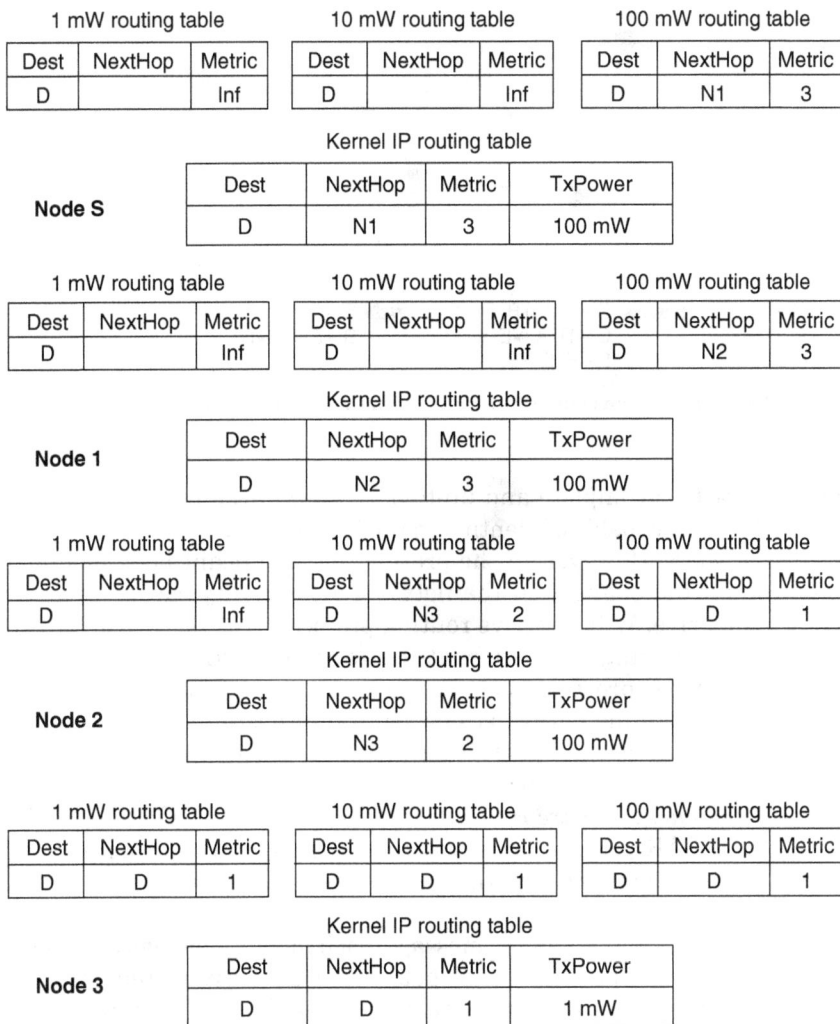

Figure 2.32 Routing tables for all power levels, and the kernel IP routing table, at all the nodes in the network of Fig. 2.31.

and [S→E→B→D]. Since these routes have some node in common, the alternate route set is not good enough. Each routing path is decomposed to its constituent links, and additional alternate routes can be constructed with improved diversity and reduced length—[S→A→B→D] and [S→E→C→D].

The APR routing method can use both proactive and reactive protocols for the routing functions. With proactive routing protocols (see section "Proactive (table-driven) routing" for discussions), each node is

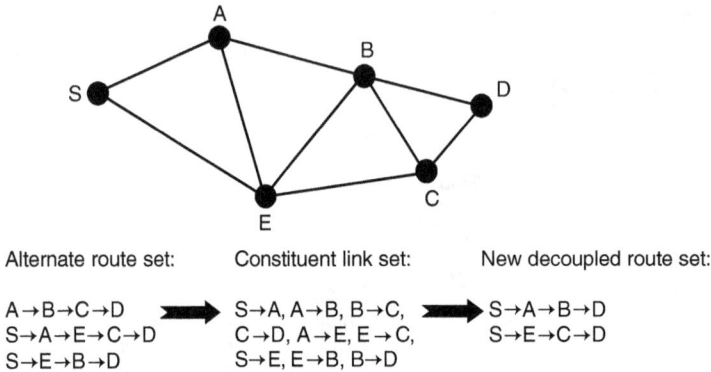

| Alternate route set: | Constituent link set: | New decoupled route set: |

A→B→C→D   ➤   S→A, A→B, B→C,   ➤   S→A→B→D
S→A→E→C→D   C→D, A→E, E→C,   S→E→C→D
S→E→B→D   S→E, E→B, B→D

Figure 2.33 Construction of alternate route set in the APR protocol.

provided with a complete and up-to-date view of the network connectivity. It is thus capable of identifying the best alternate routes that exist in the network. However, in the presence of significant node mobility, tracking all the changes in the network connectivity can be prohibitively expensive. With reactive routing protocols (see section "Reactive (on-demand) routing"), the alternate route set is constructed during the route discovery process, as a route query may produce multiple responses containing paths to the sought-after destination. During the reply phase, the cached path information is used to redirect replies along more diverse paths back to the source.

Localized energy aware routing (LEAR) protocol [WOO01]. Unlike APR, the LEAR protocol achieves a greater balance of energy consumption among all participating mobile nodes. LEAR is based on DSR for discovering routes.

On receiving a route request message, each mobile node has the choice to determine whether or not to accept and forward the route request message depending on its *remaining battery power* (E_r). If E_r is higher than a *threshold value* (Th_r), the route request message is forwarded; otherwise, the message is dropped. Eventually, the destination will receive a route request message only when all intermediate nodes along the route have good battery levels. Thus, the first arriving message is considered to follow an energy-efficient as well as a reasonably short path.

If any of the intermediate nodes along every possible path drop the route request message, the source will not receive a reply message even though a path to the node in search may in fact exist. To prevent this case from occurring, the source will resend the same route request message, but this time with a higher sequence number. When an intermediate node receives the same request message again with a larger sequence number, it adjusts (lowers) its Th_r to allow forwarding to continue. Table 2.4 describes the LEAR algorithm. In order to reduce the repeated request messages and

TABLE 2.4   The LEAR Algorithm

Node	Steps
Source node	Broadcast a route request; wait for the first arriving route reply; select the source route contained in the message; ignore all later replies
Intermediate node	On receipt of a route-request message:
	If the message is not the first trail and $E_r < Th_r$ adjust (lower) Th_r by d; If it has the route to the destination in its cache, If $E_r > Th_r$ forward DROP_ROUTE_CACHE and ignore all later requests; Else, If $E_r > Th_r$ forward (broadcast) route request and ignore all later requests; Else, forward (broadcast) DROP_ROUTE_CACHE and ignore all later requests on receipt of a ROUTE_CACHE.
	If the message is not the first trial and $E_r < Th_r$ adjust (lower) Th_r by d; If $E_r > Th_r$ forward (unicast) ROUTE_CACHE and ignore all later requests; Else, forward (unicast) DROP_ROUTE_CACHE and ignore all later requests; and send backward (unicast) CANCEL_ROUTE_CACHE
Destination node	On receipt of the first arriving route request of ROUTE_CACHE, send a route reply to the source with the source route contained in the message

to utilize the route cache, four routing-related control messages are introduced: DROP_ROUTE_REQ, ROUTE_CACHE, DROP_ROUTE_CACHE, and CANCEL_ROUTE_CACHE.

Power-aware localized routing (PLR) protocol [STOJME01a]. PLR is a localized, fully distributed energy-aware routing algorithm. Assuming that the location information of its neighbors and of the destination is available through GPS, each node selects one of its neighbors through which the overall transmission power to the destination is minimized.

Since the transmission power needed for direct communication between two nodes has super-linear dependency on distance, packet transmissions via intermediate nodes is an energy-efficient practice. For example, direct transmission from node A to node D in Fig. 2.34 may consume more energy than indirect transmission via N_i provided that $|AD|$ is larger than $(c/(a(1-2^{1-\alpha})))^{1/a}$, where the transmission and reception power between two nodes separated by a distance d is $u(d) = ad^a + c$. It is also shown that the power consumption is minimized (denoted as $v(d)$), when $(n-1)$ equally spaced intermediate nodes relay transmissions along the two end nodes, where $n = d[a(\alpha-1)/c]^{1/a}$ and $v(d) = dc[a(\alpha-1)/c]^{1/a} + da[a(a-1)/c]^{(1-a)/a}$.

Therefore, the selection of an intermediate node from among its neighbors requires the evaluation of $u(d) + v(d)$. In other words, a node (A), whether it is a source or an intermediate node, selects one of its neighbors (N_1, N_2, N_3, ...) as the next intermediate node (N_i) to the destination node (D),

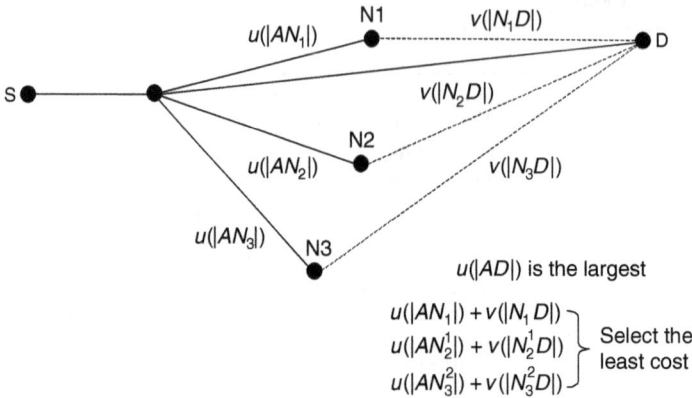

Figure 2.34   Transmission from node A to node D.

which minimizes $u(|AN_i|) + v(|N_iD|)$. Note that A to N_i is a direct transmission, while N_i to D is an indirect transmission. If the goal is to maximize the network lifetime, one only needs to generalize the cost function by including the remaining lifetime of node N_i or all N_i's neighbors.

Online max-min (OMM) routing protocol [LI01]. The data transmission sequence (or data generation rate) is not usually known in advance. Without requiring that information, OMM makes routing decisions that optimize two different metrics—minimizing power consumption and maximizing the minimal residual power in the nodes of the network. Given the power level information of all nodes and the power cost between two neighboring nodes, OMM first finds the path that minimizes the power consumption (P_{min}) by using the Dijkstra algorithm. From among the power-efficient paths (with some tolerance, which is less than zP_{min}, where $z \geq 1$), it then selects one that optimizes the second metric by iterative application of the Dijkstra algorithm with edge removals.

The algorithm steps are then as follows:

1. Find the path with the least power consumption (P_{min}) using the Dijkstra algorithm.

2. Find the path with the least power consumption in the graph. If the power consumption is greater than $z \times P_{min}$ or no path is found, then the previous shortest path is the solution, stop.

3. Find the minimal residual power fraction on that path, and let it be U_{min}.

4. Find all the edges that have a residual power fraction smaller than U_{min} and remove them from the graph.

5. Go to step (2).

The parameter z measures the trade-off between the max-min path and the *minimum power* path. When $z = 1$, the algorithm optimizes only the first metric and thus provides the minimal-power-consumed path. When $z = \infty$, it optimizes only the second metric and thus provides the max-min path. Thus, proper selection of the parameter z is important in determining the overall performance.

A perturbation method is used to compute z adaptively. First, the algorithm randomly chooses an initial value of z and estimates the lifetime of the most overloaded node. Then, z is increased by a small constant, and the lifetime is estimated again. The two estimates are compared, and the parameter z is increased or decreased accordingly. Since the two successive estimates are calculated during two different time periods, the whole process is based on the assumption that the message distributions become similar as time elapses.

A reason for inefficiency of OMM is that the protocol requires information about the power levels of all mobile nodes. In large networks, this requirement is not trivial.

Table 2.5 provides a synopsis of the main characteristics of some of the routing protocols described in this chapter.

TABLE 2.5   Synopsis of the Main Characteristics of Some Representative MANET Routing Protocols

	DSDV	AODV	RDMAR	DSR	ZRP	TORA/ IMEP	CBRP
Loop-free	Yes	Yes	Yes	Yes	Yes	No, short -lived loops	Yes
Multiple routes	No	No	Yes	Yes		Yes	Yes
Distributed	Yes	Yes	Yes	Yes	Yes	Yes	Yes
Reactive	No	Yes	Yes	Yes	Partially	Yes	Yes
Unidirectional link support	No	No	Partially	Yes	No	No	Yes
QoS support	No	No	No	No	No	No	No
Multicast	No	Yes	No	No	No	No	No
Security	No	No	No	No	No	No	No
Power conservation	No	No	No	No	No	No	No
Periodic broadcasts	Yes	Yes	No	No	Yes	Yes (IMEP)	Yes
Control flooding	Yes	Yes	No	Yes	Partially	Yes	Partially
Requires reliable data	No	No	No	No	No	Yes	No

Chapter

3

QoS-Sensitive Routing in Mobile Multimedia Ad Hoc Networks

3.1 Introduction to Multimedia Communications

Mobile computing has enjoyed an exponential growth that began in the late 1990s [GERLA95] and is expected to continue well into this new millennium. As more and more users are becoming mobile [KLEIN95], mobile communications has to evolve at a faster rate than ever before to meet the increasing demands being placed on it. Observing the growing demands of roaming users, it is predicted by the mobile computing research community that the future integrated services wireless networks will be burdened with bandwidth-intensive traffic generated by personal multimedia applications such as video-on-demand, news-on-demand, Web browsing, and traveler information systems. Multimedia services support over wireless mobile networks is the hottest telecommunications buzzword today.

Of late, the information superhighway (Internet) is being used to carry real-time data such as voice and video. The current trend of ubiquitous connectivity is a new paradigm of accessing these services via wireless links. The economics behind this trend is evident—lower cost of information delivery when compared to traditional telecommunication networks, scope for introduction of innovative user interfaces for applications like Internet telephony, and integration of basic multimedia (such as voice, video, and text) with software such as Web browsers. The demand for multimedia services over high-speed links among wireless devices, such as notebook computer, PDA, digital TV, camcorder, and digital camera, has grown in the wireless network segment too.

For multimedia traffic to be supported successfully, it is necessary to provide quality-of-service (QoS) guarantees between the end-systems. In the commercial Internet environment, QoS is a competitive mechanism to provide more distinguished services offered by competing ISPs. All ISPs are basically the same, except for the services they offer. If mechanisms exist to provide a differentiated class of service levels, so that one customer's traffic can be treated differently from another's, it certainly is possible to develop different economic models on which to base these services.

To support QoS for real-time applications, it is necessary to effectively control the total traffic that can flow into the network system. And the key to successfully addressing QoS control is *QoS routing*. QoS routing couples the coarse grain (routing) and fine grain (congestion control) resource allocation processes. The goals for QoS routing are twofold. First, to help admission control; that is, a routing protocol should not only provide route(s) between two endpoints but also compute the QoS that is supportable on these routes. The network control mechanism decides whether to accept a new connection request by examining whether the route given by the route finding scheme has sufficient QoS to adapt this new connection. Second, QoS routing schemes that consider multiple service constraints should provide a fine-grain load balance by allocating traffic on different paths, subject to their QoS requirements.

The IETF has made tremendous efforts to provide QoS on the Internet, including the *differentiated services* (DiffServ) framework, the *integrated services* (IntServ) framework, the *multiprotocol label switching* (MPLS) technology, and various studies on traffic management mechanisms, such as call admission control and resource reservation protocols. The IntServ model starts from the assumption of a best-effort service (no strict QoS requirements) and refines this to add guarantees of various kinds, typically of delay variance, bounds, and then throughput [CROWCR95].

The DiffServ model divides services into several classes with various requirements and priorities [RFC 2475]. The idea is to assign each datagram to one of these classes and, according to its class, map each datagram to a different level of service. MPLS is rooted in the IP switching and tag switching technologies that were developed to bring circuit-switching concepts to the IP's connectionless routing environment. It is a hybrid technology in that it enables very fast forwarding in the core through the use of labels that are used to identify a forwarding equivalence class, and conventional routing at the edges, where datagrams are routed on the basis of information carried in the header of the packet.

When addressing QoS in the wireless mobile segment, several unique and distinguishing issues emerge. Lower throughput, higher delay and delay variation (packet jitter), and higher bit error rates are inevitable in wireless communications networks. As different QoS-sensitive applications require different QoS guarantees (e.g., some require stringent end-to-end delay, some require a minimal transmission rate, while others with no strict delay and/or bandwidth requirements may simply require high throughput), mobile users with active calls may have to adapt to network dynamics as they move.

Besides, host mobility has a significant impact on QoS parameters. When a mobile host with an active flow moves from one location to another the data-flow path changes. As a result the packet delay may change due to changes in the path length and different congestion levels on the new path. If the new location is overcrowded, the available bandwidth in the new location may not be sufficient to provide the throughput the mobile user was receiving at the previous location. In addition, during a handoff procedure where the existing connection is torn down and a new one is established, a mobile user may suffer temporary service disruptions.

When addressing QoS in a *wireless multihop environment* these fundamentally limiting factors become all the more difficult [CORS97]. With no central authority available to monitor and control the heterogeneous peer-to-peer multihop communications, a major challenge in multihop, multimedia networks seem to be that of accounting for the network resources so that bandwidth allocations can be placed on them. In cellular (single hop) networks such accountability is made easily by the fact that all stations learn of each other's requirements through a centrally administered control station (e.g., home/foreign agent in mobile IP [RFC 2002], base station subsystem (BSS) in GSM [GSM92], and SGSN in GPRS [GOOD97]). However, because of the decentralized nature of MANETs, this approach cannot easily be extended to the multihop environment; thus rendering QoS guarantees a very complex task.

3.2 Current Notion of Quality-of-Service

Quality has always been a problematic kind of word. Everyone knows what it means—or is instantly aware of its absence in any kind of transaction—but, when it comes to providing an accurate definition, things invariably start getting more complex.

QoS in a broad sense contracts the level of user satisfaction in the services provided by a communication system. Intrinsic to the notion of QoS is an agreement or a guarantee by the network to provide a set of measurable prespecified service attributes to the user in terms of network delay, delay variance (jitter), available bandwidth, probability of packet loss (loss rate), and so on.

Many standardization bodies for telecommunications around the globe have carried out intensive studies for defining QoS standards. The International Telecommunication Union (ITU) is one of the principal standardization bodies. Within the ITU, the principal forum for discussing the studies was the Comite Consultatif International des

Telegraphes et des Telephones (CCITT), now known as the International Telecommunication Union—Telecommunications Standardization Sector (ITU-T). The principal ITU-T activity consists of producing recommendations. Many countries have their own national standardization bodies such as the British Standards Institution (BSI) in the United Kingdom and the American National Standards Institute (ANSI) in the United States. These national bodies usually specify standards from the parent organization, the International Standards Organization (ISO). The European Telecommunications Standards Institute (ETSI) is concerned with the development of telecommunications standards for Europe.

Main QoS-related standards are listed in Table 3.1.

TABLE 3.1   QoS-Related Standards

ITU-T	ISO	ESTI
ITU-T E.430: Quality of service framework	ISO/IEC JTC1 SC21 QoS basic framework	ETSI ETR 003: General aspects of quality of service and network performance
ITU-T E.600: Telephone network and ISDN/ quality of service, network management and traffic engineering	ISO/IES DIS 13236 information technology – quality of service – framework	ETSI ETR 138: Network, QoS indicators for ONP of voice telephony and ISDN
ITU-T E.800: Quality of service and dependability vocabulary		ETSI TR 22.25 V3.1.0 (1998-03) universal mobile telecommunication system (UMTS); quality of service and network performance.
ITU-T E771: Network GoS parameters and target values for circuit-switched public and land mobile services		
ITU G.281: Error performance of an international digital connection forming part of an integrated services digital network		
ITU-T I.350: General aspects of quality of service and network performance in digital networks, including ISDNs		
ITU-T I.356: Quality of service configuration and principles		

3.2.1   QoS-sensitive real-time and non-real-time applications

Real-time applications can generally be defined as those with playback characteristics—a data stream that is packetized at the source and transported through the network to its destination, where it is depacketized and played back by the receiving application. *Real-time applications* can be divided in two general classes [CLARK92]—those that are tolerant and those that are intolerant of induced *jitter* (the variance between the minimal and maximal delay). Tolerant applications can be characterized as those that can function in the face of nominal induced jitter and still produce a reasonable signal quality when played back. Examples of such tolerant real-time applications are various packetized audio- or video-streaming applications.

Intolerant applications can be characterized as those in which induced jitter and packet delay result in enough introduced distortion to effectively render the playback signal quality unacceptable or to render the application nonfunctional. In fact certain multimedia applications may become useless if their data incur more than an x seconds delay - where x can range from 100 ms to 5 s. Examples of intolerant applications are two-way telephony applications (interactive voice, as opposed to noninteractive audio playback) and circuit-emulation services.

On the other hand, *non-real-time, or elastic*, applications differ from real-time applications in that they always wait for packets to arrive before the application actually processes the data. Several types of elastic applications exist—interactive burst (e.g., Telnet), interactive bulk transfer (e.g., *file transfer protocol* (FTP)), and asynchronous bulk transfer (e.g., *simple mail transport protocol* (SMTP)). The delay sensitivity varies dramatically with each of the types, so they are often referred to as belonging to a *best-effort service class* [WC96].

The data carried by non-real-time applications, such as Web, file transfer, and electronic mail, is, for the most part, static content (e.g., as text and images). When static content is sent from one host to another, it is desirable for the content to arrive at the destination as soon as possible. Nevertheless, moderately long end-to-end delays, up to tens of seconds, are often tolerated for static content.

Notably, whereas delay is a critical parameter in real-time applications, multimedia networking applications are *loss tolerant*; occasional loss only causes occasional glitches in the audio/video playback, and often these losses can be partially or fully concealed. Thus, in terms of service requirements, multimedia applications are diametrically opposite to static-content applications—multimedia applications are delay sensitive and loss tolerant whereas the static-content applications are delay tolerant and loss intolerant.

These real-time QoS service classes can be further divided into the following general categories.

Delay and delay-sensitive services. A broad class of multimedia applications, such as interactive multimedia, Internet telephony, and video conferencing, may require stringent delay, delay jitter (or delay variation), and loss guarantees. For example, in real-time playback applications, packets arriving after the playback point will be useless, and the loss of a certain number of packets will seriously degrade the quality of voice and pictures. When the application is not merely passive, as it is in playback, but interactive, these effects can be even more insidious and, if severe, can render the application useless.

End-to-end delay and delay jitter are cumulative (or additive) attributes of all links in the path. The end-to-end delay is divided into the following components—propagation delay, transmission delay, packetization/buffering delay, congestion delay, access delay, and queuing delay. While *propagation delay* is determined by the physical distance between the source and the destination, *transmission delay* is determined by the capacity of the bottleneck link on the path. *Packetization delay* is introduced while packets are being formed and buffering delay may be introduced when they are being disassembled. Packetization delay is defined in Ref. [ETSI-QoS] as "the time taken for enough information to fill a whole packet or until enough information is available, before sending it to the network." Congestion delay arises due to the statistical sharing of the finite link bandwidth by packets belonging to multiple flows: packets arriving at a switch are buffered in a service queue until the outgoing link is available. *Congestion delay* of a packet is the time spent by it in the service queue; this delay depends on the number of flows using a link, the link capacity, and the traffic generation rate of different sources. To provide real-time services, this congestion delay must be bounded or minimized.

The *access (MAC) delay* is defined as the time between the instant at which a packet comes to the head of a station's transmission queue and the end of the packet transmission [CONTI97]. *Queuing delay* [CONTI97], on the other hand, is determined by the network load, the burstiness of the traffic source, and the service disciplines employed in the network. Typically, the transmission delay is seen only for the first packet transmitted because of the pipelining of all other packets transmitted after it, while propagation delay is only a very small fraction of the end-to-end delay because of the increased link capacity. Queuing delay is potentially the dominant component in the end-to-end delay (in a packet-switched network) and can be limited through the use of different scheduling algorithms, conforming traffic sources, and bandwidth reservations [ZHAN95].

Delay-sensitive services can be seen as equivalent to circuit-switched services, in the sense that preallocation of a certain bandwidth needs to be done. In delay-sensitive service, therefore, guaranteed resources are preallocated and maintained throughout the service lifetime. Such a service can be voice telephony over a packet-based network, which assumes that a certain amount of bandwidth is reserved for the corresponding connection. Voice traffic is sensitive to delay, which in the case of a packet-based network brings into consideration the end-to-end latency due to media transcoding and signaling adaptation.

Throughput and bandwidth-sensitive applications. Traditional best-effort applications, such as *remote procedure call* (RPC), electronic mail, ftp, and telnet, usually send messages as small as a few kilobytes. The main performance index for these applications is the *end-to-end per packet delay*. As the sophistication of networked applications has grown, so too has the amount of data transmitted. It is now not uncommon to observe applications whose data payload reaches from several hundred megabytes to several terabytes [MIGU94, GIBS95, GIBS97].

In contrast to the transmission of small messages, these new applications can consume as much network bandwidth as is available. It is the end-to-end throughput rather than the per packet end-to-end delay that is the main performance concern. *Throughput* can be defined as the maximum fraction of channel bandwidth [CONTI00, KUROSE] used by successfully transmitted messages.

Transmission of multimedia streams requires a minimum bandwidth to ensure end-to-end QoS guarantees. Bandwidth guarantees can be requested for different time intervals depending on applications. For example, assuming that an application is adaptive and its source and destination nodes have sufficiently large buffer space, the bandwidth provided by the network can vary over time, as long as the average bandwidth provided is higher than the minimum bandwidth required by the application. We will look at adaptive applications in more depth in Sec. 3.9.8.

If an application is less adaptive, the network may have to reserve more bandwidth than the amount that matches the average packet sending rate. Alternatively, a mechanism to support bandwidth renegotiations is needed [GROSS95], such that bandwidth reservations are provided at a finer time scale than per-session bandwidth guarantees (see Sec. 3.9.9).

Error and error-sensitive services. The error-sensitive services are mostly ones that assume reliable end-to-end transmission of data packets.

The wireless medium is seen as a weak link in the chain of transmission media, however, end-to-end protocols, such as TCP, or physical layer methods, such as using proper FEC (forward error correction) schemes, ensure a reliable transmission of data.

3.2.2 Examples of multimedia applications

There exist a large variety of multimedia applications. Three general classes of multimedia applications are defined as follows.

Real-time interactive audio and video.   This class of applications allows people to use audio/video to communicate with each other in real-time. Real-time interactive audio is often referred to as *Internet phone*. Internet phone can potentially provide PBX, local, and long-distance telephone service at very low cost. It can also facilitate *computer-telephone integration* (CTI), group real-time communication, directory services, caller identification, caller filtering, and the like.

Note that in a real-time interactive audio/video application, a user can speak or move at anytime. The delay from when a user speaks or moves until the action is manifested at the receiving hosts should be less than a few hundred milliseconds. For voice, delays smaller than 150 ms are not perceived by a human listener, delays between 150 and 400 ms can be acceptable, and delays exceeding 400 ms result in frustrating, if not completely unintelligible, voice conversations.

There are many Internet telephone products currently available. With real-time interactive video, also called *video conferencing*, individuals communicate visually as well as orally. During a group meeting, a user can open a window for each participant the user is interested in seeing. There are also many real-time interactive video products currently available for the Internet, including Microsoft's *Netmeeting* (*http://www.microsoft.com/windows/netmeeting/*).

One-to-many streaming of real-time audio and video.   This class of applications is similar to ordinary broadcast of radio and television, except that the transmission takes place over the Internet. These applications allow a user to receive a radio or television transmission emitted from any corner of the world. This class of applications is noninteractive; a client cannot control a server's transmission schedule. As with streaming of stored multimedia, requirements for packet delay and jitter are not as stringent as those for Internet telephony and real-time video conferencing. Delays up to tens of seconds from when the user clicks on a link until audio/video playback begins can be tolerated. Distribution of the real-time audio/video to many receivers is efficiently done either with multicast or with separate unicast streams to each of the receivers.

Streaming stored audio and video.   In this class of applications, clients request on-demand compressed audio or video files, which are stored on servers. For audio, these files can contain a professor's lectures, rock songs, symphonies, archives of famous radio broadcasts, as well as historical archival recordings. For video, these files can contain videos of professors' lectures, full-length movies, prerecorded television shows, documentaries, video archives of historical events, video recordings of sporting events, cartoons, and music video clips. At any time a client machine can request an audio/video file from a server. In most of the existing stored audio/video applications, after a delay of a few seconds the client begins to playback the audio file while it *continues* to receive the file from the server. The feature of playing back audio or video while the file is being received is called *streaming*. For streaming to work, the client side receiving the data must be able to collect the data and send it as a steady stream to the application that is processing the data and converting it to sound or pictures.

Streaming applications are very asymmetric and therefore typically withstand more delay than more symmetric conversational services. The delay from when a user makes a request (e.g., request to hear an audio file or skip 2-min forward) until the action manifests itself at the user host (e.g., user begins to hear audio file) should be on the order of 1 to 10 s for acceptable responsiveness. This also means that they tolerate more jitter in transmission. Requirements for packet delay and jitter are not as stringent as those for real-time applications such as Internet telephony and real-time video conferencing. Jitter can be easily smoothed out by buffering.

Internet video products and the accompanying media industry as a whole is clearly divided into two different target areas: (1) Web broadcast and (2) video streaming on-demand. Web broadcast providers usually target very large audiences that connect to a highly performance -optimized media server (or choose from a multitude of servers) via the actual Internet. The on-demand services are more often used by big corporations that wish to store video clips or lectures to a server connected to a higher bandwidth local Intranet—these on-demand lectures are seldom used simultaneously by more than hundreds of people. There are many streaming products for stored audio/video, including RealPlayer from RealNetworks (*http:// www.real.com/player/index. html?*) and NetShow from Microsoft (*http://www.microsoft.com /presspass/press/ 1997/Apr97/ nsmompr.asp*).

Table 3.2 summarizes the performance characteristics of various examples of QoS service types.

TABLE 3.2   Examples of UMTS Specific Quality-of-Service Values (from [RAPELI95])

Teleservice	Throughput (kb/s)	Residual error rate	Delay (ms)
Speech telephony/terrestrial	8–32	10E-4	40
Voice data band	2.4–64	10E-6	200
Program sound	128	10E-6	200
High quality audio	940	10E-5	200
Video telephony	64–384	10E-7	40–90
Short message/paging	1.2–64	10E-6	100
Electronic mail	1.2–9.6	10E-6	100
Telefax (G4)	64	10E-6	100
Broadcast/multicast	1.2–9.6	10E-6	100
Public voice announce	8–32	10E-4	90
Public A/N announce	1.2–9.6	10E-6	100
Unrestricted digital data	64–1920	10E-6	100
Database access	2.4–768	10E-6	200+
Teleshopping	2.4–768	10E-6/10E-7	90
Electronic newspaper	2.4–2000	10E-6	200
Remote control service	1.2–9.6	10E-6	100
Location and navigation	64	10E-6	100
Telewriting	32–64	10E-6	90

3.3   Routing with Quality-of-Service Constraints

Today the Internet operates in a connectionless and stateless mode. That is, the network of routers is not aware of any association between the source and the destination except on a per-packet basis. QoS is a big issue with packet services. Left to their own devices, packet networks tend to sacrifice deterministic performance for relative performance, where one packet gets priority over another; yet, it gets no guarantee. Switched services, in general, do not have this immediate technical problem, since a channel or circuit, or in the case of ATM a virtual circuit, is reserved through the network for the flow booked to cross it.

For a packet (IP) network, there is no intrinsic way of guaranteeing a performance "floor" for any particular data flow. This means that sudden surges in traffic can cause congestion, say through a peering point on the Internet, which will first cause the packets to be queued in wireless routers, and in turn cause the application to slow. If the congestion becomes really acute, a router is forced to "drop" packets when its storage buffer overflows. This causes the end-systems to request a resend of the missing packet. When random, these occurrences matter little in the context of a file transfer (for email, say) or even an Internet browsing session—it just means that whatever is supposed to happen, it will take a little longer to appear on the screen than might otherwise.

For time-sensitive applications, such as voice over IP, these are a major headache—in the worst case they can render a telephone conversation useless, since jitter (irregular arrival of packets) and latency can cause satellite-link style gaps between different ends of a conversation, and dropped packets can cause dead patches or bangs.

In this regard, in packet-based networks each packet is routed individually without any information about the state of the flow. However, QoS is meaningful only for a *flow* of packets that associates the notion of a logical connection with the *duration of the flow*.

Consequently, appropriate routers must remain aware of the logical connection and the state of the flow to ensure that adequate network resources such as link bandwidth, nodal buffers, and processing power are available for the duration of the logical connection [STOICA99].

A QoS provisioning policy should provide QoS-supporting modules such as resource reservation techniques, QoS routing protocol, and a call admission control policy. The most prominent among them is *QoS routing*. QoS routing is the process of choosing the routes to be used by the flow of a logical connection in attaining the preestablished QoS guarantees. The task of network resource identification and reservation is the other indispensable ingredient of QoS routing—to produce routes that have the necessary resources available to meet the QoS constraints imposed by the desired service.

The main objectives of QoS-based routing are [RFC 2386]:

1. *Dynamic determination of feasible paths.* QoS-based routing can determine a path, from among many possible choices, that has a good chance of accommodating the QoS of the given flow. Feasible path selection may be subject to policy constraints, such as path cost and provider selection.

2. *Optimization of resource usage.* A network state-dependent QoS-based routing scheme can aid in the efficient utilization of network resources by improving the total network throughput. Such a routing scheme can be the basis of efficient network engineering.

3. *Graceful performance degradation.* State-dependent routing can compensate for transient inadequacies in network engineering (e.g., during focused overload conditions), giving better throughput and a more graceful performance degradation as compared to a state-insensitive routing scheme.

Candidate routes for a flow with specific QoS objectives are determined by using various QoS metrics. A metric is associated with a link and determines its QoS capabilities (with respect to this metric). Metrics collectively characterize the *state* of the nodes and links that compose a path.

A typical *link-state* is an ordered tuple of its specific QoS metrics of interest, and is usually represented as follows:

link-state = < bandwidth, propagation delay, cost >

where bandwidth is the maximum residual bandwidth of that link, and "cost" here is used as a generic catch-all for other parameters such as packet loss statistics and service class (if multiple service classes are to be supported, each with its own QoS requirements).

Likewise, the *state* of a node is characterized by an ordered tuple of typical QoS metrics as follows:

node-state = < CPU bandwidth capacity, delay distribution, cost >

where the CPU bandwidth capacity is the minimum rate at which the node can place data into the link, delay distribution at a minimum includes the mean and the variance of the queuing delay, and "cost" is again a generic term for many other parameters that need be considered for different service classes for different traffic types, including service classes with multiple priorities.

It becomes evident then that the QoS routing problem is a constrained combinatorial (graph) optimization problem and shall be solved likewise. As a consequence, the computational and communication complexities of route selection criteria must also be taken into account.

It is a general practice for the node-state to be incorporated into the state of each of the links incident on it. As a result, the state of a route becomes a function of the individual states of its constituent links. Formally, the end-to-end route delay is the cumulative sum of the transmission times and queuing hold times, the nominal bandwidth of a route is the minimum of its constituent links, whereas cost is either the sum (if it is an additive quantity) or another appropriate deterministic or stochastic numerical operation of its component costs.

To minimize delay, selecting short routes is an important criterion in determining suitable routes. For QoS flows, a *feasible* route is one with sufficient available resources to satisfy certain QoS requirements but is not necessarily the shortest in length. The family of minimum-hop routing algorithms assign to each hop (i.e., link) a unity cost, thus treating equally a high-speed 5.6 Gb/s link with a low-speed 150 kb/s link, or likewise a low-quality link (due to interference, shadowing and the like) with a link with good quality.

To put this into context, consider Fig. 3.1, where the numbers next to the links represent their respective bandwidths, say in Mb/s. Suppose that the packet flow from A to E requires a bandwidth guarantee of 3 Mb/s. Shortest-path routing would select the shortest route, which is [A-D-E]. QoS routing on the other hand will select the route [A-B-C-E] over

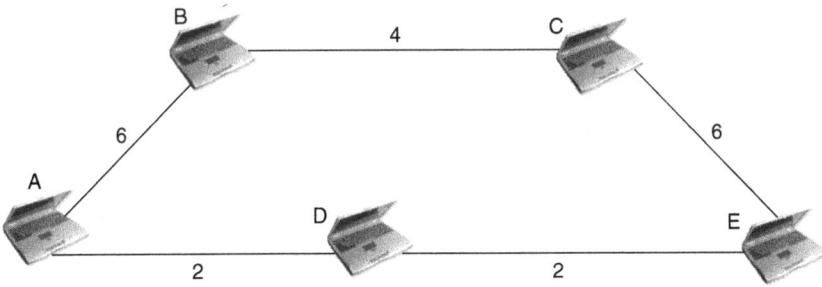

Figure 3.1   Links with fixed bandwidth in a network.

the routes [A-D-E]. This is because the latter, although with fewer hops, does not meet the bandwidth requirements of the data flow. If there are several feasible paths available, the one that leads to higher network throughput should be selected.

To define suitable link and node metrics, several considerations shall be taken into account [WC96]. First, the metrics must represent the basic network properties of interest. Such metrics include residual bandwidth, delay, and jitter. Since the flow QoS requirements are mapped onto path metrics, the metrics constitute an abstract representation of the QoS guarantees the network can support.

Second, path computation based on a metric or a combination of metrics must not be too complex so as to render them impractical. In general, path computation based on certain combinations of metrics (e.g., delay and jitter) is theoretically hard. Experiments in Ref. [WANG96] verify this conjecture. The allowable combinations of metrics must thus be determined while taking into account the complexity of computing paths based on these metrics and the QoS needs of flows. A common strategy to allow flexible combinations of metrics while at the same time reduce the path computation complexity is to utilize *sequential filtering.* In this approach, a combination of metrics is ordered in some fashion, reflecting the importance of different metrics (such as cost followed by delay). Paths based on the primary metric are computed first using a simple algorithm (e.g., shortest path) and subsets of them are eliminated based on the secondary metric and so forth until a single path is found. This is an approximation technique and trades off global optimality with path computation simplicity. The filtering technique may, however, be simpler depending on the set of metrics used. For example, with bandwidth and cost as metrics, it is possible to first eliminate the set of links that do not have the requested bandwidth and then compute the least-cost path using the remaining links.

3.4   Hurdles for Multimedia on the Internet

Routing deployed on the Internet today is focused on connectivity and typically supports only "best-effort" services. With best-effort services, all packets are typically treated equally in the network. Any congested link can induce increased packet delivery times, which, in turn, can result in generally poor performance. Traditional best-effort traffic, such as electronic mail, telnet, and RPC, constitutes mostly small messages with a payload typically less than a few tens of kilobytes of data. Users would like to have their messages arrive at their destination as quickly as possible. For this kind of low-latency traffic, it has been shown [WANG91] that the shortest-path routing may work well when the path is not congested.

The best-effort service, however, does not make any promises whatsoever about the end-to-end delay for an individual packet. Nor does the service make any promises about the variation of packet delay within a packet stream. The delay perceived by users is one of the main QoS factors in a number of existing services (such as voice transmissions). This delay must be kept almost constant and below a certain level in order to allow a normal conversation over the network. If this delay is too high, an interactive communication becomes very difficult or even impossible.

Due to the lack of any special effort to deliver packets in a timely manner, it is an extremely challenging problem to develop successful multimedia networking applications for the Internet. To date, multimedia over the Internet has achieved significant but limited success. For example, streaming store audio/video with user-interactivity delays of 5 to 10 s is now commonplace on the Internet. But during peak traffic periods, performance may be unsatisfactory, particularly when intervening links are congested links (such as congested transoceanic links).

Internet phone and real-time interactive video on the other hand has, to date, been less successful than streaming stored audio/video. Indeed, real-time interactive voice and video impose rigid constraints on packet delay and packet jitter. Real-time voice and video can work well in regions where bandwidth is plentiful, and hence delay and jitter are minimal. But again quality can deteriorate to unacceptable levels as soon as the real-time voice or video packet stream hits a moderately congested link.

The design of multimedia applications would certainly be more straightforward if there were some sort of first-class and second-class Internet services, where first-class packets are limited in number and always get priority in router queues. Such a first-class service could be satisfactory for delay-sensitive applications. But, to date, the Internet has mostly taken an egalitarian approach to packet-scheduling in router queues—all packets receive equal service; no packets, including delay-sensitive audio and video packets, get any priority in the router queues. No matter how much money you are willing to spend, you must join at the end of the line and wait your turn!

So, for the time being we have to live with the best-effort service. No matter how important or how rich we are, our packets have to wait their turn in router queues. But given this constraint, we can make several design decisions and employ a few tricks to improve the user-perceived quality of a multimedia networking application. For example, we can delay playback at the receiver by 100 ms or more in order to diminish the effects of network-induced jitter. We can timestamp packets at the sender so that the receiver knows when the packets should be played back. For stored audio/video we can prefetch data during playback when client storage and extra bandwidth is available. We can even send redundant information in order to mitigate the effects of network-induced packet loss.

Key to QoS support in the case of the Internet is QoS-based routing. QoS routing must extend the current routing paradigm in two basic ways. First, to support integrated-services traffic, multiple paths between node pairs will have to be calculated. Some of these new classes of service will require the distribution of additional routing metrics, such as delay, and available bandwidth. If any of these metrics change frequently, routing updates can become more frequent thereby consuming network bandwidth and router CPU cycles.

Second, today's "opportunistic" routing will shift traffic from one path to another as soon as a "better" path is found. The traffic will be shifted even if the existing path can meet the service requirements of the existing traffic. If routing calculation is tied to frequently changing consumable resources (e.g. available bandwidth) this change will happen more often introducing routing oscillations as traffic shifts back and forth between alternate paths. Frequently changing routes can further increase the variation in the delay and jitter experienced by the end users.

3.5   How Should the Internet Evolve to Better Support Multimedia?

Over the last few years, there has been extensive—and sometimes ferocious—debate about how the Internet should evolve in order to better accommodate multimedia traffic. At one extreme, some researchers argue that it is not necessary to make any fundamental changes to the best-effort service and the underlying Internet protocols. Instead, it is only necessary to add more bandwidth to the links. In fact, until recently, we've avoided many problems, thanks to the overprovisioning in core IP operator networks.

Opponents to this viewpoint argue that additional bandwidth can be costly, and as soon as it is put in place it will be eaten up by new bandwidth-hungry applications (e.g., high-definition video on demand).

At the other end, some researchers argue that fundamental changes should be made to the Internet so that applications can explicitly reserve end-to-end bandwidth on as-needed basis. If, for example, a user wants to make an Internet phone call from host A to host B, then the user's Internet phone application should be able to explicitly reserve bandwidth in each link along a route from host A to host B. But allowing applications to make reservations as well as the network to honor them requires fundamental changes both on the network side as well as the end-hosts running the applications. First a protocol is needed that, on behalf of applications, reserves bandwidth from the senders to their receivers. Second, scheduling policies in the router queues need be modified so that bandwidth reservations can be honored. With these new scheduling policies, all packets no longer get equal treatment; instead, those that reserve (and pay) more get more. Third, in order to honor reservations, applications need to give the network a description of the traffic that they intend to send into the network. The network must then police each application's traffic to make sure that it conforms to the description. Finally, the network must have a means of determining whether it has sufficient available bandwidth to support any new reservation request. These mechanisms, when combined, require new and complex software in the hosts and routers as well as new types of services. This QoS-based routing architecture is named integrated services architecture.

There is a conciliative approach between these two extremes—the so-called differentiated services framework. DiffServ intends to make relatively small changes at the core network and introduce simple pricing and policing schemes at the edge of the network (i.e., at the interface between the user and the user's ISP). The idea is to introduce a small number of classes, assign each datagram to one of these classes and, according to its class, map each datagram to a different level of service. A simple example of a differentiated-services Internet is as follows. By toggling a single bit in the datagram header, all IP datagrams are labeled as either first-class or second-class datagrams. In each router queue, each arriving first-class datagram jumps in front of all the second-class datagrams; in this manner, second-class datagrams do not interfere with first-class datagrams—this model behaves then as if the first-class packets have their own network! The network edge counts the number of first-class datagrams each user sends into the network each week. When a user subscribes to an Internet service, it can opt for a "platinum service" whereby the user is permitted to send a large but limited number of first-class datagrams into the network each week; first-class datagrams in excess of the limit are converted to second-class datagrams at the network edge. The network is dimensioned and the first-class service is priced so that, almost always, first-class datagrams experience insignificant delays at all router queues. In this manner, sources of audio/video can subscribe to the first-class service, and thereby receive, almost always, satisfactory service.

A user can also opt for a low-budget service, where all the datagrams are second-class datagrams. Of course, the user shall pay a higher subscription rate for a platinum service than for a low-budget service.

3.5.1 Integrated Services

Integrated services [RFC 1633] is a framework developed within the IETF to provide customized QoS guarantees to individual application sessions. Two key features lie at the heart of IntServ architecture:

- *Reserved resources.* A router is required to know the amount of its resources (buffers, link bandwidth) that are already reserved for ongoing sessions.

- *Call setup.* A session requiring QoS guarantees must first be able to reserve sufficient resources at each network router on its source-to-destination path to ensure that its end-to-end QoS requirement will be granted. This call setup (also known as call admission) process requires the participation of each router on the path. Each router must determine the local resources required by the session, consider the amounts of its resources that are already committed to other ongoing sessions, and determine whether it has sufficient resources to satisfy the per-hop QoS requirement of the session at this router without violating QoS local QoS guarantees made to already admitted sessions.

Figure 3.2 depicts the call setup process. Let us now consider the steps involved in call admission in more detail.

- *Traffic characterization and specification of the desired QoS.* For a router to determine whether or not its resources are sufficient to meet the QoS requirements of a session, that session must first declare its QoS requirement, as well as characterize the traffic for which it requires QoS guarantee. In the IntServ architecture, the so-called Rspec (R for reserved) defines the specific QoS being requested by a connection; the so-called Tspec (T for traffic) characterizes the traffic the sender will be sending into the network, or the receiver will be receiving from the network. The specific forms of the Rspec and Tspec vary, depending on the service requested, as discussed below. The Tspec and Rspec are defined in part in Refs. [RFC2210] and [RFC 2215].

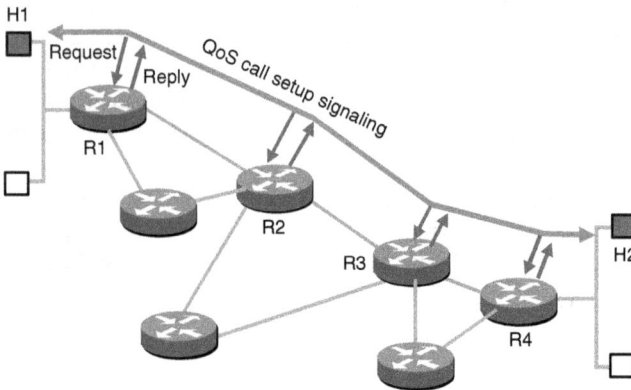

Figure 3.2 The call setup process.

- *Signaling for call setup.* A session's Tspec and Rspec must be carried to the routers where resources will be reserved for the session. On the Internet, the *resource reservation protocol* (RSVP) [RFC 2205] (see "The resource reservation protocol" in Sec. 3.5.1 for details), is currently the signaling protocol of choice. [RFC 2210] describes the use of the RSVP resource reservation protocol with the IntServ architecture.

- *Per-element call admission.* Once a router receives the Tspec and Rspec for a session requesting a QoS guarantee, it can determine whether or not it can admit the call. This call admission decision will depend on the traffic specification, the requested type of service, and the existing resource commitments already made by the router to ongoing sessions.

The IntServ architecture defines two major classes of service—*guaranteed service* and *controlled-load service.*

Guaranteed quality-of-service. The guaranteed service definition in Ref. [RFC 2212] provides a framework for delivering traffic for applications with a bandwidth guarantee and delay bound-applications that possess intolerant real-time properties. The guaranteed service only provides firm (mathematically provable) bounds on the queuing delays that a packet will experience in a router. The guaranteed service model is based on a class of rate-proportional packet scheduling algorithms [PAREK93, DEME89, CLARK92, ZHAN95, STILIAD97, GOYA95]. Under these service disciplines, packets of different connections sharing the same output link are sent in an order that ensures a weighted fair sharing of link capacity among these connections. As a result, a guaranteed rate is ensured for flows that use the guaranteed service. Packets transmitted on such flows are thus protected from either ill-behaved applications or intentional link sabotage.

To invoke the guaranteed service, an application specifies its traffic characteristics and desired performance guarantees. The network, on the other hand, reserves a certain amount of resources at each node (switch or router) on its path. The traffic specification and the QoS guarantees constitute thus a "contract" between the network and the application—once a flow is admitted into the network and the traffic source conforms to its traffic specification, the network will provide guaranteed QoS. The guaranteed service guarantees that packets will arrive within a certain delivery time and will not be discarded because of queue overflows, provided that the flow's traffic stays within the bounds of its specified traffic parameters. The guaranteed service, however, does not control the minimal or average delay of traffic nor does it control or minimize jitter—it only controls the maximum queuing delay.

To a first approximation, a source's traffic characterization is given by a leaky bucket (see "Leaky-bucket implementation" in Sec. 3.9.10) with parameters (r, b) and the requested service is characterized by a transmission rate R at which packets will be transmitted. In essence, a session requesting guaranteed service requires that the bits in its packet be guaranteed a forwarding rate of R bits per second. Given that traffic is specified using a leaky bucket characterization, and a guaranteed rate of R is being requested, it is also possible to bound the maximum queuing delay at the router. Recall that with a leaky bucket traffic characterization, the amount of traffic (in bits) generated over any interval of length t is bounded by $rt + b$.

Controlled-load network service.   A session receiving controlled-load service will receive "a QoS closely approximating the QoS that the same flow would receive from an unloaded network element" [RFC 2211]. Assuming that the network is functioning correctly (i.e., routing and forwarding), applications may assume that a very high percentage of transmitted packets will be delivered successfully and that any latency introduced into the network will not greatly exceed the minimum delay experienced by any successfully transmitted packet.

Even under congestion, network nodes offering controlled-load service are expected to provide flows with low delay and low packet loss. To ensure that this set of conditions is met, the application requesting the controlled-load service provides the network with an estimation of the traffic it will generate using the TOKEN-BUCKETTSPEC—the Tspec or traffic specification. In turn, each node handling the controlled-load service request ensures that sufficient resources are available to accommodate the request.

To limit the number of flows receiving the service, the controlled-load service requires applications to make explicit requests. Such requests can be placed using a reservation setup protocol, such as the RSVP, or some other means. Each network node that receives a request for service can either accept or reject the request. In this vein, a flow can receive a guaranteed average rate with no per-packet delay guarantees.

Traffic control.   The integrated services model defines four mechanisms that comprise the traffic-control functions at Layer 3 (the router) and above:

1. *Packet scheduler.*   The scheduling mechanism may be represented as an abstract queuing implementation and may be implementation specific. The packet scheduler also assumes the function of traffic policing.

2. *Packet classifier.*   This maps each incoming packet to a specific class so that these classes may be acted on individually to deliver traffic differentiation.

3. *Admission control.*   Just as the manager of a restaurant should not accept reservations for more tables than the restaurant has, the admission policy—in the context of the IETF integrated services architecture—determines whether a network node has sufficient available resources to supply the requested QoS. Admission control, in the most primitive sense, is the simple practice of determining whether a flow can be granted the requested QoS without affecting other established flows in the network.

4. *Resource reservation.*   A resource reservation protocol (in this case, RSVP) is necessary to set up flow state in the requesting end-systems as well as each router along the end-to-end flow-transit path.

Figure 3.3 shows a reference model illustrating the relationship of these functions.

The resource reservation protocol.   The resource reservation protocol [RFC 2205] is a signaling protocol that allows applications to reserve resources. RSVP is referred to as a *signaling protocol* in that it allows hosts to establish and tear down reservations for data flows. The term *signaling protocol* comes from the jargon of the circuit-switched telephony community. The ability to request and reserve *per-flow* resources makes it possible for the IntServ framework to provide QoS guarantees to individual flows.

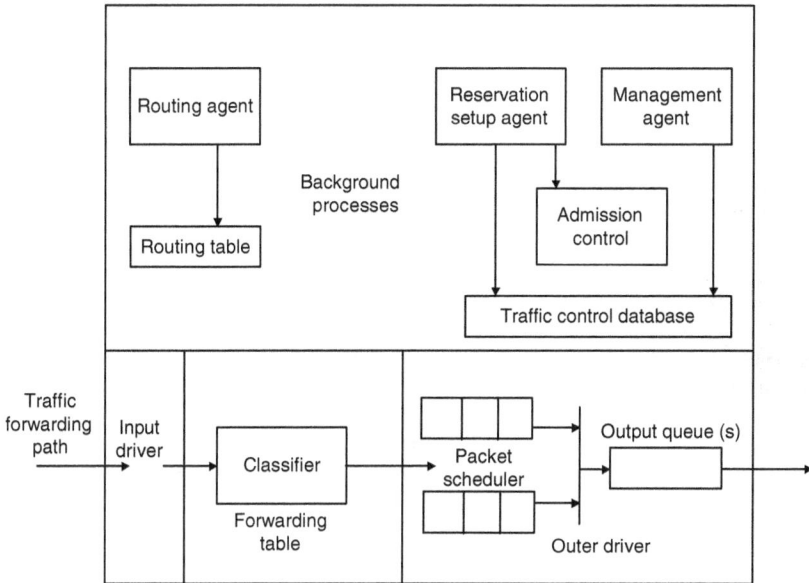

Figure 3.3 Implementation reference model.

When people talk about *resources*, at least in the context of the Internet, they usually mean link bandwidth and router buffers. To keep the discussion concrete and focused we shall assume that the word *resource* is synonymous with *bandwidth*.

RSVP requests generally result in resources being reserved and soft-state timers maintained for each flow in each router in the transit path. A soft state is maintained by periodic refresh messages sent along the data path to maintain the reservation and path state. In the absence of these periodic messages, which typically are sent every 30 s, the state is deleted as it times out. This soft state is necessary because RSVP is essentially a QoS reservation protocol and does not associate the reservation with a specific static path through the network. As such, it is entirely possible that the path can change. Reservation state shall therefore be refreshed periodically. By this, it is meant that each reservation for bandwidth stored in a router has an associated timer. If a reservation's timer expires, then the reservation is removed. If a receiver desires to maintain a reservation, it must periodically refresh the reservation by sending reservation messages.

The senders must also refresh the path state by periodically sending *path* messages. When a route changes, the next path message initializes the path state on the new route whereas future reservation messages will establish reservation states on the new route. The state on the old segments of the route will time-out.

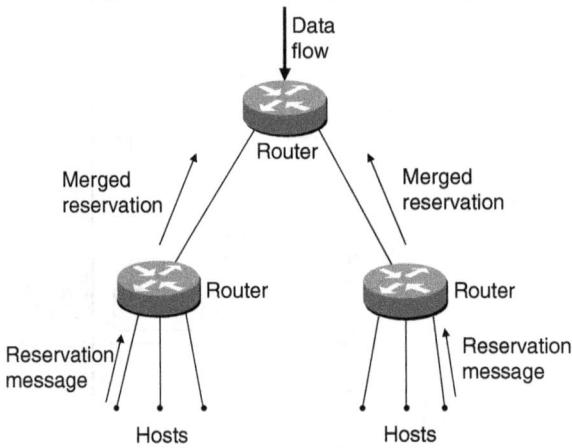

Figure 3.4   RSVP: multicast- and receiver-oriented.

The two principal characteristics of RSVP are:

1. It provides *reservations for bandwidth in multicast trees* (unicast is handled as a special case).

2. It is *receiver-oriented*, i.e., the receiver of a data flow initiates and maintains the resource reservation used for that flow.

These two characteristics are illustrated in Fig. 3.4.

The above diagram shows a multicast tree with data flowing from a single data source to six hosts. Although data originate from the sender (source of data call), the reservation messages originate from the receivers (sink of data call). When a router forwards a reservation message upstream toward the sender, the router may merge the reservation message with other reservation messages arriving from downstream.

Figure 3.5 shows a simplified version of RSVP in hosts and routers..

Reservation (Resv) messages.   The Resv message contains information about the reservation style, the appropriate flowspec object, and the filter spec that identify the sender(s). The pairing of the flowspec and the filter spec is referred to as the flow descriptor. The flowspec is used to set parameters in a node's packet-scheduling process, and the filter spec is used to set parameters in the packet-classifier process. Data that does not match any of the filter specs is treated as best-effort traffic.

As we mentioned in the previous section, various bits of information must be communicated between the receiver(s) and intermediate

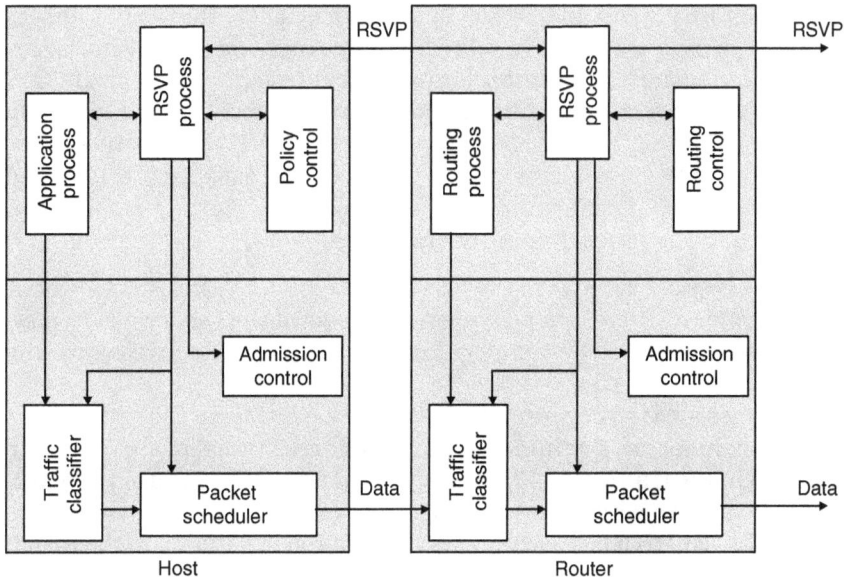

Figure 3.5 RSVP in hosts and routers.

nodes to appropriately invoke QoS control services. Among the data types that need to be communicated between applications and nodes is the information generated by each receiver that describes the QoS control service desired, a description of the traffic flow to which the resource reservation should apply (receiver Tspec), and the necessary parameters required to invoke the QoS service (receiver Rspec). This information is contained in the flowspec (flow specification) object carried in the Resv messages. The Resv message is generated by the receiver and is transported back upstream toward the sender, creating and maintaining the reservation state at each node along the traffic path, following the reverse path on which path messages were previously sent. This is the same path data packets will sub-sequently use.

The information contained in the flowspec object may be modified at any intermediate node in the traffic path because of reservation merging and other factors.

The format of the flowspec is different depending on whether the sender is requesting controlled-load or guaranteed service. When a receiver requests controlled-load service, only a Tspec is contained in the flowspec. When requesting guaranteed service, both a Tspec and an Rspec are contained in the flowspec object. This restriction may be removed in future revisions of the RSVP specification.

At each RSVP-capable router in the transit path, the sender Tspecs arriving in path messages and the flowspecs arriving in Resv messages are used to request the appropriate resources from the appropriate QoS control service. State merging, message forwarding, and error handling proceed according to the rules defined in the RSVP specification. Also, the merged flowspec objects arriving at each RSVP sender are delivered to the application, informing the sender of the merged reservation request and the properties of the data path.

Path messages.   Similar to the RSVP *reservation messages*, which originate at the receivers and flow upstream toward the senders, *path messages* originate at the senders and flow downstream toward the receivers.

Path messages store path information in each node in the traffic path, which includes at a minimum the IP address of each previous hop (PHOP) in the traffic path. The IP address of the PHOP is used to determine the path in which the subsequent Resv messages will be forwarded.

The RSVP path message contains information in addition to the PHOP address, which characterizes the sender's traffic. These additional information elements are called the sender template, the sender Tspec, and the Adspec.

The sender template contains information called a filter specification (filter spec), which uniquely identifies the sender's flow from other flows present in the same RSVP session on the same link. The Tspec parameter characterizes the traffic the sender expects to generate; it is transported along the intermediate network nodes and received by the intended receiver(s). The sender Tspec is not modified by the intermediate nodes.

Path messages may contain additional fragments of information contained in an Adspec. When an Adspec is received in a path message, it is passed to the local traffic control process, which updates the Adspec with resource information and then passes it back to the RSVP process to be forwarded to the next downstream hop. Receivers need the information carried within the Adspec in order to choose a QoS control service and determine the appropriate reservation parameters. The Adspec also allows the receiver to determine whether a non-RSVP-capable node lies in the transit path or whether a specific QoS control service is available at each router in the transit path. The Adspec also provides default or service-specific information for the characterization parameters for the guaranteed service class.

Additional message types.   Besides the Resv and path messages, the remaining RSVP message types include path and reservation errors (PathErr and ResvErr), path and reservation torn down (path tear and ResvTear), and confirmation for a requested reservation (ResvConf).

The PathErr and ResvErr messages are simply sent upstream to the sender that created the error and do not modify the path state in the nodes through which they pass. A PathErr message indicates an error in the processing of path messages and is sent back to the sender. ResvErr messages indicate an error in the processing of Resv messages and are sent to the receiver(s).

RSVP tear-down messages remove path or reservation state from nodes as soon as they are received. It is not always necessary to explicitly tear down an old reservation because the reservation eventually times out if periodic refresh messages are not received after a certain period of time. PathTear messages are generated explicitly by senders or by the time-out of path state in any node along the traffic path and are sent to all receivers. An explicit PathTear message is forwarded downstream from the node that generated it; this message deletes path state and reservation state that may rely on it at each node in the traffic path. A ResvTear message is generated explicitly by receivers or any node in which the reservation state has timed out and is sent to all pertinent senders. Basically, a ResvTear message has an effect opposite to that of a Resv message.

Reservation styles.   A reservation request contains a set of options that are collectively called the *reservation style*. The reservation style characterizes how reservations should be treated in accordance with the sender(s) requests.

One option is concerned with the treatment of reservations for different senders within the same RSVP session. The option has two modes—establish a distinct reservation for each upstream sender or establish a shared reservation used for all packets of specified senders.

Another option controls the selection of the senders. This option also has two modes—an explicit list of all selected senders or a wildcard specification that implicitly selects all senders for the session. In an explicit sender-selection reservation, each filter spec must match exactly one sender. In a wildcard sender selection, no filter spec is needed.

Three reservation styles are defined—wildcard-filter style, fixed-filter style, and shared-explicit style.

- *Wildcard-filter style.*   A receiver uses the wildcard-filter style in its reservation message to instruct the network to receive all flows from all upstream senders in the session. Its bandwidth reservation is to be shared among the senders.

- *Fixed-filter style.*   A receiver uses the fixed-filter style in its reservation message to specify a list of senders from which it wants to receive a data flow along with a bandwidth reservation for each of these senders. These reservations are distinct, i.e., they are not to be shared.

■ *Shared-explicit style.* A receiver uses the shared-explicit style in its reservation message to specify a list of senders from which it wants to receive a data flow along with a single bandwidth reservation. This reservation is to be shared among all the senders in the list.

Shared reservations, created by the wildcard filter and the shared-explicit styles, are appropriate for a multicast session whose sources are unlikely to transmit simultaneously. Packetized audio is an example of an application suitable for shared reservations; because a limited number of people talk at once, each receiver might issue a wildcard-filter or a shared-explicit reservation request for twice the bandwidth required for one sender (to allow for overspeaking). On the other hand, the fixed-filter reservation, which creates distinct reservations for the flows from different senders, is appropriate for applications like video teleconferencing.

Merging. The concept of merging is necessary for the interaction of multicast traffic and RSVP. Merging of RSVP reservations is required because of the method multicast uses for delivering packets—basically replicating packets that must be delivered to different next-hop nodes. At each replication point, RSVP must merge reservation requests and compute the maximum of their flowspecs.

Flowspecs are merged when Resv messages, each originating from different RSVP receivers and initially traversing diverse traffic paths, converge at a merge-point node and are merged prior to being forwarded to the next RSVP node in the traffic path (Fig. 3.6). The largest flowspec from all merged flowspecs—the one that requests the most stringent QoS reservation state—is used to define the single merged flowspec, which is forwarded to the next-hop node. Because flowspecs are opaque data elements to RSVP, the methods for comparing them are defined outside the base RSVP specification.

As mentioned earlier, different reservation styles cannot be merged, because they are fundamentally incompatible. You can find specific

Sender Selection	Reservations	
	Distinct	Shared
Explicit	Fixed – Filter (FF) Style	Shared – Explicit (SE) Style
Wildcard	(No style defined)	Wildcard – Filter (WF) Style

Figure 3.6 Flowspec merging.

ordering and merging guidelines for message parameters within the scope of the controlled-load and guaranteed service classes in Refs. [RFC 2211] and [RFC 2212], respectively.

RSVP in operation.   Let us first describe RSVP in the context of a concrete one-to-many multicast example. Suppose there is a source that is transmitting a video stream through the Internet. This session has been assigned a multicast address, and the source stamps all its outgoing packets with this multicast address. Also suppose that an underlying multicast routing protocol has established a multicast tree from the sender to four receivers as shown in Fig. 3.7; the numbers next to the receivers are the rates at which the receivers want to receive data. Let us also assume that the video is layered encoded to accommodate this heterogeneity of receiver rates.

RSVP operates as follows.   Each receiver sends a reservation message upstream into the multicast tree. This reservation message specifies the rate at which the receiver would like to receive the data from the source. When the reservation message reaches a router, the router adjusts its packet scheduler to accommodate the reservation. It then sends the reservation upstream. The amount of bandwidth reserved

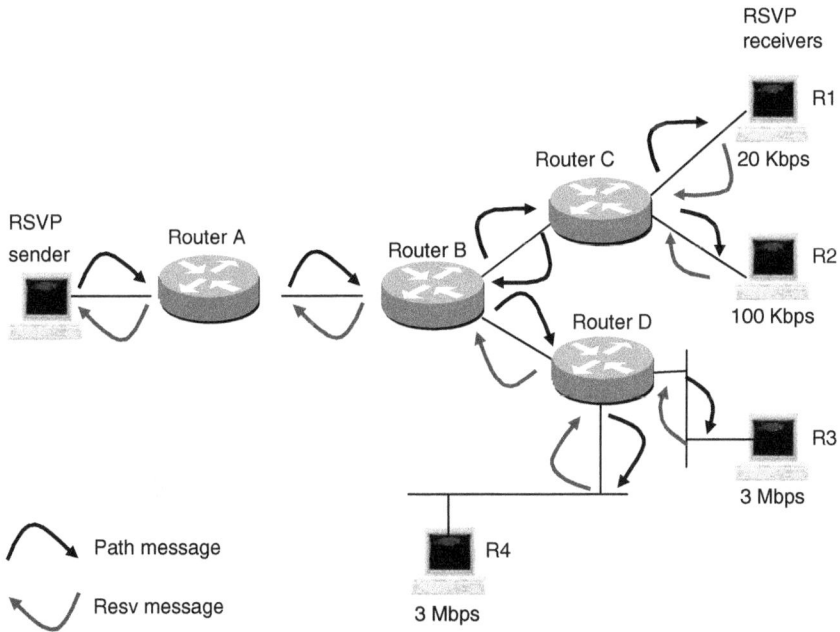

Figure 3.7 An RSVP example.

upstream from the router depends on the bandwidths reserved downstream. In the example in Fig. 3.7, receivers R1, R2, R3, and R4 reserve 20 kb/s, 100 kb/s, 3 Mb/s, and 3 Mb/s, respectively. Thus router D's downstream receivers request a maximum of 3 Mb/s. For this one-to-many transmission, router D sends a reservation message to router B requesting that router B reserve 3 Mb/s on the link between the two routers. Note that only 3 Mb/s is reserved and not 3 + 3 = 6 Mb/s; this is because receivers R3 and R4 are receiving the same video stream, so their reservations may be merged. Similarly, router C requests that router B reserve 100 kb/s on the link between routers B and C; the layered encoding ensures that receiver R1's 20 kb/s stream is included in the 100 Mb/s stream. Once router B receives the reservation message from its downstream routers and passes the reservations to its schedulers, it sends a new reservation message to its upstream router, router A. This message reserves 3 Mb/s of bandwidth on the link from router A to router B, which is again the maximum of the downstream reservations.

Call admission.   Whenever a router receives a new reservation message, it must first determine if its downstream links on the multicast tree can accommodate the reservation. This *admission* test is performed whenever a router receives a reservation message. If the admission test fails, the router rejects the reservation and returns an error message to the appropriate receiver(s). (RSVP does not define the admission test; it simply assumes that the routers perform such a test and that RSVP can interact with the test.)

In case a reservation request fails an admission test, a reservation error must be reported to all the concerned receivers (recall that a reservation request may embody a number of requests merged together). These reservation errors are reported within *ResvError messages*. The receivers can then reduce the amount of resource that they requested at their previous attempt and try reserving again. The RSVP standard provides mechanisms to allow the backtracking of the reservations when insufficient resources are available. Unfortunately, these mechanisms add significant complexity to the RSVP protocol.

Furthermore, RSVP suffers from the so-called *killer-reservation problem*, whereby a receiver requests over and over again a large reservation, each time getting its reservation rejected due to lack of sufficient resources. Because this large reservation may have been merged with smaller reservations downstream, the large reservation may be excluding smaller reservations from being established.

To solve this thorny problem, RSVP uses the ResvError messages to establish additional state in routers, called *blockade state*. Blockade state in a router modifies the merging procedure to omit the offending reservation from the merge, allowing a smaller request to be forwarded and established. Yet, the blockade state adds further complexity to the RSVP protocol and its implementation.

Inefficiencies of IntServ architecture.   The main source of inefficiencies of the IntServ *per-flow reservation* model includes

- *Scalability.*   The per-flow resource reservation in RSVP indicates the need for a router to process resource reservations and to maintain per-flow state for *each* flow passing though the router. With recent measurements [THOMSO97] suggesting that even for an OC-3 speed link approximately 256,000 source-destination pairs might be seen in 1 min in a backbone router, per-flow reservation processing represents a considerable overhead (protocol signaling plus processing plus storage) in large networks.

- *Flexible service models.*   The IntServ framework provides for a small number of prespecified service classes. This particular set of service classes does not allow for more qualitative or relative definitions of service distinctions (e.g., "Service class A will receive preferred treatment over service class B."). These more qualitative definitions might better fit our intuitive notion of service distinction (e.g., first class versus coach class in air travel; *platinum* versus *gold* versus *standard* credit cards).

- *Fair share of network resources.*   It is important to understand that the allocation of network resources is accomplished on a flow-by-flow basis, and that although each flow is subject to admission control criteria, many flows share the available resources on the network, which is described as link sharing. With link sharing, the aggregate bandwidth in the network is shared by various types of traffic. These types of traffic generally can be different network protocols (e.g., IP, IPX, SNA), different services within the same protocol suite (e.g., Telenet, FTP, SMTP), or simply different traffic flows that are segregated and classified by the sender. It is important that different traffic types do not unfairly utilize more than their fair share of network resources as that could result in a disruption of other traffic.

- *Better-than-best-effort service to applications, without the need for host RSVP signaling.*   Few hosts on the Internet are able to generate RSVP signaling or express the Rspec and Tspec in the detail needed by the IntServ model.

3.5.2   Differentiated services

Differentiated services [DIFFSERV-ID] evolved from the integrated services standard. It eliminates much of the complexity introduced from IntServ's use of RSVP to set up and tear down QoS reservations,

```
0                                    7
┌──┬───┬──┬──┬──┬──┬──┬──┐
│  │   │  │  │  │  │  │  │
│  │ DSCP │  │  │ CU  │
│  │   │  │  │  │  │  │  │
└──┴───┴──┴──┴──┴──┴──┴──┘
```

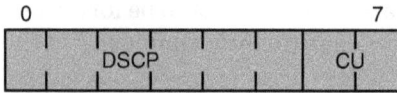

Figure 3.8   Structure of the DS field in the IPv4 and IPv6 headers.

and instead provides an architecture for providing *scalable* and *flexible* service differentiation. Service differentiation refers to the ability of the architecture to handle different classes of traffic in different ways within the Internet. (The differentiated services architecture is flexible in the sense that it does *not* define specific services or service classes (as is the case with IntServ for instance). Instead, the differentiated services architecture provides the functional components with which such services can be structured.) The need for *scalability* arises from the fact that hundreds of thousands of simultaneous source-destination traffic flows may be present at a backbone router of the Internet.

Traffic classification and conditioning.   In the differentiated services architecture, a packet's mark is carried within the so-called *differentiated services* (DS) field in the IPv4 or IPv6 packet header. The definition of the DS field is intended to supersede the earlier definitions of the IPv4 type-of-service field [RFC 791] and the IPv6 traffic class field [IPV6-ID]. The structure of this 8-bit field is shown in Fig. 3.8.

The 6-bit *differentiated service code point* (DSCP) subfield determines the so-called per-hop behavior (see "Per-Hop Behavior" in [DF1] Sec. 3.5.2) that the packet will receive within the network. The 2-bit CU subfield of the DS field is currently unused.

Packets arriving at the edge router (see Fig. 3.9) are first classified. The classifier selects a packet based on the values of its header fields (e.g., source address, destination address, source port, destination port, protocol ID) and steers the packet to the appropriate marking function. The DS field value is then set accordingly at the marker. For example, in Fig. 3.9, packets being sent from H1 to H3 might be marked at R1, while packets being sent from H2 to H4 might be marked at R2. The mark that a packet receives identifies the class of traffic to which it belongs. Different classes of traffic will then receive different services within the core network. The RFC defining the differentiated service architecture [RFC 2475] uses the term *behavior aggregate* rather than *class of traffic*.

Once packets are marked, they are then forwarded along their route to the destination. At each subsequent DS-capable router, these marked packets then receive the service associated with their marks. Even this simple marking scheme can be used to support different classes of service within the Internet. For example, all packets coming from a certain set of source IP addresses (e.g., those IP addresses that have paid for

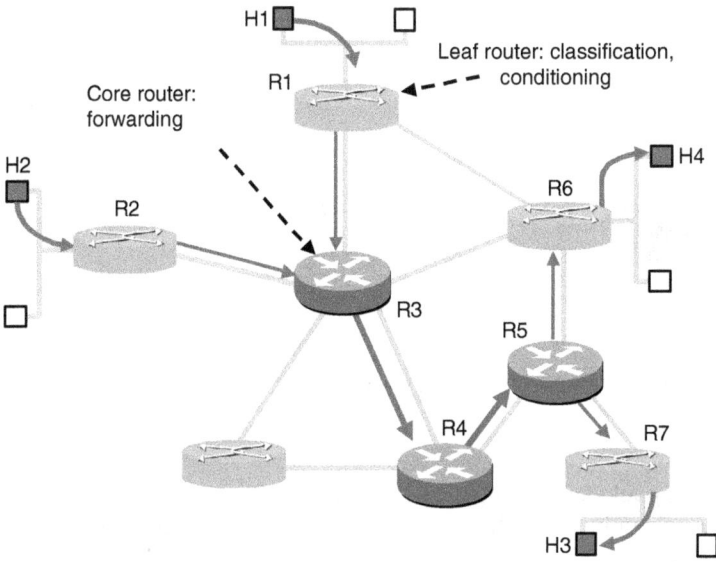

Figure 3.9   A simple DiffServ network example.

an expensive priority service within their ISP) could be marked on entry to the ISP and then receive a specific forwarding service (e.g., a higher-priority forwarding) at all subsequent DS-capable routers.

In certain cases it might be desirable to limit the rate at which packets bearing a given marking are injected into the network. For example, an end user might negotiate a contract with its ISP to receive high-priority service, but at the same time agree to limit the maximum rate at which it would send packets into the network. That is, the end user agrees that its packet-sending rate would be within some declared *traffic profile*. The traffic profile might contain a limit on the peak rate, as well as the burstiness of the packet flow (see "Leaky-bucket implementation" in Sec. 3.9.10 for details on the leaky bucket mechanism).

As long as the user sends packets into the network in a way that conforms to the negotiated traffic profile, the packets receive their priority marking. On the other hand, if the traffic profile is violated, the out-of-profile packets might be marked differently, might be shaped (e.g., delayed so that a maximum rate constraint would be observed), or might be dropped at the network edge. The role of the *metering function*, shown in Fig. 3.10, is to compare the incoming packet flow with the negotiated traffic profile and to determine whether a packet is within the negotiated traffic profile. The actual decision about whether to immediately re-mark, forward, delay, or drop a packet is *not* specified in the DiffServ architecture. The DiffServ architecture only provides the framework

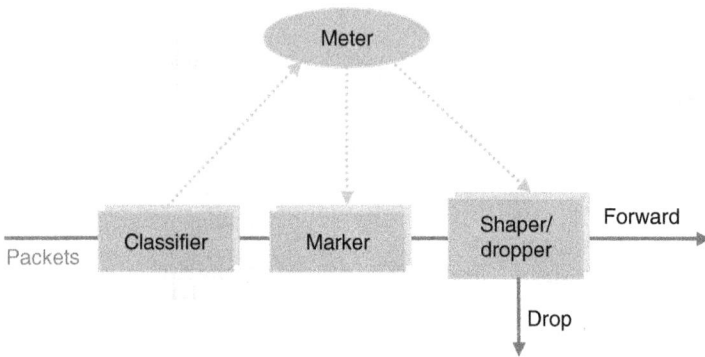

Figure 3.10   Logical view of packet classification and traffic conditioning at the edge router.

for performing packet marking and shaping/dropping; it does *not* mandate any specific policy for which marking and conditioning (shaping or dropping) is actually to be done.

With access control for QoS definition and management at the network edge, providers can deliver service levels with suitable granularity. CBQ-based architecture of access points is an example of how access QoS can be implemented in a DiffServ environment.

Class-based queuing: an example of DiffServ in practice [STEPH99].   Class-based queuing (CBQ) is an open, nonproprietary traffic classification and queuing technology that provides QoS and bandwidth management in IP networks. In CBQ, several output queues can be defined each associated with a different service preference. You also can define the amount of queued traffic, measured in bytes that should be drained from each queue on each pass in the servicing rotation. This servicing algorithm is an attempt to provide some semblance of fairness by prioritizing queuing services for certain types of traffic, while not allowing any one class of traffic to monopolize system resources and bandwidth.

The configuration in Fig. 3.11, for example, has created three buffers—high, medium, and low. The router could be configured to service 300 bytes from the high-priority queue, 200 bytes from the medium-priority queue, and then 100 bytes from the low-priority queue on each rotation. High-priority classes are always first in line, but the extra bandwidth is also allocated fairly among a range of other traffic classes. This automated redistribution of idle bandwidth is fundamental to achieving high multiplexing efficiency and performance.

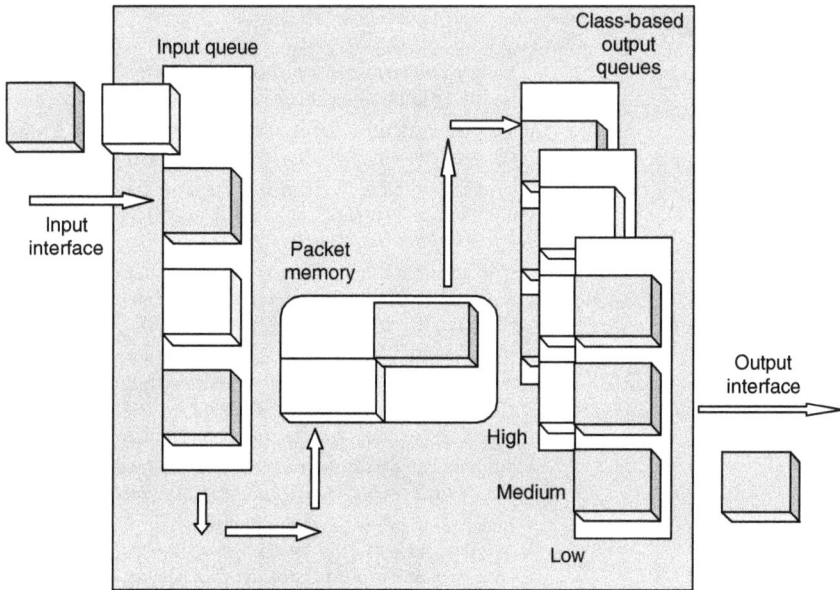

Figure 3.11   Class-based queuing.

After traffic in each queue is processed, packets continue to be serviced until the byte count exceeds the configured threshold or the current queue is empty. In this fashion, traffic that has been categorized and classified to be queued into the various queues has a reasonable chance of being transmitted without inducing noticeable amounts of latency thereby allowing the system to avoid buffer starvation. CBQ was also designed with the concept that certain classes of traffic, or applications, may need minimal queuing latency to function properly; CBQ provides the mechanisms to configure how much traffic can be drained off each queue in a servicing rotation, providing a method to ensure that a specific class does not sit in the outbound queue for too long.

With CBQ as a model for access QoS in a DiffServ network, enterprise network managers can eventually establish business policies for allocating Internet bandwidth to different classes of users and applications. CBQ can classify enterprise traffic and define service levels by a range of parameters based on information from Layers 3 to 7 of the user traffic. For example, traffic can be classified by IP address range, host address, protocol, port, application, or any combination of these parameters.

QoS requirement	CBQ
Traffic classification	Enterprise customers can classify traffic flows based on internal business priorities
Traffic shaping	The network edge device explicitly rate shapes traffic to meet customer or business-defined policies
Policing	Traffic is subject to trusted control at the network edge to limit traffic flows that would violate the agreed service contracts
Prioritization	It is possible to prioritize the use of bandwidth to ensure bandwidth is available for the most important time-sensitive or business-critical traffic flows
Measurement	Enterprises and network providers can accurately measure bandwidth usage to ensure that it is effectively allocated and delivered
Provisioning	The customer controls bandwidth allocation to support the dynamic provisioning of access bandwidth according to changing business policies
Service level management and monitoring	Local traffic flows are mapped to the service levels offered by the network provider to ensure consistent and predictable service

Per-hop behavior. So far, DS classification focused on the edge functions in the differentiated services architecture. The second key component of the DS architecture involves the per-hop behavior (i.e., packet forwarding function) performed by DS-capable routers. The per-hop behavior (PHB) is defined as "a description of the externally observable forwarding behavior of a DS node applied to a particular DS behavior aggregate" [RFC 2475]. Digging a little deeper into this definition, we can see several important considerations embedded within.

A PHB can result in different classes of traffic (i.e., traffic with different DS field values) receiving different performance (i.e., different externally observable forwarding behavior).

While a PHB defines differences in performance (behavior) among classes, it does not mandate any particular mechanism for achieving these behaviors. As long as the externally observable performance criteria are met, any implementation mechanism and any buffer/bandwidth allocation policy can be used. For example, a PHB would not require that a particular packet queuing discipline, e.g., a priority queue versus a weighted-fair-queuing queue versus a first-come-first-served queue, be used to achieve a particular behavior. The PHB is the "end," to which resource allocation and implementation mechanisms are the "means."

Differences in performance must be observable, and hence measurable.

An example of a simple PHB is one that guarantees that a given class of marked packets receive at least x percent of the outgoing link bandwidth over some interval of time. Another per-hop behavior might specify that one class of traffic will always receive strict priority over another class of traffic, i.e., if a high priority packet and a low priority packet are present in a router's queue at the same time, the high priority packet will always leave first. Note that while a priority queuing discipline might be a natural choice for implementing this second PHB, any queuing discipline that implements the required observable behavior is acceptable.

Two "classic" PHBs are the *expedited forwarding* (EF) [EX-ID] and the *Assured forwarding* (AF) PHB [AS-ID]:

The expedited forwarding PHB specifies that the departure rate of a class of traffic from a router must equal or exceed a configured rate. That is, during any interval of time, the class of traffic can be guaranteed to receive enough bandwidth so that the output rate of the traffic equals or exceeds this minimum configured rate. Note that the EF per-hop behavior implies some form of isolation among traffic classes, as the guarantee is made *independently* of the traffic intensity of any other classes that are arriving to a router. Thus, even if the other classes of traffic are overwhelming router and link resources, a share of those resources must still be made available to the class to ensure that it receives its minimum rate guarantee. EF thus provides a class with the simple *abstraction* of a link with a minimum guaranteed link bandwidth.

EF is designed for implementing a low-latency, low-loss and/or low-jitter service class, such as that desired for high-quality VoIP conversations. EF can be implemented with a strict priority over other service classes to achieve low jitter, though other approaches are allowed by the standard that may yield somewhat different service characteristics.

The *assured forwarding* PHB is more complex. AF divides traffic into four classes, where each AF class is guaranteed to be provided with some minimum amount of bandwidth and buffering. Within each class, packets are further partitioned into one of three "drop preference" categories. When congestion occurs within an AF class, a router can then discard (drop) packets based on their drop preference values. See [AS-ID] for details. By varying the amount of resources allocated to each class, an ISP can provide different levels of performance to the different AF traffic classes.

Concluding remarks. DiffServ eliminates the need for RSVP and instead makes a fresh start with the existing type-of-service (ToS) field found in the header of IPv4 packets. The original intent of the IP's designers was to make the ToS field support capabilities much like QoS—allowing applications to specify high or low delay, reliability, and throughput requirements—but was never used in a well-defined, global manner. But since the field is a standard element of IP headers, DiffServ is reasonably compatible with existing IP gear and can be implemented to a large extent by software/firmware upgrades.

DiffServ takes the ToS field (8 bits), renames it the DS (DiffServ) byte, and restructures it to define IP QoS parameters. The DS byte determines how routers handle a packet. Instead of trying to identify and manage IP traffic flows as RSVP must do, DiffServ brings QoS consideration to the normal IP practice of packet-by-packet, hop-by-hop routing. Forwarding decisions are made according to parameters defined by the DS byte, and are manifested in a range of standard PHBs at each router. Each PHB effectively determines the QoS treatment a packet receives at that router hop. A DiffServ domain is a logical IP network where PHB definitions are applied consistently across router hops. For example, the default PHB in a DiffServ network is traditional best-effort service, using first-in first-out (FIFO) queuing. Other PHBs are defined for higher classes of service, including expedited forwarding and assured forwarding.

The standard describes how to mark IP packets so that they receive a particular PHB by using what is called the differentiated services code point, a 6-bit field in the IP packet header that allows for 64 possible PHBs. It also describes the characteristics of the standard PHBs (and their associated DSCPs) and the DSCPs that may be used for customized local or experimental use within a DiffServ domain. Best-effort is used in the event that no other PHB is specified for the traffic. It may also serve as a PHB for demoted traffic (i.e., traffic that arrives in excess of metering profiles). The standard, however, only requires two of the eight class selector PHBs to have forwarding behaviors that are actually different from each other. It is also worth noting that neither AF nor EF are required for a router to be DiffServ compliant; without these though, there is little hope for multivendor or multiservice provider compatibility.

DiffServ standardizes PHB classification, marking and ranking procedures, but it leaves equipment vendors and service providers free to develop PHB flavors of their own. Router vendors often define parameters and capabilities they think will furnish effective QoS controls, and service providers can devise combinations of PHB to differentiate their qualities of service.

The main underlying principle of the DiffServ standard is that for every PHB, or at least every service class implemented using a set of

PHBs, there must be separately managed queuing resources available in the router. This is essential for the predictable operation of each service relative to every other, so that a queue overflow in one service class or PHB does not adversely affect the others. This has the consequence that each logical interface on the system must support many possible queues, and therefore queuing resources must be managed effectively.

3.5.3 Multiprotocol label switching

Multiprotocol label switching is rooted in the IP switching and tag switching technologies that were developed to bring circuit-switching concepts to IP's connectionless routing environment. It is a hybrid technology in that it enables very fast forwarding in the core and conventional routing at the edges.

In conventional IP routing, the next hop for a packet is determined based on the information carried in the header of the packet and the result of running a network layer routing algorithm. Choosing the next hop at the routers is thus a combination of two functions.

1. Dividing the whole set of possible packets into *forwarding equivalence classes* (FEC).

2. Mapping each FEC to a next hop.

In MPLS routing, packets are assigned a "label" at the entry to the backbone and are switched by a simple label lookup. "A label is a short fixed-length locally significant identifier that is used to identify a forwarding equivalence class (Fig. 3.12). The label represents the FEC to which that packet is assigned" [MPLS-ID]. The label is "locally significant" in the sense that two routers agree upon using a particular label

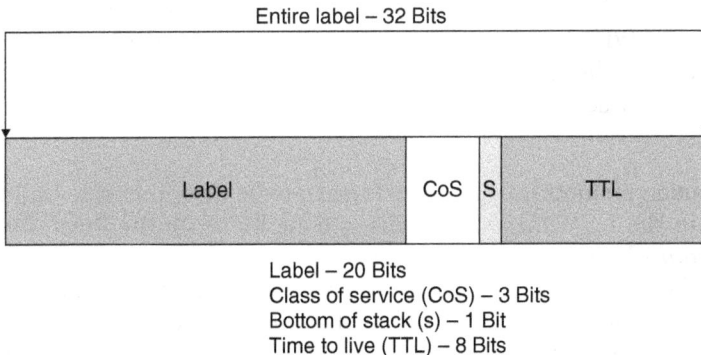

Entire label – 32 Bits

Label	CoS	S	TTL

Label – 20 Bits
Class of service (CoS) – 3 Bits
Bottom of stack (s) – 1 Bit
Time to live (TTL) – 8 Bits

Figure 3.12 CoS bits carried in the MPLS label.

to signify a particular FEC, between themselves. The label itself is a function of the packet's class of service and the FEC. The same label can be used to distinguish a different FEC by another pair of routers. In the same spirit, the same FEC can be assigned a different label by other routers. A router which uses MPLS is called a *label switching router* (LSR). The path taken by a packet after being switched by LSRs through an MPLS domain is called a *label-switched path* (LSP).

The packets are stripped of the labels at the egress router and may be routed in the conventional fashion, thereafter, before reaching their final destination.

If routers R1 and R2 agree to bind an FEC F to label L for traffic going from R1 to R2, then R1 is called the upstream LSR and R2 is the downstream LSR. If R1 and R2 are not adjacent routers, then R1 may receive packets labeled L from R2 and R3.

In MPLS, the final decision for the binding of a label L to an FEC F is made by the downstream LSR with respect to that binding. (i.e., who will be receiving the traffic). The downstream LSR then announces the binding to the upstream LSR. Thus labels are distributed in a bottom-up fashion. This distribution is accomplished with the label distribution protocols [MPLS-WP].

Route selection in MPLS.   Route selection refers to the problem of mapping an FEC to an LSP. An LSP can be chosen for a particular FEC in two ways.

1. *Hop-by-hop routing.*   The route is chosen at each LSR in the same manner as in conventional IP forwarding.

2. *Explicit routing.*   Here the boundary ingress LSR specifies the particular LSP that a packet will take through the MPLS cloud. This is done by explicitly declaring all the LSPs along the path. This is almost like TCP/IP source routing although it has several advantages over TCP/IP source routing. If the whole LSP is declared then it is "strictly explicitly routed," whereas for a partial LSP stated, it is said to be "loosely explicitly routed."

Label distribution protocol (LDP).   Labels have to be distributed to build and maintain the *LSR-databases*. This is done by using the *label distribution protocol*. This protocol belongs to the general family of *hard-state* protocols: when an FEC is bound to one label and this information is *flooded* using the *LDP*, that bind stays until a new call tears down the bond. Conceptually, the labels are bound to a prefix or set of prefixes. The association and prefix-to-tag binding are done by an MPLS network edge node. An MPLS edge node exchanges routing information with

In label	Prefix	Out label	Out interface
	199.1.0.0/24	6	0
	128.10.0.0/16	7	0

In label	Prefix	Out label	Out interface
1	199.1.0.0/24	5	4
2	128.10.0.0/16	8	4

In label	Prefix	Out label	Out interface
6	199.1.0.0/24	1	1
7	128.10.0.0/16	2	1

In label	Prefix	Out label	Out interface
5	199.1.0.0/24		0
8	128.10.0.0/16		0

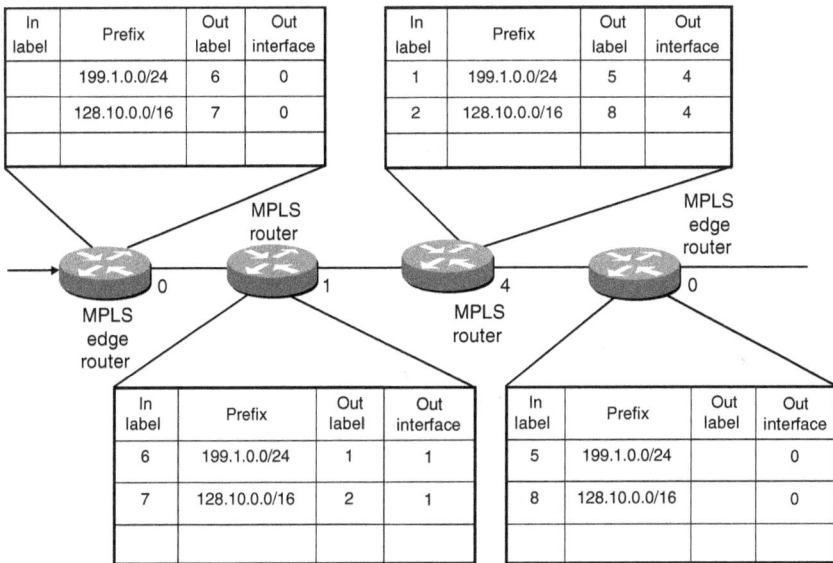

Figure 3.13 The label-swapping process.

non-MPLS-capable nodes, associates and binds prefixes learned via Layer 3 routing to MPLS labels, and distributes the labels to MPLS peers (Fig. 3.13).

The constraint route label distribution protocol. The ability of MPLS to define explicit routes has been augmented by support for QoS. QoS support allows the introduction of QoS-enabled services. There are two methods of QoS support in MPLS. The first uses traffic engineering extensions to the RSVP signaling protocol (RSVP-TE) [XIAO99], and the second utilizes extensions to the LDP called constraint route label distribution protocol (CR-LDP) [MPLS-ID].

CR-LDP is an extension of QoS-based routing which takes into account the viability of a route, certain QoS requirements, and other constraints of the network, such as policy. The goals are twofold.

1. Select routes which can meet the QoS requirements.

2. Increase utilization of network and load distribution—a longer and less congested path may be better for QoS demanding traffic than the shortest and highly congested path.

CR-LDP is a signaling protocol for the establishment of explicitly-routed paths with QoS attributes attached to them. CR-LDP specifically modifies the original LDP design to allow for traffic specification parameters

0				1		2		3

0 1 2 3 4 5 6 7 8 9 0 1 2 3 4 5 6 7 8 9 0 1 2 3 4 5 6 7 8 9 0 1

0	0	Type	0 × 0810	Length	24
Flags		Frequency	Reserved	Weight	
Peak data rate (PDR)					
Peak burst size (PBS)					
Committed data rate (CDR)					
Committed burst size (CBS)					
Excess burst size (EBS)					

Type

A fourteen-bit field carrying the value of the traffic parameters TLV type = 0 × 0810.

Length

Specifies the length of the value field in bytes = 24.

Flags

The flags field is shown below:

Res	F6	F5	F4	F3	F2	F1

Res: These bits are reserved—zero on transmission, ignored on receipt
F1: Corresponds to the PDR
F2: Corresponds to the PBS
F3: Corresponds to the CDR
F4: Corresponds to the CBS
F5: Corresponds to the EBS
F6: Corresponds to the weight

Each flag Fi is a negotiable flag corresponding to a traffic parameter. The negotiable flag value zero denotes "not negotiable" and value one denotes "negotiable."

Figure 3.14 TLV Traffic parameters.

(TLVs—type, length, and value). This protocol adds new fields to the original LDP protocol (see Fig. 3.14)—*committed information rate* (CIR), *peak information rate* (PIR), *committed burst size* (CBS), *peak burst size* (PBS), and *excess burst size* (EBS).

Both PDR and CDR are measured over specified time intervals as defined by the PBS and CBS. In addition to CBS, an LSP could burst at its CDR up to a level specified by an EBS.

From the user's perspective, these traffic-related parameters are useful for signaling to the MPLS domain of the user-desired values (e.g., a specific CDR). To the network these parameters are used to execute local admission control at each of the label-switched routers along the path.

If specified by the user, a simple token bucket algorithm [KUROSE02] could be used to monitor both PDR and CDR. PDR could be monitored, for instance, by a token bucket with token rate equal to PDR and maximum token size equal to PBS. With respect to CDR, since two burst sizes are specified—CBS and EBS—two token buckets, C and E, are proposed. The C and E token buckets are incremented at the CDR rate. The C bucket, when full, overflows into E.

For the applications that tolerate some amount of delay variations, the service frequency can be assigned the value "frequent." This implies that the available rate should average at least the CDR measured over any time interval equal to or longer than a small number of shortest packet times at CDR. Hence, packets are served after a delay that is bounded by a few packet times measured at CDR. The number of packets is a configurable parameter and is set to meet the service delay requirement. Other services with no delay requirements can be assigned the value "unspecified." This implies that the CDR may be provided at any granularity. Thus, CDR might be guaranteed over a longer timescale than that offered by very frequent or frequent assignments.

CR-LDP allows for resource sharing among a number of CR-LSPs competing for the same pool of resources. The weight parameter determines the CR-LSP relative priority when assigning excess nodal bandwidth or under congestion conditions. The implementation of a particular fairness criterion is domain-specific.

Renegotiation allows the network to offer CR-LSP a set of revised parameters and weight more suitable for the current network state. Parameter negotiation is allowed only if indicated by the appropriate flag. Flags are provided for PDR, PBS, CDR, CBS, EBS, and weight. Negotiation is only allowed in the direction in which the parameter values are lowered from those defined in the original signaling message. If modified, the node must overwrite the modified parameters before forwarding the message to the next node. A node might elect to change its reservation level based on the new parameter values carried in the traffic parameters TLV in the label mapping message.

CR-LDP versus RSVP. Although the two protocols have similar flows for reservation setup—sending an end-to-end request and replying with an end-to-end response, the way they operate is different, and the detailed function they offer is also not consistent. Each protocol has its champions and detractors, and the specifications are still under development.

RSVP includes an optional function (Adspec) whereby the available resources on a link can be reported on the path message. This allows the destination host to know what resources are available, and modify the flowspec on the Resv accordingly. Unfortunately, not only does this function require that all the routers on the path support the option, but it also has an obvious window where resources reported on a path message may already have been used by another path by the time the Resv is received.

CR-LDP, on the other hand, carries the full traffic parameters on a LABEL_REQUEST. This allows each hop to perform traffic control on the forward portion of path setup. The traffic parameters can be negotiated as the setup progresses, and the final values are passed back on the LABEL_MAPPING allowing the admission control and resource reservation to be updated at each QoS path. This approach ensures that a path will not be set up on a route where there are insufficient resources. CR-LDP offers a slightly tighter approach to traffic control especially in heavily used networks, but individual RSVP implementations can provide a solution that is almost as good. A problem with CR-LDP, however, is that it does not allow explicit sharing of the allocated network resources.

3.6   QoS in Infrastructured Wireless Mobile Networks

In wireline (ISPN-like) networks, consisting of static hosts, QoS guarantees are provided by reserving sufficient bandwidth at each link along the path from the sender to the receiver so that the congestion delay at each switch is bounded. In a mobile computing network, a geographical area is divided into several cells. Each cell is served by a *base station* connected to the fixed network. A mobile host maintains connectivity with the fixed network through the base station of the cell in which it is currently located. When a mobile host moves from one location to another, the delivery delay of a packet is affected in two ways. The first factor is the propagation delay, which may change due to the change in the length of the path from the sender to the receiver. Second, the congestion delay at the switches along the new path may be different. Hence, when a mobile host initiates a session with a certain QoS guarantee and reserves link bandwidth from the sender to its current location, the QoS guarantee is valid only in that location; when the mobile host moves to a new location, the QoS guarantee is not valid in the new location.

There are two broad ways to provide service guarantees to mobile hosts. The first approach is location dependent—the QoS is guaranteed specific to a location, *i.e.,* QoS guarantee is maintained as long as the mobile host stays at the location from where the session was initiated.

To obtain such a service the mobile host makes a reservation for a certain QoS from its current location. As soon as it moves to a new location, the QoS guarantee is no longer valid. The application has to renegotiate its desired QoS at the new location. If sufficient resources are not available along the new data-flow path, the mobile host may suffer service degradation.

The second approach is location independent—the mobile host receives the same QoS guarantee at all locations (given by a *mobility profile*[1]) where it may move during the lifetime of the session. For example, suppose at time t_1, the mobile host is at cell C_1 where it initiates a session with a certain QoS guarantee, and at time t_2 is located in cell C_2 where $t_1 < t_2$. The user will receive the same QoS guarantee at time t_2 as long as C_2 is within its mobility profile.

To obtain mobility-independent QoS, a mobile host must make reservations along the data-flow path from the sender to each location it may visit. Evidently, this leads to very low utilization of resources because, although data is physically flowing to all locations, data is being used only at the current location of the mobile. The utilization of the network resources can be significantly improved if the reserved resources of unused data flows could be used by other flows. However, the users utilizing these unused resources may suffer service degradation when the original reservers start using these resources. For example, suppose a mobile host M_1 has placed reservation at the two cells C_1 and C_2 and is currently located at C_1 (hence using the reserved resources at C_1). In this situation, another host M_2 which is located in C_2 *can* use the resource reserved by M_1 in C_2. But when M_1 moves into C_2, M_2 must release the resources reserved for M_1 at C_2.

3.7   Hurdles for Multimedia in Mobile Multihop Wireless Networks

Unlike existing (2G and 3G) wireless networks that comprise a wired as well as a wireless segment, wireless mobile ad hoc networks (MANETs) consist of mobile nodes interconnected by wireless multihop communication links with no fixed network infrastructure or administrative support. The topology of such networks changes dynamically as mobile nodes join or depart the network or radio links between nodes become unusable due to time-varying link capacity and limited transmission power [XIAO00].

Routing is one of the key issues for supporting QoS-demanding applications in this rather unstable and resource limited wireless

---

[1]The *mobility profile* of a mobile host is the set of locations the mobile is expected to visit.

networking environment. Successful QoS routing includes the necessary knowledge of the network state and algorithms for feasible route selection and resource reservation. Network state information includes radio-link quality and resource availability. In real-time applications it is beneficial for the data source to know, prior to call setup, not only the path to the destination but also the available resources on such path (for instance, the average available/achievable data rate). This is important for many reasons: (1) if bandwidth is not available, the call is dropped without congesting the network unnecessarily (i.e., call admission control); (2) if bandwidth and/or channel quality is inadequate for full rate transmission, the source may still put the call through by reducing the rate or using a more robust coding scheme; (3) if path bandwidth/quality deteriorates during the call, the source node can then modify or drop the call.

However, the lack of infrastructure in MANETs implies a lack of, or imprecise, network state information maintained at each node. The lack of sufficient knowledge of network state information linked with the underlying network dynamics makes QoS guarantees in MANETs a complex and difficult task.

Among other issues, key to supporting QoS for real-time applications in wireless mobile networks is the design of a flexible and robust routing and *medium access control* (MAC) protocols. Wireless stations use some form of multiple access technique with suitable collision avoidance and "hidden terminal" mitigation for accessing radio resources. The higher the number of nodes contending for radio resources, the greater is the delay (random variable) in accessing the radio channel for transmitting a packet.

Besides, enough reserve of radio channel capacity must be available to ensure an upper bound on end-to-end delay as part of QoS. The decentralized nature of wireless ad hoc networks makes it difficult to dynamically assign a central controller for monitoring and maintaining the connection state and reservations. Because of this, best-effort distributed MAC controllers are still used in wireless ad hoc networks.

It is not just the availability of enough resources that affects the QoS. Equally essential is the connectivity relationship among nodes. A principal objective of network engineering, as emphasized earlier, is the minimization of routing updates. Topology updates occur when a new node joins the network or an existing node is detected to have become unavailable with respect to a particular flow. One naturally expects that such topology updates should not affect the QoS for the rest of the nodes as long as the topology of the rest of the network remains unchanged. However, updates consume network bandwidth and wireless spectrum is a scarce and expensive resource. Since radios are operating using their batteries, the processing capability of nodes (CPU capacity) is limited, which further imposes low processing overheads of nodes. Mobiles should thus use the available spectrum efficiently as well as try to keep the transmission/reception overhead at minimum levels.

Route stability is another critical consideration for QoS in a wireless mobile ad hoc network. Route stability follows directly when the distribution of the mobile nodes does not change much relative to one another during the time interval of interest. Frequently changing routes could increase the delay jitter experienced by the users. To accommodate real-time traffic needs such as those of voice or live video, both the overall delay and the delay variance must be kept under a certain bound. This objective is extremely difficult to attain in wireless networks because of the involuntary network state changes as nodes join in or depart, traffic loads vary, and link quality swings dramatically [CORS97].

3.8   Hard QoS Guarantees in MANETs:
Fact or Fiction?

We have repeatedly noted that the policies, protocols, and algorithms in an ad hoc network with QoS support must be QoS-preserving. But, how badly do the rapid topology changes militate against the QoS guarantees? Let τ_c and T_u denote the interval between two consecutive topology change events and the time it takes to detect the change, complete the calculation, and propagate the topology updates resulting from the last topology change, respectively, to all pertinent nodes. Recall that an ad hoc network is combinatorially stable only if $T_u < \tau_c$. If the just-computed feasible route ceases to exist during the corresponding topology update, the QoS guarantee becomes meaningless.

Maintaining bounds on delay jitter may also become impractical even in a combinatorially stable network if τ_c remains "close" to T_u. It may be necessary to investigate more rigorous criteria for different *degrees* of combinatorial stability for different QoS constraints. Since combinatorial stability is governed by random processes arising from random changes in the topology and link traffic intensity, making the network QoS-robust for a particular flow and its associated QoS constraints is clearly impossible as a deterministic objective for an arbitrary ad hoc network.

This is why no MANET-tailored QoS routing algorithm to date can offer hard QoS guarantees, and why it may not be possible to do so in the future. Any such guarantee at best could only be statistical in nature, where QoS robustness is specified as a probability bound for QoS violation during a topology update, the duration of which does not exceed a fixed upper bound. This is also why performance studies on ad hoc networks with QoS support are meaningful only for combinatorially stable networks. The smaller the value of T_u, the smaller is the probability of QoS violation, assuming that resources remain available for use whenever necessary.

It becomes obvious that considerable additional work is necessary to understand better both the specific conditions and extents under which various QoS objectives could be satisfied for dynamic ad hoc networks in the real world. It is in this latter context that the use of admission control, multiple classes of service with possibly different priorities, preemptive priority of control messages, and segregation of dedicated resources for QoS-robust ad hoc networking offer promising areas of investigation.

3.9 A Reference Framework for QoS Routing in Wireless Mobile Networks

Overview

As described above, the goal of QoS routing is to select paths for flows with QoS requirements in such a manner as to increase the likelihood that the network will be capable of supporting and maintaining them. The reference framework proposed in Ref. [AGGE04] for QoS routing (QoSR) in a multihop mobile network is described in this section.

This model can be briefly structured as follows:

When a mobile user requests a new connection with QoS bounds, the network conducts an admission control test and tentatively reserves resources during the forward pass to the destination via a suitable loop-free route. The route is provided by some routing algorithm. During the reverse pass, the network allocates resources for it. Resources include bandwidth, buffer space, and schedulability.

Because nodes are highly mobile, node mobility implies that reservation has to be performed at all places a mobile node may visit during the lifetime of a data flow connection. This is a significant problem since the movement of a mobile node is unpredictable to a large extent. To reduce the transient behavior of connections to a mobile on handoff or when network conditions change, the QoSR framework shall initiate resource adaptation and degradation to satisfy the QoS bounds of all the existing connections.

3.9.1 Functional specification

The QoSR reference framework is illustrated in Fig. 3.15. To keep our discussions consistent, let us focus on a single QoS metric, the link bandwidth. The goal of a routing bandwidth-sensitive service is to deliver a stream of bits with at least the minimum bandwidth fraction of the requested bandwidth that would minimize the number of bandwidth changes and would prefer the shortest paths (minimize end-to-end delay).

As illustrated, the QoSR reference architecture consists of four horizontal layers and a virtual vertical layer. The four horizontal layers are

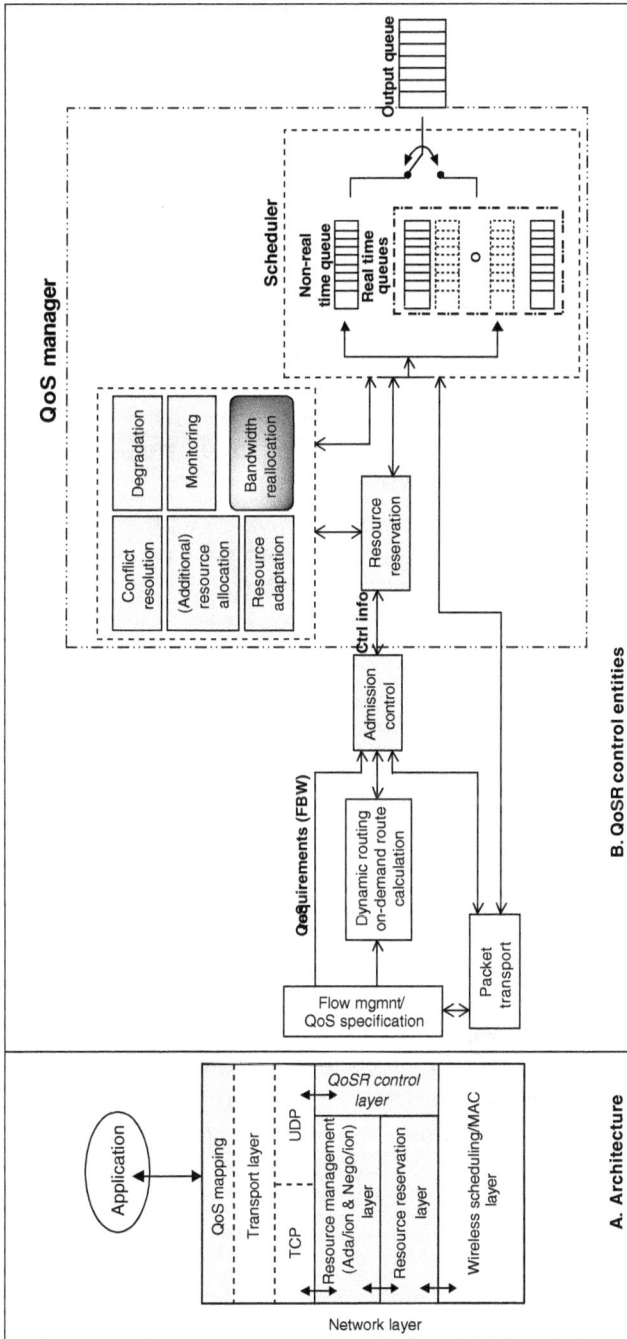

Figure 3.15 A framework for link-layer QoS support in wireless multimedia systems.

the scheduling/MAC layer [HA98, LU97a], the resource reservation layer, the resource adaptation layer [LU96], and the transport layer [DWYE98].

During the call setup period, the higher layers are required to provide the following input parameters to the lower layer modules: (1) *maximum bandwidth desired* (HBW), and (2) *minimum bandwidth required* (LBW). In the following text, the bandwidth window [LBW, HBW] shall be referred to as flow bandwidth window (FBW). The HBW and LBW parameters serve as upper and lower bounds during the bandwidth assignment as well as during the resource reservation, adaptation, and negotiation phases.

In order to provide adaptive QoS services, various resource management algorithms interact in the following sequence of events.

1. The application notifies the network that it wishes to set up a flow between two endpoints, and provides the flow specifications and *resource specifications* (FBW).

2. The network performs *admission control* along the computed route and decides whether to accept or reject the request based on the input requirement profile (FBW) and the available bandwidth on its constituent links. The admission controller is described in Sec. 3.9.6.

3. Once the admission controller at each radio along the multihop path accepts the call, it passes the admission decision and the requirement profile of the user to the packet sorter module (this module does not appear in Fig. 3.15 as it is part of the scheduler module). The packet sorter sorts and classifies traffic packets of admitted flows. Its functionality is detailed in Sec. 3.9.10.

4. The network performs *resource reservation* for guaranteed flows.

5. The network performs *resource adaptation* among the adaptable flows in order to resolve resource conflicts and also to allocate additional resources, if any.

There is a close interaction between the admission control, resource reservation, and resource adaptation algorithms. There is also a close interaction between the transport protocol, resource reservation, and packet scheduling algorithms. In particular, admission control determines if a new flow can be admitted without violating the minimum resource bounds of ongoing flows. Resource reservation is then performed for the new flow. For the network to deliver quantitatively specified QoS (e.g., a bound on delay) to a particular flow, it is usually necessary to set aside certain resources such as a share of the bandwidth. The function of resource reservation is to ensure that a flow will get its requested resources once it is admitted into the network.

Resource adaptation is performed on a set of adaptable flows in order to increase the allocated resources of these flows and hence the revenue of the network. One of the critical tasks of resource adaptation in a mobile environment is to prevent frequent adaptations due to the dynamics of resources and mobility of flows, while still optimizing the utilization and revenue of the network.

This sequel highlights the specifics of the building blocks of the QoSR reference model.

3.9.2 Application (service) model and QoS bounds

Bandwidth guarantee is one of the most critical requirements for real-time applications. Since future application bit rates are not known in advance, the problem of network QoS representation of application requirements is of an online nature. In this view, the reference service model shall present a slightly general formulation of the application requirements. To this end, it is a general practice to express the requested bandwidth in terms of time slots (e.g., see Ref. [LIN99]). Given the application bit rate and the number of bits per time slot, the number of time slots per second required to accommodate a specified bit rate can be derived. Using this framework for bandwidth representation, any future application could be (abstractly) mapped on this.

Furthermore, a simplistic approach for QoS specification is to allow users to specify a constant bandwidth requirement per session. Having a constant bandwidth allocation has many advantages from both the user and network perspectives. Once established, it is completely predictable, and enables a simple pricing model that depends on the total bandwidth consumption, i.e., the product of the bandwidth allocation and the duration of the session. However, the required bandwidth is known in advance only for very few tasks (e.g., real-time voice). Even video communication involves a variable requirement of bandwidth (due to compression), and in other applications with bursty nature of traffic the required bandwidth may change dramatically over time, usually in an unpredictable manner. In order to accommodate such situations it is reasonable to require the network to accommodate dynamic modification of the bandwidth allocation (see Refs. [GROSS95, CIDO95, AFEK96] for trends in applying dynamic bandwidth allocation).

With these in mind, the basic service model shall be extended to support *adaptive services* [LU97]. Service adaptation is an application-specific process. Some applications may be incapable of adapting while others may adapt discretely (e.g., scalable profiles of Moving Picture Expert Group, version 2 (MPEG2)) or continuously (e.g., dynamic rate-shaped applications [ANGIN98]). The time scale over which applications can adapt is also application specific. For example, greedy data applications (e.g., image downloads) may want to take advantage of any change in available bandwidth at any time. In contrast, adaptive continuous media applications (such as audio and video) may prefer to follow trends (via some low-pass filtering scheme) in available bandwidth based on slower adaptation time scales, preferring some level of "stable" service delivery rather than responding to every instantaneous change in bandwidth availability. Adaptive applications therefore should manage the adaptation process and dictate the time scales and semantics of their adaptation process. For example, as mentioned previously, the resource specification for rate is given by a [LBW, HBW] bound. When the network admits a flow, it guarantees a stable rate of *at least* the LBW.

Ideally, it is desirable to always provide the maximum required bandwidth for each connection. However, due to user mobility or the dynamically fluctuating channel and network conditions, this will not always be possible. For example, suppose a connection is provided the maximum bandwidth within a cell. When the user moves into an adjacent cell, the new cell may not have enough available bandwidth to maintain the maximum bandwidth for that connection. If the cell has only bandwidth b e (LBW, HBW) available, the connection need not be dropped but its quality (bandwidth) instead be lowered to b. The minimum required bandwidth can be considered as the bandwidth required to support the lowest-level quality of the connection the user can "live with." The upper bound on the other hand is determined by several factors such as how much the user is willing to pay for a better quality service. Hence LBW can be used for admitting/rejecting each new/handed-off connection.

3.9.3 QoS mapping

The application-to-network QoS mapping control block performs the function of automatic translation between representations of QoS at different system levels (i.e., application, operating system, transport and network, and the like) and thus relieves the user from maintaining intelligence in terms of lower-level specifications. The QoS requirements information distilled from the application service specification leverages the operation of the lower building blocks of the QoSR reference model.

Given the diversity of applications and underlying networks and operating systems, how feasible is it to formulate a generic framework for QoS mapping? The typical technique used today is through some form of application profile characterization. The scope of mapping considered in most schemes is still fairly simplistic and static in nature—focusing on a certain application, operating system and network, one may assume that the whole field of QoS mapping is still in its infancy.

The primary difficulty in formulating good QoS mapping stems from the fact that we do not yet understand the perceptual effect of QoS. Another point is related to the spectrum of application addressed from the existing profile characterization systems. In fact, various QoS mapping schemes in the literature address only continuous media services avoiding other communications services such as transactions. The interaction problem among the many layers in a system complicates further the issue of QoS mapping since the dynamics are usually nonlinear and difficult to characterize.

Several QoS mapping algorithms exist in the literature [KNOC97, ISSA97] following different approaches for translating between application QoS specification and network QoS.

3.9.4 Routing layer

The routing protocol is responsible for finding a loop-free path for a flow or aggregate of flows with the required QoS level [CHEN99, CHEN97, IWATA99, LIN97, RFC 31401, RAMAN98, SINHA99, HSU98]. Proactive, reactive, or hybrid self-organized routing structures (see Chap. 2) can well be used. From our earlier discussions on the topic (see Sec. 3.6.3), we note that, given the requirements of efficient network resource control, distributed clustering techniques facilitate the development of important features in wireless multihop environments such as channel access, routing, power control, virtual circuit support, and bandwidth allocation.

3.9.5 QoS manager

The heart of the resource management and control architecture is the QoS manager. From the functional point of view the QoS manager is a control plane component primarily responsible for:

1. Creating and maintaining the reservation states of QoS connections

2. Allocating and managing network buffers for these connections

3. Scheduling different classes of packets

4. Radio resource usage monitoring

5. Call degradation or reducing bandwidth allocation to degradable applications in the face of scarcity of available radio channels

6. Bandwidth reassignment to maximize utilization of available channel resources

The QoS manager cooperates with other components in the data plane, such as socket (transport) layer and network interface layer, to coordinate the management of network resources.

In the following section, we describe in detail the architectural and functional building blocks of the QoS manager as a part of the resource management phase.

3.9.6   Call admission block

Reservation protocols provide a general facility for creating and maintaining distributed reservation states across a mesh of multicast or unicast delivery paths. Given that the network's resources are finite, the network cannot grant all resource reservation requests. In order to maintain the network load at a level where as many QoS commitments as possible can be met, the network must employ an admission control algorithm that determines which requests to grant and which to deny, thereby maintaining a relatively high utilization of network resources [CLARK92, JAMI95].

To this avail, the main responsibility of the *call admission control* (CAC) is to check whether a new connection can be established with respect to the service parameters, or whether the request has to be rejected to protect the existing connections from a violation of the negotiated QoS. The decision whether sufficient resources are available for a new flow or an aggregate of flows at the requested QoS, without violating the existing QoS commitments to existing connections, greatly relies on resource management policies and resource availability.

Since most of the traffic sources have different characteristics, such as variable bit rate, mean bit rate, sustainable bit rate, peak bit rate, burstiness, and duration of the peak bit rate, the task of the CAC is to estimate the equivalent bit rate of all data sources, based on the traffic characteristics of the potential new source and the currently connected sources and decide whether the new and existing connections can coexist.

Recall, however, that hosts are mobile radios. To assess the equivalent capacity C_{new} of all connections at a wireless station including the new one, all stations that share the same radio spectrum within their receiver sensitivity have to be considered. If the newly evaluated equivalent capacity C_{new} is below the total capacity C_{total}, the new connection is accepted. A new virtual end-to-end connection will be accepted if this procedure is successful on every link involved in the multihop path. Once admission testing has been successfully completed, local resources are immediately reserved. If insufficient resources are available, the flow is rejected [PNNI-SPEC]. If this is the case, a procedure similar to ResvErr message triggering in RSVP protocol is followed, where a notification message is sent to the source(s) of the connection.

The call admission algorithm is presented as a flowchart in Fig. 3.16. For real-time users, admission is primarily based on the availability of bandwidth. Call admission and resource reservation are initiated in

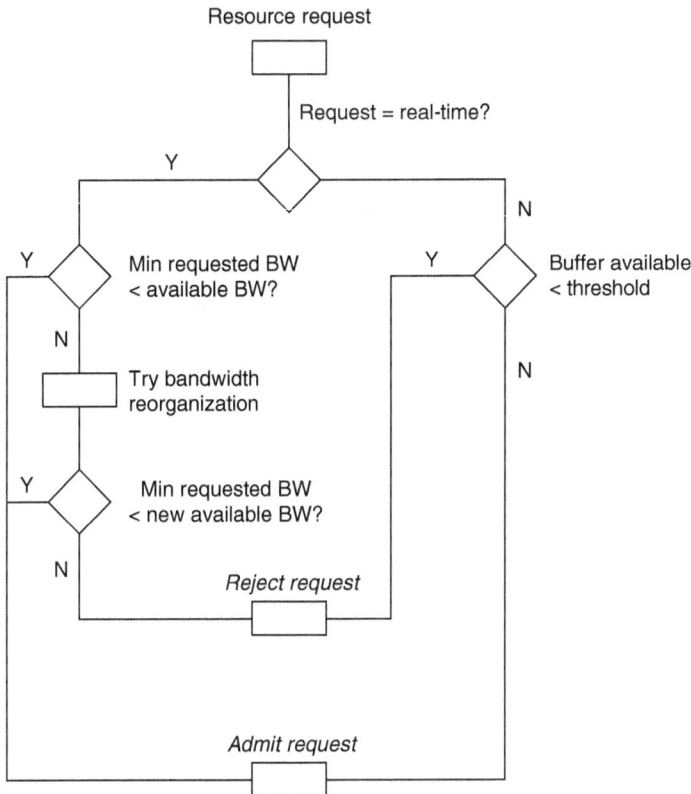

Figure 3.16 Flowchart of the call admission algorithm.

each hop. If only this reservation is successful, the call is forwarded to the next hop along the path. The rule applied here is "all or nothing," in that even if a call/admission reservation on a single hop along the path fails, the request resumes with failure. For non-real-time users, the admission is primarily based on the availability of buffer space in the *non-real-time packet queue* (NRTQ). This criterion may be set against a queue length threshold, in order to prevent queue overflow once the new call is admitted.

In the same line of thought, one may reasonably expect that all packet exchanges are not treated with equal priority in a QoS network. The exchange of control packets, for instance, should receive higher priority than user data packets in a QoS network. If so, control packets should receive preemptive priority over user data packets. Second, the QoS policy may allow different priorities to exist even among different flows of user packets. In heavy traffic situations, accommodating packets with preemptive priorities or guaranteeing QoS for lesser priority traffic may be difficult or even impossible. In such cases, the admission control policy needs to address whether the QoS guarantee for flows could be preserved and what actions to take in case they are not.

Notably, admission control decisions could also be leveraged by additional sophisticated adaptive/online optimization mechanisms, as for example to account for channel conditions as well. To this extent, the QoS monitoring module, which is responsible for monitoring the QoS parameters that affect system-wide performance, such as interference levels, number of hand-off drops, and bit error rate performance, could provide system QoS feedback to the call admission controller to help it make certain policy-based admission decisions. For fundamental discussions on this issue, Pravin's thesis in Ref. [BHAGW95] is an excellent resource.

3.9.7   Bandwidth reservation

As described earlier, in order to provide a stable QoS, it is not sufficient to simply select a route that can provide the correct resources. These resources must be reserved to ensure that they will not be shared or "stolen" by another session [WANG01]. Guaranteed services then require routers to make a reservation and, for each reservation request, routers must make admission control decisions.

A general problem related to resource reservation though is how to determine what resources have already been allocated to a flow. The availability of this information during path computation/selection improves the chances of finding a QoS path. It is necessary to consider the total resources already booked for a flow along a path in relation to available resources, to determine whether or not the flow should be routed on the path.

Consequently, the amount of the remaining network resources available to accommodate the QoS requests will have to be recalculated then and propagated to all pertinent nodes. The propagation of the new state of the network is part of the topology update information. In doing so, the resource reservation protocol interacts with the QoS routing protocol to establish a path through the network in the first instance; then, based on admission control decisions at each node, end-to-end resources are allocated. Reservation can be done per-flow or per-class. The protocol maintains per-flow/per-class specific state information of traffic and updates the traffic control database.

In the QoSR reference model, applications can access the services provided by the QoS manager directly through the resource reservation interface. For example, in order to set up a new QoS connection, an application can use the resource reservation interface to communicate the endpoints of the connection and the traffic specification for the flow. The resource reservation communicates this information to the QoS manager for creating the local reservation. In addition, on the arrival of a new flow request, the reservation protocol sets up a connection along the path that is selected by the routing mechanism. Similarly, the QoS manager is also invoked when the application decides to modify the reservation level or to remove the reservation altogether.

Thus, to create and maintain local reservation states for different flows a resource reservation signaling protocol is used prior to the flow of data. Similarly, an associated reservation is removed after a data connection is terminated.

For the resource reservation functions one could directly apply the RSVP signaling protocol. However, the reservation mechanism in RSVP presents several limitations when used in a mobile wireless environment. This is due to the following reasons:

- RSVP does not have any provision for passive reservation.

- In RSVP, reservation can be initiated from a location only when the sender or the receiver is present at that location. Thus, in RSVP, a mobile host cannot make advance reservation from a location where it is not currently present. For a mobile host, *path* or *Resv* messages must originate from the locations where it wants to make advance reservation.

- In RSVP, the sender IP address and port number is used to identify the senders in the filter-spec. A *path* message carries the IP address of its origin in the sender-template. This ensures that the *path* message is properly routed to the destination by the routing protocols in which routing decision depends on the source address of the packet. As a consequence, if *path* messages originate from several locations in the mobility specification (MSPEC) of a mobile sender, a receiver or an intermediate router cannot determine the identity of the mobile host from the sender-template object of the message. As a result, *Resv* message forwarding for the different reservation styles becomes difficult.

- RSVP does not support reservations over IP-IP tunnels, even though the routers in the tunnel may be RSVP capable.

There exist a number of resource reservation protocols in the literature [RFC 2205, RFC 1190, SCHELEN97, HAFID97, LEE98, GERLA95]. Among these, a protocol of particular interest is the *mobile resource reservation protocol* (MRSVP) [MRSVP].

The mobile resource reservation protocol (MRSVP) Mobile resource reservation protocol considers a network architecture in which a mobile host can make advance resource reservations along the data-flow

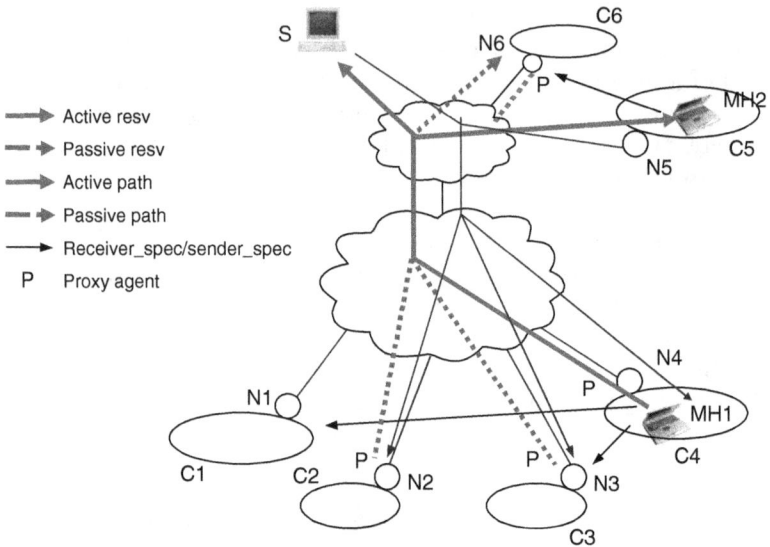

Figure 3.17   Overview of MRSVP.

paths to and from the locations it may visit during the lifetime of the connection.

The reservation model of RSVP is used. The set of locations from which the mobile host makes advance reservations is called MSPEC. Authors of MRSVP assume that the mobile host has acquired its MSPEC either from the network or from its mobility profile, when it initiates a reservation. The MSPEC of a mobile host can be changed dynamically while the flow is in transit. In such a case, resources will be reserved at the newly added locations of the MSPEC only if enough resources are available on the data-flow paths to/from these locations.

MRSVP supports *active* and *passive* reservations (Fig. 3.17). For a mobile sender, it makes an active reservation from the current location of the mobile host and it makes passive reservations from the other locations in its MSPEC. Similarly, for a mobile receiver it makes an active reservation to its current location and passive reservations to the other locations in its MSPEC.

Similar to mobile-IP protocol [MIP] that requires home and foreign agents to aid in routing, MRSVP requires *proxy agents* to make reservations from the locations in the MSPEC of the sender to the locations in the MSPEC of the receiver. The proxy agent at the current location of a mobile host is called the *local proxy agent*; the proxy agents at the other locations in its MSPEC are called *remote proxy agents*. The remote proxy agents make passive reservations on behalf of the mobile host.

The local proxy agent of a mobile host acts as a normal router for the mobile host and an active reservation is set up from the sender to the mobile host (or from the mobile host to the sender) via its local proxy agent.

If the mobile host is the sender of the flow, the paths of active reservations from the current location of the mobile host and the paths of passive reservations from its proxy agents are determined by the routing mechanism of the network. When the mobile host is a receiver, the paths of active and passive reservations to its current location and the proxy agents depend on the flow destination as follows:

1. *The mobile host joins a multicast flow.*   In this case the mobile host directs the proxy agents to join the multicast group. The data paths are then set up along the multicast routes.

2. *The mobile host initiates a unicast flow.*   In this case the paths may be set up by unicast routing or by multicast routing.

In MRSVP there are two types of *path* messages as well as two types of *Resv* messages. These are:

1. *Active path* message.   Carries a sender-Tspec for active reservation.

2. *Passive path* message.   Carries a sender- Tspec for passive reservation.

3. *Active Resv* message.   Carries a flowspec for active reservation. In addition, it may carry a FLOWSPEC for passive reservation when an active and a passive reservation are merged.

4. *Passive Resv* message.   Carries a flowspec of only passive reservation.

A sender host periodically sends active *path* messages to flow destination. In addition, if the sender is mobile, its proxy agents will send passive *path* messages. After the routes of active and passive reservations are set up, the mobile host and the proxy agents will start receiving the *path* messages. On receiving a *path* message the mobile host will send a *Resv* message for active reservation. If a proxy agent receives *path* messages for a multicast group, for which it is acting as a proxy agent, or for a mobile host from which it has received a request for acting as a proxy, it will make a passive reservation on the downstream link to which the mobile host will attach when it arrives in its subnet, and then send a *Resv* message to make a passive reservation. *Resv* messages for active reservations are converted to *Resv* messages for passive reservation when they are forwarded toward nodes which contain only proxy agents of mobile senders and no active sender (Fig. 3.17).

In addition to the messages present in RSVP, some additional messages are required in MRSVP. These are:

1. *Join-group.*   This message is sent by a mobile receiver to its remote proxy agents to request them to join a multicast group. It contains the multicast address of the group to join.

2. *Receiver spec.*   This message is used by a mobile receiver to send the flowspec and the flow identification (i.e. the SESSION object) to its remote proxy agents.

3. *Sender-spec.*   A mobile sender uses this message to send its sender-Tspec, ADspec and the destination address of a flow to a proxy agent.

4. *Receiver-Mspec.*   This message is used by a mobile host to send its MSPEC to the appropriate node that sets up the routes of active and passive reservations. It contains the addresses of proxy agents of the locations in the MSPEC of the mobile host.

5. *Sender-Mspec.*   This message is used by a mobile sender to send its MSPEC to a proxy agent that sets up the routes of active and passive reservations for the mobile sender.

6. *Forward-Mspec.*   This message is used by a mobile sender to forward the MSPEC of a mobile receiver to its local proxy agent.

7. *Terminate.*   This message is used by the mobile host to request its remote proxy agents to terminate reservation.

Currently, RSVP and integrated services do not provide any support for passive reservations. MRSVP considers two approaches in handling the active and passive messages. In the first method, the proxy agents play a more significant role in processing active and passive messages and no change is necessary in the RSVP message processing and forwarding roles except at the proxy agents and the mobile hosts. In the second approach, the RSVP message processing and forwarding rules need to be augmented at all nodes, but this ensures better utilization of network resources.

3.9.8 Resource adaptation

In a mobile radio environment, resource changes occur frequently as a result of wireless channel dynamics and user mobility. The motivation behind resource adaptation is to maintain an acceptable user presentation quality when resources fluctuate in a media-shared wireless network. Multimedia applications are capable of being adaptive in their operations and exhibit flexibility in dealing with fluctuations in network conditions, such as by changing between coding techniques. This is a particularly valuable feature to be exploited when a mobile node traverses dissimilar access systems during the course of a session.

QoS adaptation algorithms can, for example, trade temporal and spatial quality with available bandwidth or manipulate the playout time of continuous media in response to variations in delay [MOGHE97, SISAL97, OTT97, OH97, REVEL97].

Resource adaptation may be triggered by changes in the measured QoS over the wireless link on increase or decrease or resource availability, on handoff, or when the application initiates QoS renegotiation. A mobile radio, for instance, may move from lightly loaded areas (e.g. cluster or service area) to heavily loaded ones. In practice this can occur when, during a call handover process, the available bandwidth in the new area may be lower than that granted to the active flow in the previously visited areas (this is often referred to as *resource conflict* [QMA98]). The problem of resource conflict arises when the network cannot accept a new connection or hand over an ongoing connection without reducing allocated resources to currently ongoing connections (within prenegotiated QoS bounds).

Rate adaptation is of paramount importance. The goal of rate adaptation is two-fold: (1) on resource increase, allocate additional rate to a subset of ongoing flows, and (2) on resource decrease, reclaim rate from ongoing flows. With the goal of maximizing the network resource utilization rate when sporadic bandwidth shortage or link congestions exist, the system should manage to successfully resolve resource conflict occurrences so that existing QoS connections with a granted bandwidth higher than their LBW share could enter a lower rate. If this can be achieved, the system can potentially accommodate news calls, thus minimizing the average blocking rate of the system.

Adaptation for bandwidth-sensitive services is realized in the following two cases:

- A real-time connection that has been accepted but was not granted its maximum desired share (HBW) during the call setup time.

- When a real-time call request arrives and finds all channels occupied, it may, under certain circumstances, force one (or more) ongoing non-real-time call(s) to be temporarily buffered so that the released channel can be used to admit the real-time request.

Notably, each adaptation attempt has a certain cost to each flow that needs to adapt. This cost is often expressed in terms of either short or long interruptions and/or additional signaling to account for the coordination and management of these changes. Thus, a critical task of the rate adaptation algorithm is to increase revenue through improved network utilization but at the same time avoid repeated adaptations due to resource changes.

In addition, an inherent problem is ensuring that the new declared rate (*during* the adaptation phase) can be supported at all links of the data path. Applications with very tight bandwidth constraints may want to know the resulting end-to-end available bandwidth prior to their adaptation requests so as to choose an appropriate service level. This is an extremely tedious procedure that involves both extra signaling and intermittent service interruptions.

Moreover, since the resolution of resource conflicts readjusts the resources allocated to each connection, it should do so without violating the prenegotiated bounds of the existing connections. Besides, the existing calls can be degraded if and only if these are degradable and adjustable applications, as mentioned in Sec. 3.9.9. Whether an application can be degraded or not is declared during the connection set-up process. At this point, the degradation algorithm is conducted.

3.9.9 Degradation policies

When a node attempts to find a QoS route for a service with a QoS class q, the route discovery fails if there is no feasible QoS route available to accommodate q. In this case, the call request should either be rejected, or, if the call is degradable, should be accepted but with lower QoS than q. In the latter case, the admission control policy should investigate whether the requesting node could negotiate with the destination for a lower QoS q'. If such negotiations are allowed, then the policy will also specify how many, how often, and for how many different "values" of q'. The default option, as considered in Ref. [CHEN99], would be to switch to the "best-effort" service only.

To implement the degradation policy, the packet sorter (see next section for details) marks packets of flows that are degradable and the scheduler provides treatment accordingly. For example, degradable flows may not be scheduled at times of bandwidth shortage, or may even be discarded. For bandwidth-sensitive services one type of degradable behavior is considered—the bandwidth degradation.[2]

Bandwidth degradation implies that an application occupying multiple channels releases some of them to enter a degraded mode. To put this into context, when available bandwidth increases on a link, due to call termination for instance, the nodes associated with this link could allocate additional bandwidth to their supporting QoS flows, which were not granted their maximum resource share (HBW) in a previous negotiation attempt.

In its general context, if the average rate specified in the requirement profile equals the minimum rate for the application, then the application

---

[2]In an integrated system, other types of degradation may also be imposed, such as delay degradation. Delay degradation implies an increase in the delay jitter or delay variance of a session. Typically, non-real-time applications are expected to exhibit a large amount of delay degradation if the number of higher priority real-time applications grow in the system.

is not degradable. For example, some forms of compressed video may exhibit this behavior. Bandwidth degradation may be precipitated by various circumstances, such as when the mobile user faces sudden resource constraint while trying to hand over to a heavily loaded area.

Furthermore, the fundamental principles underlying the graceful degradation phase assume that the network is responsible for informing the application the extent to which the call shall be degraded. This insulates applications from having to understand, or even know about, the detailed nature of link heterogeneity or how the different services along the path compose. This insulation is critical for enabling applications to function relatively undisturbed while the network infrastructure evolves.

To enforce degradation policies on real-time traffic, the degradation control block being part of the QoS manager interacts with the resource reservation signaling protocol of the local system. As a result, when a QoS path enters the degradable mode, the state information associated with the QoS path is altered to reflect the changes. The application source, however, is free to either accept the suggested change from the network QoS-degraded mode or instead to act autonomously, such as to use an alternate path with sufficiently more resources than the existing one to accommodate the call.

Furthermore, since non-real-time traffic packets can sustain much longer delays, real-time traffic usually has preemptive priority over the former when there is a scarcity of available channels in the system. The preempted non-real-time traffic is buffered for future scheduling when free channels become available. Hence, in the case of an overloaded system, a suitable QoS parameter is the average buffer occupancy or queue length for non-real-time packets. The following assumptions are made:

- Real-time packets are never buffered and always allocated a channel if one is available or being used by non-real-time traffic.

- Each channel has a queue used only by the arriving non-real-time packets being serviced by that channel. If the channel is servicing a real-time call, its queue remains empty.

- When a non-real-time call is preempted from a channel, the system distributes the incoming packets (for the queue associated with the channel) with equal probability among the queues servicing other ongoing calls of similar type.

3.9.10 Packet sorter and shaper

The packet sorter and shaper modules operate between the IP and MAC layers. The packet sorter is responsible for differentiating real-time and non-real-time (best-effort) packets.

More specifically, the packet sorter has the following main functionalities:

1. Identify traffic packets entering the QoS sublayer based on a flow id.

2. Sort the incoming packets based on the packet-type information in the requirement profile.

3. Segment the packets and append the flow id to the segmented packets.

4. Mark the packets by setting a certain bit pattern in their header to distinguish them further, so that the scheduler (and also the QoS manager) can treat them differentially while allocating radio resources (e.g., time slot) or while faced with resource constraints. One example of this service differentiation is the priority of real-time packets over non-real-time. This is illustrated in Fig. 3.15 where the packet sorter interfaces with a two-level priority queue in the system—the *real-time packet queue* (RTQ) and the *non-real-time packet queue* (NRTQ), the former having a higher scheduling priority over the latter. Note that both RTQ and NRTQ are abstractions for actual physical buffer implementations, of which there can be multiple instances under each type. For example, the system could implement a multilevel queue for real-time traffic, based on delay jitter or reliability requirements for further differentiating the real-time traffic packets.

   Hence, in the case of buffer overflow, the non-real-time packets are discarded first. Another example might be to allocate smaller rate time slots to packets generated by degradable applications and allocate larger time slots to nondegradable ones. This concept leads to more efficient bandwidth usage.

5. Route the packets into the appropriate queue.

Traffic shaping provides a mechanism to control the amount and volume of traffic being sent into the network and the rate at which the traffic is being sent. This is required so that all packets entering the network and belonging to an existing reserved connection are complying with their initial FBW. It may also be necessary to identify traffic flows with a granularity that allows the traffic-shaping control mechanism to separate traffic into individual flows and shape them differently.

Two elegant methods for shaping traffic exist—a *leaky-bucket implementation* and a *token-bucket implementation*. Both these schemes have distinctly different properties and are used for distinctly different purposes. Discussions of each scheme follow.

Leaky-bucket implementation.   The leaky-bucket method (Fig. 3.18) is used to control the rate at which traffic enters the network. A leaky

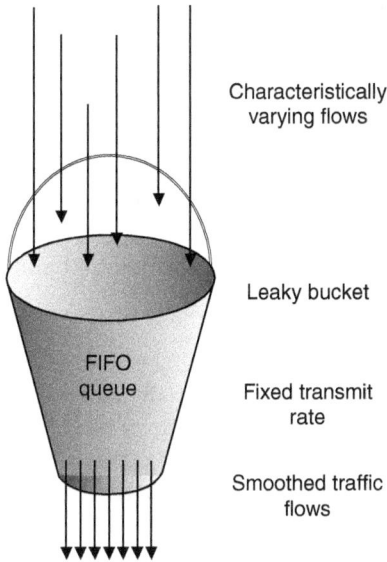

Characteristically
varying flows

Leaky bucket

FIFO
queue

Fixed transmit
rate

Smoothed traffic
flows

Figure 3.18   A simple leaky bucket.

bucket provides a mechanism by which bursty traffic can be shaped to present a steady stream of traffic to the network, as opposed to traffic with erratic bursts of low- and high-volume flows. An appropriate analogy for the leaky bucket is a scenario in which four lanes of automobile traffic converge into a single lane. A regulated admission interval into the single lane of traffic flow helps the traffic move. The benefit of this approach is that traffic flow into the major traffic arteries (the network) is predictable and controlled. The major liability is that when the volume of traffic is far greater than the bucket size, in conjunction with the drainage-time interval, traffic backs up in the bucket beyond bucket capacity and is discarded.

The size (depth) of the bucket and the transmit rate are generally user-configurable and measured in bytes. The leaky-bucket control mechanism uses a measure of time to indicate when traffic in a FIFO queue can be transmitted to control the rate at which traffic is leaked into the network. It is possible for the bucket to fill up and subsequent flows to be discarded. This is a very simple method to control and shape the rate at which traffic is transmitted to the network. Besides, this is a fairly straightforward implementation.

The important concept to bear in mind here is that this type of traffic shaping has an important and subtle significance in controlling network resources in the core of the network. Essentially, you could use traffic shaping as a mechanism for traffic flows to conform to a threshold that the network administrator has calculated arbitrarily. This is especially useful if the administrator has oversubscribed the network capacity. However, although this is an effective method of shaping traffic into flows with a fixed rate of admission into the network, it is ineffective in providing a mechanism that provides traffic shaping for variable rates of admission.

It is also arguable whether the leaky-bucket implementation makes use of the available network resources efficiently. Because the leak rate is a fixed parameter, there may exist a number of instances when the traffic volume is very low and large portions of network resources (bandwidth) are not being used.

Token-bucket implementation.   Another method of providing traffic shaping and ingress rate control is the token bucket (Fig. 3.19). The token bucket differs from the leaky bucket substantially. Whereas the leaky bucket fills with traffic and steadily transmits traffic at a continuous fixed rate when traffic is present, traffic does not actually transit the token bucket. The token bucket is a control mechanism that dictates when traffic can be transmitted based on the presence of tokens in the bucket. The token bucket also makes more efficient use of the available network resources by allowing flows to burst up to a configurable burst threshold.

The token bucket contains tokens, each of which can represent a unit of bytes. The administrator specifies how many tokens are needed to transmit however many number of bytes; when tokens are present, a flow is allowed to transmit traffic. If there are no tokens in the bucket,

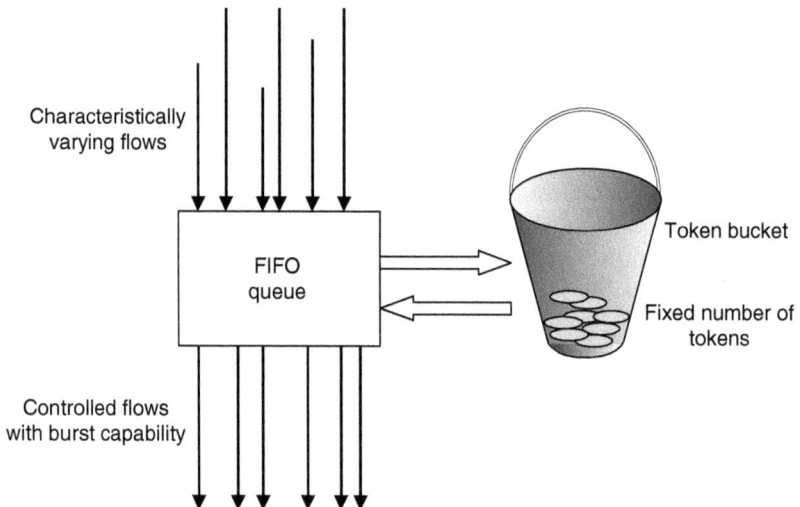

Figure 3.19   A simple token bucket.

a flow cannot transmit its packets. Therefore, a flow can transmit traffic up to its peak burst rate if there are adequate tokens in the bucket and if the burst threshold is configured appropriately.

The token bucket is similar in some respects to the leaky bucket, but the primary difference is that the token bucket allows bursty traffic to continue transmitting while there are tokens in the bucket, up to a user-configurable threshold, thereby accommodating traffic flows with bursty characteristics. An appropriate analogy for the way a token bucket operates would be that of a toll plaza on the interstate highway system. Vehicles (packets) are permitted to pass as long as they pay the toll. The packets must be admitted by the toll plaza operator (the control and timing mechanisms), and the money for the toll (tokens) is controlled by the toll plaza operator, not the occupants of the vehicles (packets).

3.9.11 Scheduler

The requirements of multimedia traffic are faced with the limitations of a highly error-prone radio channel. Because of the bursty error characteristic of radio channels, not all wireless terminals experience a good channel at the same time. Instead, the components at the link layer attempt to average out this channel variability and assure a fair channel resource assignment to mobile terminals.

In wireless networking environments, a crucial issue is how to handle the time-variable nature of a wireless channel and match this impediment with the application's traffic requirements so as to manage the scheduling of traffic with respect to the state of the channel. Various research efforts on *channel-state dependent scheduling* have considered guaranteeing a (possibly weighted) fair share of the radio resources to every terminal [LU97, EUGENE98, LU00, FRAGOU98, VAIDYA00] as the goal of fair queuing. The scheduler assures that terminals with an intermittently bad channel state are compensated for once the channel state is good again. Some of this research has also considered the integration with MAC protocols acting as a channel-state feedback.

An alternative to channel-state-dependent scheduling is *prediction-based scheduling* of traffic flows. Being able to predict the channel quality for a practically relevant time frame allows adapting transmission parameters, taking into account the particular requirements of a particular traffic flow. This is essentially a scheduling decision between multiple flows with a number of degrees of freedom. In particular, the quality of the channel prediction is an important factor to adequately support packet traffic in general and TCP/IP traffic in particular.

Another important aspect of QoS is that, regardless of the mechanism used, some traffic must be given a predictable service characteristic.

It is not always important that some traffic be given preferential treatment over other types of traffic, but instead the characteristics of the network remain predictable. These (behavioral) characteristics rely on several issues, such as end-to-end response time (or RTT), latency, queuing delay, available bandwidth, or other criteria. Some of the characteristics may be more predictable than others, depending on the application, the type of traffic, the queuing and buffering characteristics of the net services, and the architectural design of the net.

The Havanna scheme [GOMEZ99] presents an integrated solution to deliver QoS over a wireless channel. It consists of a predictor, a compensator, and an adaptor. The first predicts the channel state, which is used to decide on transmitting or not; the compensator uses a credit system to compensate flows for bad channel periods; and the latter consists of a buffer controller and a regulator, working together to deal with long periods of bad channel state (sort of channel state aware flow control). The performance greatly depends on the channel prediction method, which requires a contention-free medium access. The wireless IP framework [STERNAD01, FALAH01] is a centralized concept based on accurate prediction of the channel state for up to 10 ms. The timeslots are allocated according to the target error rates of the flows, taking channel state and adaptive modulation and coding into account. The chosen scheduling algorithm provides compensation on a (scheduling-) round per round basis, but not over longer timeframes.

3.9.12   Medium access control

The MAC layer is generally responsible for monitoring channel quality and accounting for available resources. MAC protocols have been extensively studied in the early 1970s, mainly for single shared channel applications. Examples include the ALOHA protocol, the CSMA-family, and CSMA/CD. The ALOHA protocol [ABRAMS70, ABRAMS85] is the earliest version of wireless MAC protocol. In ALOHA, every user is allowed to transmit its outgoing data packet whenever it is ready. A collision occurs when more than two users are transmitting at the same time. The channel utilization of ALOHA is very low due to the high probability of collisions, especially in the case of high network loads. In the Slotted ALOHA protocol, the channel is divided into time slots. Transmission is allowed to start at the beginning of a time slot only. Slotted ALOHA (S-ALOHA) [ABRAMS70, ABRAMS85] maintains higher channel utilization from ALOHA by avoiding collisions in the middle of the data transmission. An average channel utilization of 36 percent is achieved. Slotted ALOHA, however, introduces a somewhat longer access delay and more complexity due to the need for slot synchronization. ALOHA and slotted-ALOHA are particularly applicable in network environments with long propagation delay, such as satellite communication. In network with shorter end-to-end propagation delay, much better channel utilization is possible.

In the *carrier sense multiple access* (CSMA) family protocols [KLEIN75, METCAL76, LAM80, ROM90], users sense the channel for carrier before they transmit. The idea behind the CSMA-paradigm is to reserve the transmission channel at the originator (source) by carrier sensing. In principle, CSMA-based systems have lower probability of collision. This is primarily attributed to the fact that when users sense the other node's transmissions, they defer accessing the shared channel.

In a wireless ad hoc network where not all users in the network can hear each other, the hidden and exposed terminal problems are exhibited (see "The hidden and exposed terminal problem" in Sec. 2.7.1).

Various protocols have been proposed to avoid the hidden and exposed terminal problems in wireless ad hoc networks [BHARGH94, KARN90, ALWAN96, IWATA99, LIN97, LIN99, TANG99, TANG99a, GERLA95, ZHU98]. Some notable examples are the *multihop access collision avoidance* (MACA) [KARN90], the *MACA by invitation* (MACA-BI) [TALUC97], the *power-aware multiaccess protocol with signaling* (PAMAS) [RAGHAV98], and the *dual busy tone multiple access* (DBTMA) [HAAS98] protocols.

In multiple access collision avoidance [KARN90], a node wishing to transmit a data packet to a neighbor first sends a request-to-send (RTS) frame to the neighbor. All nodes that receive the RTS are not allowed to transmit. On reception of the RTS, the neighbor that the RTS was sent to replies with a clear-to-send (CTS) frame. Also, any node that hears the CTS transmission is prevented from using the channel. Hence, the RTS-CTS message exchange clearly alleviates the hidden terminal problem present in wireless networks. Due to this scheme, data frames are, at least in theory, delivered collision-free. As a result, collisions can only affect control packets. In this case, the IEEE exponential back off is used to resolve MAC contentions for control packets. In practice, however, collisions may still affect data frames. Nodes that do not properly receive a CTS frame are eligible to use the medium and their transmissions might overlap with those of the source. Hence, this MAC scheme will only decrease the probability of data collisions and the data remains vulnerable to corruption. From the standpoint of energy consumption, corrupted CTS frames will result in idle time energy losses for those neighbors that successfully receive the CTS frame. The neighborhood-wide energy loss is proportional to the number of neighbors of the sending station. On the downside also, MACA does not use link -layer positive or even negative ACKs, but rather end-to-end ACKs. It is worthy of notice, however, that the transmission of an RTS frame considerably reduces the energy costs of data collisions. RTS frames result in a favorable reduction in the consumed energy in the event of a collision, as compared with the consumed energy in addition to the larger delay due to collision time, if it were the actual data frame for which the RTS is being sent. These delay and energy savings will be achievable in most cases. The use of the RTS frame is nullified whenever the size of data is comparable to that of the RTS frame. A threshold is used to specify the size of the data frames for which an RTS frame ought to be sent.

In MACA by invitation [TALUC97], data must be foreseen beforehand by the receivers. Hence, MACA-BI may only be used for periodic and not unpredictable (bursty) traffic. Incorrect data predictions would cause all the neighboring nodes to waste an amount of energy that is proportional to the size of the ready-to-receive (RTR) frame, which is sent by the receiver in place of the CTS frame. The sender then responds with the actual data, and the use of the RTS frame is rendered obsolete. In a wireless ad hoc environment, in which nodes are allowed to move freely at all times, anticipation of a source's transmissions would be extremely difficult due to the unpredictable patterns of contention for medium access by nodes neighboring the source. Therefore, it would still not help the receiver to know the transmission schedule of the sender. Complex application-level and contention prediction schemes would be required.

The IEEE 802.11 wireless LAN standard [IEEE-STDRD] employs the idea of collision avoidance mechanism through the CSMA/CA (carrier sense multiple access with collision avoidance) protocol, which relies on immediate positive acknowledgments to the sender and possible retransmission of unacknowledged packets. The probabilistic nature of CSMA/CA is known to be unable to support real-time-oriented traffic. Besides the decentralized control, by means of the *distributed coordination function* (DCF), IEEE 802.11 optionally defines a centralized access mode (*point coordination function*, PCF) based on polling to support prioritized calls. Since the IEEE 802.11 centralized mode can neither be applied in multihop networks nor be operated simultaneously in neighboring cells, where it also requires an access point with specialized access functions implemented, it imposes severe constraints on the operation of wireless local area networks. IEEE 802.11 MAC standard [IEEE-STDRD] is designed with a reservation system similar to MACA in mind. The IEEE 802.11 protocol is discussed in detail in Sec. 2.7.1.

As mentioned above, single channel MAC schemes solve the hidden node and exposed node problem by exchanging RTS/CTS control packets. As the traffic load increases, the performance gets worse due to increased collisions of radio signals. Because of the possible large size of the ad hoc networks, much larger than the transmission range of a single transmitter, single shared channels do not perform well in this environment.

To overcome this problem, multiple channel MAC schemes have been proposed. Separating channels functionally, multiple channel MAC approaches reduce the chances of collisions and contentions, and result in performance enhancements, i.e., shorter access delay and higher throughput.

In Ref. [RAGHAV98], a power-aware multiaccess protocol with signaling [SINGH98a] is introduced for power conservation in ad hoc networks. PAMAS is based on MACA with a separate signaling channel. A distinguished feature of PAMAS is that it considers the limited battery energy of mobile. PAMAS conserves battery power by intelligently powering off the mobiles that do not perform transmitting or receiving packets. In addition to the primary data channel, PAMAS uses a secondary channel, called the *signaling channel*, to carry control traffic. The signaling channel is used for exchanging RTS/CTS packets, busy tones, and probe control packets. The separate signaling channel in Ref. [RAGHAV98] enables a host to determine when and for how long to power its antenna off.

The basic operation of the protocol is based on the observations that, due to the broadcast nature of radio transmission, a host's energy is often wasted on overhearing packets that are not meant for it. In this scheme, a current receiver can force a neighbor to defer using the medium by sending a busy tone over the signaling channel. This can be used to force the sender of an RTS to defer. All nodes that successfully receive the busy tone are prevented from using the medium for the period of data transmission over the primary channel. Even if the RTS is detected by the receiver, its CTS transmission will be corrupted by that of the busy tone since they will both be sent simultaneously over the control channel.

In PAMAS therefore, battery power is conserved by judiciously turning off radios of hosts when no transmission/reception is possible. Since most of the consumed energy in a wireless ad hoc network is attributed to the idle time, PAMAS will obtain high energy-conservation gains only in networks with high traffic loads. In networks with low to modest traffic, the energy-based performance of PAMAS will be close to that of other CSMA protocols. Therefore, nodes running PAMAS will still consume most of their battery capacity while in their idle states. As Fig. 3.20 shows, node C powers off its radio while A is transmitting to B. Let us assume now that during its sleep period, a new transmission is established between D and E. In this case, when C wakes up it does not know how long this transfer will last so as to go back to sleep. According to PAMAS, node C—on waking up—performs a binary probe to determine the length of the longest remaining transfer. To achieve this

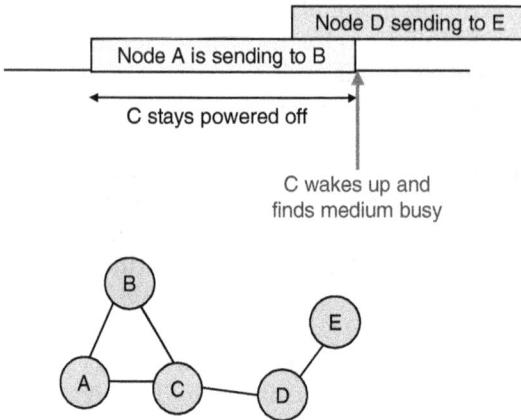

Figure 3.20 PAMAS in operation.

- C transmits a probe packet with parameter L
- All nodes with a connection that is going to resume within the interval [L/2,L] respond
- Depending on whether node C see silence, collision, or a unique response it assumes different actions

In *dual busy tone multiple access* (DBTMA) protocol [HAAS98] the single common channel is split into two subchannels—a *data channel* and a *control channel*. Data packets are transmitted on the control channel, while control packets (RTS, CTS, and the like) are transmitted on the data channel. Additionally, two out-of-band busy tones are used to inform the neighboring nodes of any ongoing transmission—rBT, the receive busy tone, which shows that a node is receiving on the data channel, and tBT, the transmit busy tone, which shows that a node is transmitting on the data channel.

When a node becomes ready for transmission, it first senses the channel. If no rBT tone is heard, the node can be sure that its transmission will not interfere with the reception of any other node in its vicinity. It then transmits the RTS message and waits for the CTS message from the receiver. The receiver, on receipt of the RTS message, checks the tBT tone. If this is absent, the receiver knows that the transmitter's transmission will not interfere with the transmission of any other node in its vicinity. The receiver then issues the CTS and raises the rBT to prevent any other nodes in the area to transmit. The receiver, on receipt of the CTS message, raises the tBT to indicate that it intends to transmit and prevent the other nodes in its vicinity to accept a connection request. When the connection is terminated, the busy signals are turned off. As the busy tones are maintained continuously during the connection, a node migrating into the vicinity of the communicating nodes will still be able to learn the status of the channel and refrain from accessing it.

A source of inefficiency of DBTMA is the use of probing that wastes battery capacity by forcing the wireless nodes to continuously sense the medium for the transmit and receive busy tone signals. A more energy-efficient MAC protocol would consider turning off the transceiver during standby time to save power.

3.9.13   Channel allocation

Once a resource request is successfully admitted along the data path, the system needs to allocate the resources (channels) to the real-time flow. Efficient channel assignment during the call setup is vital for minimizing the blocking/dropping rates of the present as well as future calls. In the following section, we elaborate on the channel allocation method proposed in Ref. [AGGE04] for bandwidth-sensitive services in cluster-based ad hoc wireless networks. The method aims at providing a differentiated service treatment to real-time *bandwidth-sensitive* and non-real-time *bandwidth-tolerant* multimedia traffic flows at the link-level using novel techniques for end-to-end path bandwidth maximization. Time-division multiple access (TDMA) within a cluster is chosen whereas code division multiple access (CDMA) is overlaid on top of the TDMA infrastructure. To reduce the effect of intercluster interference, separate codes are assigned to different clusters.

3.9.14   End-to-end path bandwidth calculation

Let us assume that the transmission time scale is organized in frames, each containing a fixed number of time slots. Similar to [BAGR95], each frame is divided into two phases—the control phase and the data phase. The size of each slot in the control phase is much smaller than the one in the data phase. The control phase is used to perform all the control functions, such as slot and frame synchronization, power measurement, code assignment, QoS path setup, slots request, and routing-protocol specific control message exchange. The system's requirements dictate the number of the data slots per frame [RAPP96] (e.g., see Fig. 3.21).

Each node takes turn to broadcast its information to all its neighbors in a predefined slot (the way this is resolved is described later), such that the network control functions can be performed in a distributed fashion. In general, it is assumed that the information can be heard by all its adjacent nodes. In a noisy environment, where the information may not always be heard perfectly at the adjacent nodes, an acknowledgment scheme is performed in which each node has to acknowledge for the last

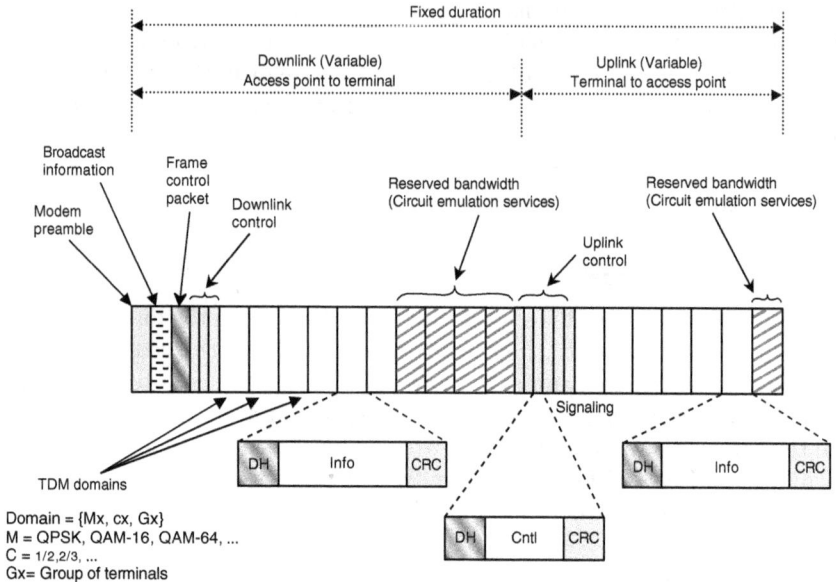

Figure 3.21 Frame/slot configuration example (from the IST ADAMAS project— *http://adamas.intranet.gr/*).

information in its control slot. By exploiting this approach, there may be one frame delay for the data transmission after sending the data slot reservation.

Ideally, at the end of the control phase, each node has learned the channel reservation status of the data phase. This information will help one to schedule free slots, verify the failure of reserved slots, and drop expired real-time packets. The data phase must support both guaranteed and best-effort traffic. Since real-time traffic needs guaranteed bandwidth during its active period, bandwidth must be preallocated to the real-time connection in the data phase before actual data transmission.

Because only adjacent nodes of the same cluster and neighbor gateways can hear the reservation information, the free slots recorded at every node in different clusters may be different. The set of the common free slots between two adjacent nodes will be referred to as the *link bandwidth*. A node that knows its link *available bandwidth* (ABW) to some other node can only determine whether a request can be satisfied through this link or not. Consider the example shown in Fig. 3.22 in which node A in cluster C_1 intends to establish a real-time call to C in cluster C_2 with a *bandwidth request* (FBW) [2,2]. Let us assume a TDMA frame with six time slots. The admission controller at node A would

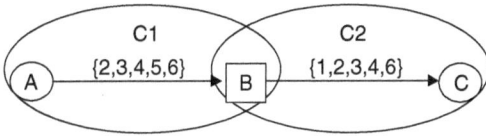

Figure 3.22 Example of end-to-end bandwidth calculation.

reject the request, as its ABW to node B is smaller than the minimum requested BW, which is 2.

The *path bandwidth* (also called end-to-end bandwidth) between two nodes, which may not be necessarily adjacent, is the minimum link bandwidth of the links that comprise the path. If the destination and source nodes of a flow are adjacent, the path bandwidth is equal to the link bandwidth; the traffic flow in this case is called *intracluster traffic*. If the destination and source nodes of a flow belong to different clusters, the traffic flow is called *intercluster traffic*.

The calculation of link bandwidth differs from node to node, depending on whether a node functions as a gateway or is merely an end-host. If a node B serves as a gateway between two clusters with codes C_1 and C_2, respectively, the available slots for reception/transmission at B are determined by:

$$ABW(B) = ITS_B(C_1) \cap ITS_B(C_2)$$

where $ITS_i(X)$ denotes the set of time slots when a node i is idle (i.e, not transmitting or receiving) in cluster X.

Thus the link bandwidth between a node A in C_1, which does not function as a gateway, and the node B, which is GW, is determined by

$$LinkBW(A, B) = ITS_A(C_1) \cap ITS_B(C_1) \cap ITS_B(C_2)$$

and if B is not a gateway (that is, intracluster traffic), their link bandwidth would be

$$LinkBW(A, B) = ITS_A(C_1) \cap ITS_B(C_1)$$

Therefore, the link bandwidth can take different calculation forms or, else, different resource constraint levels, depending on whether either the end nodes of the link are gateways or, simply, end hosts.

Furthermore, to calculate the end-to-end bandwidth (PBW) a careful analysis of the transmission and reception planes of the end nodes of a link is required. Let us consider the example illustrated in the figure below.

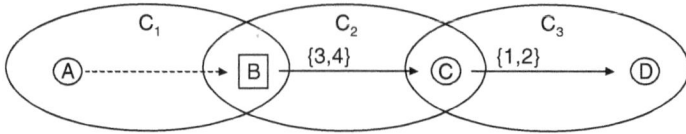

Node A in cluster C_1 intends to send real-time traffic to D in cluster C_3 through cluster C_2. Node B is the gateway node between clusters $C_1 - C_2$ and node C is the gateway node between clusters C_2 and C_3. Let us assume a TDMA frame with six time slots again and that gateways B and C are busy (either transmitting or receiving) in slots {3,4} and {1,2}, respectively; the set of busy slots of a node i are denoted by UBW(i). It is evident that the busy slots of a GW between two clusters C_a and C_b, is the union of the busy slots where this GW is transmitting to or receiving from *both* C_a and C_b. It is obvious also that the relationship between ABW and UBW for a node is: ABW = {slots per TDMA frame} – UBW

Referring to the above example, node A can use any of the slots {1,2,5,6} to transmit to B in C_1, say {1,2}. If so, the new UBW of B turns out to be UBW(B) = {1,2,3,4}. B in turn can use slots from the set UBW(B) – UBW(C) to transmit in C_2, and the procedure continues until the request either reaches D where the end-to-end path bandwidth is determined (in this example, PBW(A,D) = 2), or gets rejected along the path due to the lack of resources. The latter case may arise if during the propagation of a resource request, a LinkBW value is smaller than the minimum requested bandwidth. Failure to satisfy a resource request results in the blocking and/or dropping of the call. As mentioned earlier, call dropping manifests not only during periods of high system load but also during the handover of connections from lightly loaded to heavily loaded clusters. Thus one of the primary objectives of a handover as well as of a channel assignment algorithm is to maintain the blocking/dropping rates at minimum levels.

We next illustrate two cases where call blocking/dropping occurs. These shall form the basis for our discussions in the following sections.

The preceding calculation of path bandwidth between node A and D, PBW(A,D) involved only the transmission/reception planes of the end nodes and gateways that comprise the path; that is only the intercluster traffic of the *same* path. When we calculate end-to-end path bandwidth, additional constraints are imposed, which we need to take into account.

Case I. Assume the UBW size of A and B are the same as in Fig. 3.23 (that is, 2). Let us further assume that there are internal transmissions in cluster C_1 and C_2 as illustrated in Fig. 3.23. These internal transmissions are placed from nodes M and N to N and K. respectively, in C_1 and X and Y to Y and Z respectively, in C_2.

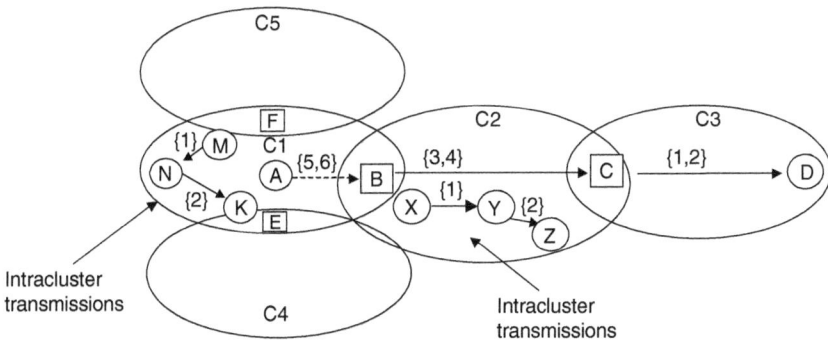

Figure 3.23 Example for channel allocation state.

It turns out that A can transmit to B only in slots {5,6} in C_1, instead of {1,2,5,6} as illustrated earlier. For B's transmission to gateway C, however, even though the LinkBW(B,C) seems to be {1,2} and thus the size of ABW(B,C) is 2, it turns out to be:

$$\text{LinkBW}(B,C) = \text{Slots\_Per\_Frame} - \text{UBW}(B) - \text{IntraCluster\_TXs}(C_1) - \text{UBW}(C_1)$$

$$\Downarrow \qquad\qquad \Downarrow \qquad\qquad\qquad \Downarrow$$

$$= \{1,2,3,4,5,6\} - \{3,4,5,6\} \quad - \quad \{1,2\} = 0$$

This scenario illustrates the impact of internal (intracluster) transmissions on the end-to-end path bandwidth. Even though there is ABW at all nodes along the path (i.e., A,B,C,D), it turns out that the link from B to C, link ABW (LinkBW) is zero.

Case II. The same problem is exhibited in intercluster flows. Assume the slot utilization as in Fig. 3.23 again. As illustrated above, node A can transmit to B in slots {5,6} in cluster C_1. This, however, is true only if an additional constraint is satisfied—if there is no intercluster traffic that flows through cluster C_1 such that it makes use of the available resource spectrum of C_1. To illustrate this with an example let us assume that another flow is routed through the gateways E and F of clusters (C_1 and C_4) and (C_1 and C_5), respectively, and also that gateway E transmits to F at slot {5} and F as slot {6}, both in cluster C_1. It is easy then to calculate that LinkBW(A,B) = ∅. The latter case effectively results in the blocking of not only A's flows but also of any inter- or intracluster traffic through cluster C_1, since there is no capacity spectrum left so that new transmissions can effectively be placed on C_1.

Again, although nodes do have ABW at their disposition, still cluster C_1 is effectively blocked as a result of the present channel configuration assignment.

From these observations, the following general observations are made:

- To compute the available bandwidth for a path in a time-slotted network, one not only needs to know the available bandwidth on the links along the path, but also needs to determine the scheduling of the free slots.

- Call blocking/dropping rates increase if time slot assignment is not performed efficiently. Efficiency here refers to the ability of the resource (time slot) assignment algorithm to select a time slot from the pool of available slots in a way that is minimizing the blocking/dropping probability of present as well as of future calls. Case II illustrates a common example where the system cannot place the call even though there is ABW in both clusters C_1 and C_2. As demonstrated in the slot assignment phase, key to efficient resource assignment within a cluster C is to account for both the UBW of the cluster C as well as of C's neighbor clusters.

- No matter how efficiently a routing protocol and a channel assignment algorithm perform, if the network load exceeds its capacity the ABW of links and clusters starts saturating thus leading to higher call dropping/blocking rates.

Finally, although it seems obvious, it is worth mentioning again that the bandwidth computation is part of the call admission control and resource reservation mechanisms during the call set-up process. As illustrated in "The resource reservation protocol" in Sec. 3.5.1, the resource reservation propagates the resource requirements, the call admission control calculates the ABW on each individual link, and based on its calculations it determines whether the real-time flow can be accepted or not. If the minimum bandwidth criteria are satisfied, the request is forwarded hop-by-hop toward its destination. Reservations can be placed on the reverse path (RSVP-style) or on both reverse and forward paths (CR-LDP-style).

3.9.15 Slot assignment phase

Once the resource request is successfully admitted along the path, the system needs to allocate the resources (slots) to the real-time flow hop-by-hop along the data path. As we argued above, efficient assignment of the available slots during the call setup is vital for minimizing the blocking/dropping rates of the present as well as of future calls. In the following section, we describe the *minimum blocking and bandwidth reassignment channel assignment* (MBCA/BRCA) methods proposed in

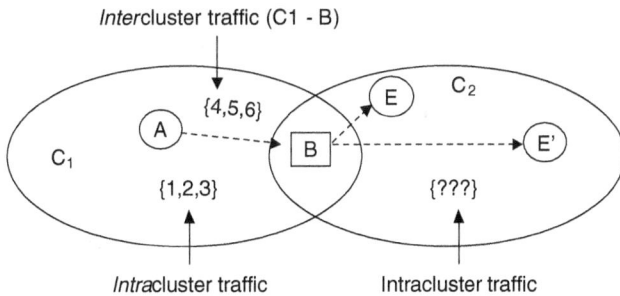

Figure 3.24 Intracluster traffic slot-configuration scenario.

Ref. [AGGE04] for efficient channel assignment in clustered wireless ad hoc networks.

The minimum blocking channel assignment.   To illustrate how the MBCA slot assignment algorithm operates, we use the communications scenario depicted in Fig. 3.24. As described earlier, for a given path, the end nodes (i.e., source/destination) and the intermediate nodes (i.e., gateways) have all different levels of resource constraints. End nodes are primarily constrained by their cluster's traffic as well as the intercluster traffic of their gateways. Gateway nodes are constrained not only by the intra- and intercluster traffic of their cluster's flows, but also by the intracluster traffic of their cluster's gateways.

For the correct operation of the MBCA mechanism, it is assumed that clusterhead nodes are aware of the UBW (i.e., transmission/reception scheduling) of their cluster's gateways as well as the intracluster UBW of their own cluster and of their gateways' clusters. That is, in Fig. 3.24 the clusterhead in C_1 is aware of the intracluster traffic in C_1, which is $UBW_{intra}(C_1) = \{1,2,3\}$, the intercluster traffic in C_1, which is $UBW_{inter}(B) = \{4,5,6\}$, and the intercluster traffic of C_1's gateway B, which is $UBW_{inter\_gw}(C_1) = \{4,5,6\}$. Clusterhead nodes in different clusters could be informed about each other's intracluster traffic through the exchange of a dedicated signaling that is leveraged by the initiation of new or the termination of existing connections.

The MBCA time slot assignment algorithm distinguishes between intra- and intercluster traffic flows.

Rule I: Channel assignment for intracluster flows.   The following rule captures the resource assignment for intracluster flows: *the scheduling of intracluster traffic within a cluster C should favor the time slots which are not used in C's neighboring clusters.*

The rationale behind this approach is based on the fact that the used slots of a cluster C cannot be used by intercluster traffic that flows through C, thus increasing the *common* available bandwidth spectrum between the two neighbor clusters. The following example illustrates the effectiveness of this approach.

Assume that C_1, C_2 are two neighbor clusters and node B is their elected gateway. A node or GW in C_1 is to forward a request to B which in turn is to forward the request either to an end node (E) or to a GW (say, E') in C_2.

Let us further assume that time slots {1,2,3} are used for intracluster traffic in C_1 and also that slots {4,5,6} are free slots in C_1. The objective is to devise a flexible mechanism to assign time slots for intracluster traffic in C_2 such that the ABW of link B-E is maximized.

It is clear from Fig. 3.24 that any intercluster traffic that flows to C_2 through C_1, will be assigned any time slots but {1,2,3}.

One can easily observe then that if time slots {1,2,3} are used for intracluster assignment in C_2, the intercluster traffic $C_1 - C_2$ (through B) will then make use of the common ABW between $C_1 - C_2$, which is {4,5,6}. That is, if slot {4} is used from C_1 to B then B can use any slot from the set of their common ABW, which is {5,6}. Thus the ABW ($C_1 - B$) is 2.

If, however, the proposed rule is applied and slots {4,5,6} which are the unused slots in C_1 are preferred for intracluster traffic in C_2, then the common ABW size is 3. This is clear as the link $C_1 - B$ can use slots {4,5,6} whereas the link B-E can use slots {1,2,3}. To put this into context, an application with LBW = 3 cannot be accommodated in the first case, whereas in the second case the system can exploit the channel states of current and neighbor clusters to achieve optimum channel allocation. The pseudocode for implementing the MBCA's intracluster channel assignment method is illustrated in Fig. 3.25.

Rule II: Channel assignment for intercluster flows. The following rules capture the time slot assignment on a gateway G for routing intercluster traffic, such that time slots are favored if:

(a) The gateways of G's neighbor cluster are busy in these time slots.

(b) Intracluster traffic in G's neighbor clusters takes place on either of these time slots.

(c) The gateways of G's present cluster, except those involved in the forwarding of the flow, use either of these time slots.

The effectiveness of these rules is illustrated again with an example. Consider the three clusters C_1, C_2, and C_3 in Fig. 3.26, with their associated busy set of slots for intracluster traffic.

ALGORITHM FOR INTRACLUSTER CHANNEL ASSIGNMENT

/* Calculate the Group of Free Slots of Cluster C, S_F^C */

for(s = 1; s <= SLOTSperFrame; s++)   **if** ($s \notin (S_{U_{intra}}^C \cup S_{U_{inter}}^C)$) $S_F^C \leftarrow s$

BW_Reassign = OFF;   */* Flag to trigger BRCA Algorithm (See Below)*/*

if (S_F^C != NULL) {   */* If the Group of Free Slots is non-zero, proceed*/*

   / Calculate the Group of Free Slots of C's Neighbour Clusters, S_F^N ,
   for each Neighbour Cluster of C, $N \in C^D$ */*

   for(s = 1; s <= SLOTSperFrame; s++)   **if** ($s \notin S_{U_{intra}}^N$) $S_F^N \leftarrow s$
   / Calculate the Intersection of S_F^C and S_F^N, for each Neighbour
   Cluster of C, $N \in C^D$*/*

   if(S_F^N != NULL) $S_{P_{intra}}^C = S_F^C \cap S_F^N$

   else { */* There is no intra-cluster traffic placed in C's neighbour clusters */*

       /* Pick in random a slot from the pool of C's free slots */

$$S_{P_{intra}}^C = drand(S_F^C)$$
   }

   if($S_{P_{intra}}^C$ = NULL)   **BW_Reassign = ON;**

} **else BW_Reassign = ON;**   */* There are no free slots in cluster C */*

If (BW_Reassign == ON) { */* Run the BRCA Algorithm */*

   if (**BW_Reassign**(C, C^D) == TRUE) {   */* Success */*

      / Bandwidth Reassignment Channel Allocation Method has
      successfully created free common bandwidth between C and its
      neighbour clusters, C^D Communication is successfully established.

      /

$$S_{P_{intra}}^C = S_{BW\_Reassign}^{C,C^D}$$

   } **else** */* Call Blocked/Dropped :: BRCA Method failed to create free
            common bandwidth */*

}

Figure 3.25 Pseudocode of MBCA's intra- and intercluster channel assignment methods.

Let us assume that node A from cluster C_1 intends to establish a connection to D from C_3 through gateways B and C. The objective is to efficiently allocate available common slots for the routing of the flow through the three clusters. Let us concentrate on the transmission from gateway B to C and determine according to the rules (a,b,c) the effectiveness of the proposed time slot selection.

ALGORITHM FOR INTERCLUSTER CHANNEL ASSIGNMENT

/* Calculate the Free Slots of G's present and neighbour clusters, C and L, S_F^C, and S_F^C, resp, */

for(s = 1; s <= SLOTSperFrame; s++) **if** ($s \notin (S_{U_{intra}}^{C,L} \cup S_{U_{inter}}^{C,L})$ && $N_G(s)==0$) $S_F^C \leftarrow s$

BW_Reassign = OFF;   /* Flag to trigger BRCA Algorithm (See Below)*/

if (S_F^C != NULL && S_F^L != NULL) { /* If the Group of Free Slots is non-zero, proceed*/
/* For each neighbour cluster of C, $N \in C^D$, calculate the busy slots of its Gateways, S_U^{GW} */

    for(s = 1; s <= SLOTSperFrame; s++)   **if** ($N_{GW^D(C^D)}(s)$ == ON) $S_U^{GW}(N) \leftarrow s$

    /* Calculate the time slots used for Intra-cluster traffic in C's neighbour clusters, $N \in C^D$ */

    for(s = 1; s <= SLOTSperFrame; s++)   **if** ($s \in S_{U_{intra}}^N$) $S_U^N \leftarrow s$

    /* Calculate the busy slots, $S_U^{GW}(C)$, of C's Gateways GW^D; exclude these in the present flow */

    for(s = 1; s <= SLOTSperFrame; s++) **if** ($GW \notin GW_F$ && $N_{GW}(s)$ == ON) $S_U^{GW}(C) \leftarrow s$

    /* * * * * * * * **Main Algorithmic Decision (Rules <a, b, c>)** * * * * * * * * * */

    if(S_U^N && $S_U^{GW}(N)$ && $S_U^{GW}(C)$!= NULL) $S_{P_{inter}}^C = S_U^{GW}(C) \mathrm{I} S_U^{GW}(N) \mathrm{I} S_U^N$

    else {

        if(S_U^N && $S_U^{GW}(N)$ && $S_U^{GW}(C)$ == NULL) {
        /*MBCA can not be applied; pick a slot in random */

$$S_{P_{inter}}^C = drand[(S_F^C \mathrm{I} S_F^{C^D}]$$

        } **else** {

        if($S_U^{GW}(C)$!= NULL && S_U^N != NULL) $S_{P_{inter}}^C = S_U^{GW}(C) \mathrm{I} S_U^N$

        if($S_U^{GW}(C)$!= NULL && $S_U^{GW}(N)$!= NULL) $S_{P_{inter}}^C = S_U^{GW}(N) \mathrm{I} S_U^N$

        if($S_U^{GW}(N)$!= NULL && S_U^N != NULL) $S_{P_{inter}}^C = S_U^{GW}(C) \mathrm{I} S_U^{GW}(N)$

        }

    }

}

if($S_{P_{inter}}^C$ == NULL)   BW_Reassign = ON;

} **else** BW_Reassign = ON;   /* There are no free slots in either cluster C or L*/

If(BW_Reassign == ON) { /* Run the *Bandwidth Reassignment Channel Allocation* Algorithm */

    if (BW_Reassign(C, C^D) == TRUE) $S_{P_{intra}}^C = S_{BW\_Reassign}^{C,C^D}$ */ Success */

    else /* Call Blocked/Dropped :: BRCA failed to create free common bandwidth */

}

Figure 3.25 (*Continued*).

According to Rule II.a, gateway B should favor the slots where *all* the gateways of C_1 are busy. That is simply because these gateways in C_1 cannot use the same slots for any future intercluster transmission to B (through C_1 of course); thus similar to Rule I, gateway B increase its ABW with C_1.

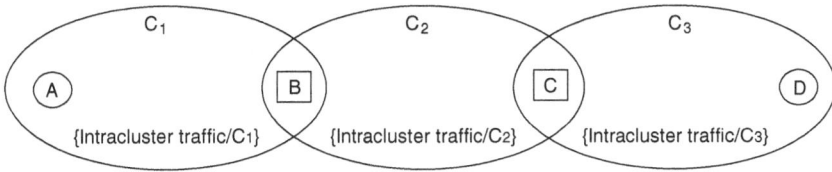

Figure 3.26   Intration of Rule II—culster configuration example.

Rule II.b is a complementary to Rule II.a. According to Rule II.b, gateway B should favor the slots which are used for intracluster traffic in C_1. The rationale behind this is that intercluster traffic from any gateway of C_1 to B be could not be placed on these slots; thus, by selecting these slots for transmission, B again maximizes its ABW with C_1.

Note here that Rules II.a and II.b are in contrast to Rule I, which does not favor the used slots in neighbor clusters.

Concluding, according to Rule II.c, gateway B should favor the slots where the gateways of its cluster (that is cluster C_2 in Fig. 3.26), except the next gateway of the present flow (i.e., C) and itself, are busy with inter- or intracluster traffic to *other* clusters. The reason is that for any future traffic that is routed through B in cluster C_2, B cannot use these slots to transmit to these gateways and, likewise, these gateways cannot use these slots to receive from B. As a result, the probability of blocking/dropping future calls through cluster in C_2 is provisionally minimized. The pseudocode for implementing the MBCA's intercluster channel assignment method is illustrated in Fig. 3.25.

The bandwidth reassignment channel allocation (BRCA) algorithm.   In view of rapidly-fluctuating wireless link bandwidths, a time slot assignment mechanism should be further leveraged to account for the dynamics of the network topology and timeliness of applications. The termination or hand over of ongoing connections within a cluster C, for example, results in a change of the intra- as well as of intercluster traffic in C. In this regard the intracluster time slot assignment (Rule I) could be triggered *proactively* in order to account for these changes and reassign new resources (if needed) to ongoing connections such that the new channel allocations adapt to the new available channel spectrum of C as well as of C's neighbor clusters.

The above process can also be triggered *reactively* when a bandwidth request arrives for forwarding and the call admission algorithm cannot accommodate the requested bandwidth. If so, the call admission algorithm triggers the so-called *bandwidth reassignment* algorithm with the hope to create common bandwidth between the current and the next cluster along the data path.

Let us assume that GW A attempts to transmit to GW B through cluster C_1 but there is no common available bandwidth.

The bandwidth reassignment process comprises the following three steps:

Step 1: Intracluster time slot reassignment. If there is a slot i that is used in C_1 for intracluster communication and i is not used by A and B, then exchange i with a time slot from the pool of free slots of C_i. If the procedure is successful, then communication can be established between A and B as A can use i for transmission and B can use i for reception and the phase end; else continue to step 2.

Step 2: Intercluster time slot reassignment at GW A. If there is a slot i where A is busy but B is free try to free this time slot. To achieve this GW A first finds out the communicating node, say X, at time slot i. Then A examines if the intersection of its ABW with X's ABW is nonzero. If so, A and X swap the used slot with the free one. This slot is used for communicating with B, since time slot i is free at B as well (see Rule II). If the phase is not successful, then try Step 3.

Step 3: Intercluster time slot reassignment at GW B. The semantics of this phase are similar to those in Step 2 with the only difference that it is gateway B that examines the intersections of its ABW with the ABWs of the communicating nodes at time slots where A is free.

Using this adaptive QoS concept, it is possible to utilize bandwidth more efficiently, and thus increase the network's reward/revenue, while reducing the number of new connection drops/blocks.

The main source of inefficiency of the bandwidth reassignment process is that, under high loads, a high number of bandwidth allocation changes may be triggered. This is an important parameter that could greatly influence the cost of ongoing sessions. The reason for frequent bandwidth allocation changes being a problem is: (1) it takes time to setup the modified bandwidth allocation and (2) in today's routers (and switches) it would normally require the invocation of software in every router on the session path, which would increase the response time even more and consume resources at the router. Thus, it is reasonable to assume that when using the bandwidth reassignment process an additional goal would be to minimize the number of changes.

Furthermore, operating the bandwidth reassignment mechanism in proactive mode could result in better accuracy of Rules I and II but however at the expense of higher signaling exchange between cluster-heads. Reactive bandwidth reassignment, on the other hand, trades off inaccuracy of Rule II, which inherently results in higher interruption rates of ongoing connections, with reduced signaling overhead.

Multihop Relaying: Stepping Stone to Systems Beyond 3G

4.1 Mobile Communications Evolves to Mobile Multimedia Ubiquitous Communications (3GSM01)

Mobile communications is the global success story of the late twentieth and early twenty-first centuries. The *global system for mobile communication* (GSM) is undoubtedly the largest technology worldwide. Second-generation (2G) telecommunication systems, such as the GSM, enabled voice traffic to go wireless—the number of mobile phones already exceeds the number of landline phones whereas, according to the GSM MoU[1] Association, mobile phone penetration exceeds 70 percent in countries with the most advanced wireless markets. In late 2002 there were close to half-a-million mobile subscribers worldwide whereas in the first quarter of 2004 GSM subscribers reached up to one billion globally [CNDN]. This success was built largely on voice communication as well as on short message service (SMS). SMS is a service enabling the sending of short messages of 160-octets length. Figure 4.1 illustrates the huge traffic exchange worldwide using text messaging.

The boom in SMS messaging has clearly shown that there is a strong and rapidly growing market for person-to-person text messaging. Multimedia messaging service (MMS), introduced in 2002, is a logical continuation of SMS that combines text, pictures, and voice. MMS appears to be well suited for the mass-market consumer (see Fig. 4.2) and probably takes over the baton from SMS in leading the way forward with mobile data [TEL-INTER02v36]. The emergence of MMS could

---

[1]Memorandum of understanding.

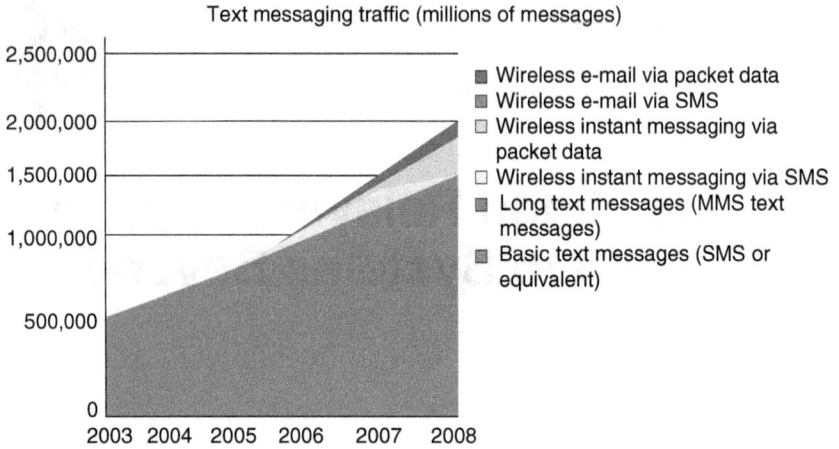

Figure 4.1   Worldwide text messaging by service.

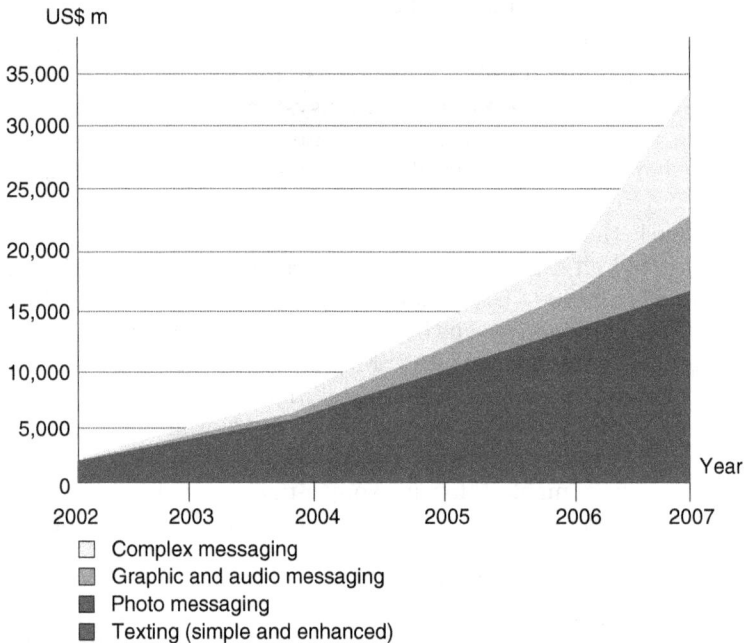

Figure 4.2   Global person-to-person MMS revenue by application.

mean that this is the case, although it's important to note that SMS still has a lot of life left in it in its own.

Meanwhile, as Internet users and content proliferated, the incredible penetration of personal communications systems as well as the increasing number of Internet connections has asked the industry to come up with solutions to combine the multimedia capacities of the World Wide Web with the flexibility of wireless communications [3GSM01]. The convergence of the trends for the next-generation—mobile services, groundbreaking mobile techniques and technological innovations—made the vision of mobile multimedia a reality in a stronger way than anyone would have predicted. This has been fuelled by the expectation of the marketplace to deliver advanced broadband multimedia and information-intensive services on the move [IMT-2000, 3GSM01]. The realization of this mobile multimedia vision requires technological evolution linked with a portfolio of innovative applications, attractive services, and access to information that is relevant to people on the move, to the lifestyles they adopt, and to the circumstances they find themselves in [3GSM01]. In this way, people using a lightweight and convenient pocket communicator will be able to experience audio and video clips, send electronic postcards to each other, and tap into entertainment and service databases, personalized according to their preferences, location, or situation [3GSM02].

Facing the challenge of multimedia handling on the move, second-generation systems are proven limited. Mobile operators moved toward a new generation of technologies—the third-generation (3G) mobile communication systems. 3G, or else *universal mobile telecommunication systems* (UMTS), systems are designed with the goal to complete the worldwide globalization process of mobile communication and provide the foundation for new services with high-rate and high-quality data.

The first step toward 3G is the *general packet radio service* (GPRS) technology [3GSM01]. GPRS has brought a whole new meaning to the mobile world with the introduction of packet switching technology for the first time to mobile wireless communications systems. The GPRS technology in particular introduces the Internet protocol (IP) into the circuit-switched world of GSM, thereby enabling instant access to new applications and services to end users as well as new charging and tariff schemes that are more appropriate for mobile services. In this stylized always-on GPRS usage model, users are charged only for actual (packet) data transmitted rather than the online time, as is the case with current circuit-switched data services.

Possible applications for GPRS range from communication tools in a laptop PC (electronic mail, file transfer, and Web browsing) to special applications with relatively low transmission needs (telemetry, road and railway traffic control, taxi and vehicle dispatch, dynamic road guidance, and monetary transactions). Since 2001, commercial GPRS services have already enabled hundreds of operators in 170 countries (in Western Europe, 55 out of the 65 GSM operators, and around the world there are some 134 operators in 40 countries with GPRS services launched or pending) to give their customers (500 million by May 2003) their first taste of 3G-type multimedia services [3GSM01].

With these new multimedia applications and services, traffic levels began to rise sharply, increasing on one hand service revenues of operators, and on the other driving up air-time usage. Operators were thus soon forced to increase network capacity and improve data speeds. Beyond GPRS, which is nominated as a 2.5G technology, operators rolled out new radio technologies, such as the *enhanced data for GSM evolution* (EDGE). EDGE delivers a huge advance in high-rate service capabilities and an "always on, always connected" experience. EDGE, in particular, is one of the International Telecommunications Union's (ITU's) 3G radio standards (IMT-SC) offering operators, which are currently using TDMA-based radio interfaces such as GSM and IS-136, the chance to enable higher (than 2G) data rates using their own existing frequency bands. For operators with EDGE-compliant base stations (for the 800, 900, 1800, and 1900 MHz bands) the upgrade is simply a matter of installing EDGE transceivers and making a software upgrade. In this regard, mobile professionals see EDGE as a low-risk technology evolution from a mature worldwide GSM base to 3G mobile multimedia networks, offering operators the chance to "accelerate" their existing GSM networks for only a small incremental investment. Combined with GPRS, EDGE results in a much improved utilization of the radio network and approximately a three times increase in the overall network capacity, enabling a finer utilization of their resource (frequency) allocations.

In parallel with EDGE, operators deployed a new air interface for 3G networks that is based on the *wideband code division multiple access* (W-CDMA) technology. W-CDMA is considered to be the main third-generation air interface worldwide and is already deployed in Europe and Asia, including Japan and Korea, in the same frequency band. W-CDMA can deliver greater bandwidth and higher data speeds than GPRS and EDGE are capable of, and enhance and extend the multimedia services and applications initially enabled by GPRS. With such broad capabilities at hand, it will be tempting for operators as well as service providers to establish synergies and use the 2.5 and 3G systems to support every new application (and many old ones too—see Fig. 4.3).

Vending machines calling for supplies, cars sending service requirements to the garage, cordless phones, communicating with personal data assistants (PDAs), W-LANs and the like could all be supported directly by 2.5/3G.

Meanwhile, operators deploying W-CDMA could use EDGE for providing complementary nationwide and hot spot support to their 3G

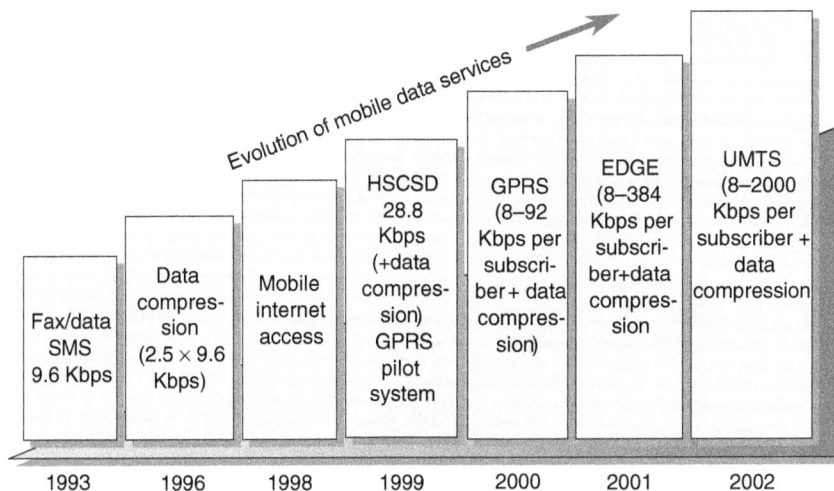

Fax/data SMS 9.6 Kbps	Data compression (2.5 × 9.6 Kbps)	Mobile internet access	HSCSD 28.8 Kbps (+data compression) GPRS pilot system	GPRS (8–92 Kbps per subscriber + data compression)	EDGE (8–384 Kbps per subscriber+data compression	UMTS (8–2000 Kbps per subscriber + data compression
1993	1996	1998	1999	2000	2001	2002

Evolution of mobile data services

Figure 4.3 Evolution of the GSM system toward the third generation.

(W-CDMA) coverage. Its appeal appears to be widening by two equally significant factors—EDGE can work on the existing GSM spectrum, and requires "minimal tweaking" to GPRS networks. In regions where 3G licenses are yet to be awarded, or where operators are not committed to a specific 3G rollout timetable, or even fail to get a 3G license, EDGE is gaining traction. 3G, with its extra bandwidth and enhanced services, would enjoy faster and more comprehensive rollout in high-density, high-use countries or areas where demand and capacity constraints could help to justify it economically [3GSM01]. In this context, EDGE technology can mesh smoothly with W-CDMA, providing coverage in rural areas, than in major cities where the number of subscribers does not justify deploying 3G, but where there is demand for high-speed data services. By deploying EDGE in rural areas as a solution for high-speed data, operators are able to offer seamless high-speed 3G-like services on a nationwide basis—albeit using two different technologies.

At the end of 2005, *mobile network operator* (MNO) professionals estimate that there will be 1.6 billion users, of which more than 1 billion will be 3G mobile Internet users. By 2010, the UMTS Forum foresees 28 percent of the world's 2.25 billion mobile cellular subscribers using 3G systems. This is a conservative estimate, taking into account slower network build-out and service commercialization in emerging and developing economies. Based on this conservative analysis approach, forecasts predict that the total service-provider-retained revenues for 3G services in 2010 will reach US$322 billion. Of these revenues, 66 percent will come from 3G-enabled data services. The cumulative revenue potential for mobile services providers between now and 2010 is over US$1 trillion—a true mass-market success.

Now that the first 3G systems are being built, it is natural to start asking what could be the next big step. Commentaries and predictions regarding wireless broadband communications are currently cultivating visions of the systems beyond the third generation (B3G). Maximum data rates up to 1 Gb/s, the combination of several available, evolving, and emerging access technologies into a common platform and their seamless interworking combined with adaptive multimode terminals will be the key characteristics of beyond third-generation mobile wireless systems. These systems will be flexible and reconfigurable by network management schemes, adaptive frequency allocation, and self optimizing networks as far as possible to support the needs of the different players and a variety of terminals in future communication systems.

4.2   The Evolution of Wireless Mobile Technologies

Since the early 1980s, when the first-generation mobile systems were introduced, mobile communications has experienced enormous growth. The world's first cellular systems such as the *advanced mobile phone system* (AMPS) [YOUNG79], *total access communications system* (TACS), and *nordic mobile telephony* (NMT) were pretty basic using radio analog transmission and allowing the transmission of speech and only speech. Within a few years, cellular systems began to hit a capacity ceiling with millions of new subscribers signed up for service, demanding more and more airtime. The limited amount of available radio spectrum, however, could not accommodate the ever growing needs of customers and, as a consequence, dropped calls and network busy signals became common in many areas.

In this regard, in the late 1980s, the industry developed a new set of digital wireless technologies that allowed more advance services such as low-to-medium bit-rate data transmission. First-generation systems evolved then to 2G and further to 2.5G.

New advanced data services are driving network traffic even higher and the continued growth in voice as markets are migrating from fixed to mobile has challenged the industry to come up with new ways to fulfill the requirements of the twenty-first century mobile multimedia information society. This challenge called for new wireless communications technologies capable of supporting a wide range of services from voice to low, high, and advanced data-rate services to deliver mobile multimedia such as wireless e-mail, Web, digital picture receiving/sending, and assisted-GPS (global position system) position location applications. 2.5G systems evolved then to third generation, and beyond.

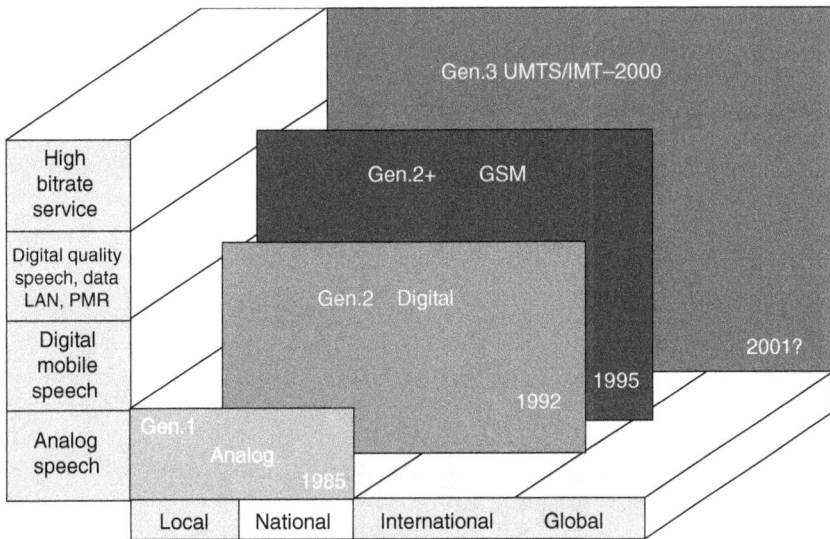

Figure 4.4 Evolution of wireless systems capacities.

With the advent of 3G systems, person-to-person communication is enhanced with high-quality images and video whereas access to information and services on public and private networks is enhanced by higher data rates and new flexible communication capabilities, not provided by the 2G systems [UMTS-RPRT].

Figure 4.4 illustrates this evolution.

These three generations of wireless communications networks are summarized as follows (see Table 4.1 for details):

- In the first generation, the radio interface was analogical.

- In the second generation, the radio interface became numerical (digital systems). For the second generation of cellular networks (Fig. 4.5):
  - Europe has developed the GSM network.
  - The United States has developed the IS-95 (Interim Standard 95) [TIA-IS95] and IS-136 (Interim Standard 136) networks.
  - Japan has developed the *pacific digital cellular* (PDC) networks.

    These systems have enabled voice communications to go wireless in many of the leading markets, and customers are increasingly finding value also in other services such as text messaging and access to data networks. An excellent overview of the various cellular systems and standards can be found in Refs. [RAPP96, KETCHUM95].

- Third-generation communications systems integrate multimedia applications [DIXIT01, FRODIG01]. For 3G networks:

TABLE 4.1 Comparison of Various Cellular Standards

Name	AMPS	GSM/DCS - 1900	IS-136 USDC	IS-95	Cdma2000	W-CDMA/ UTRA
Generation	1	2	2	2	3	3
Year introduced and origin	1983	1992/1994	1996	1993	2002	2002
	United States	Germany	United States	United States	United States	Europe
Region of coverage	United States	Europe, India, United States (PCS)	United States	United States, Hong-Kong, Middle—East Korea	United States	Europe
Frequency Band		Cellular/PCS	Cellular /PCS	Cellular/PCS	PCS	
Uplink (MHz)	824–819	890–195/ 1850–1910	824–849/ 1850–1910	824–849/ 1850–1910	1850–1910	1920–1980
Downlink (MHz)	869–894	935–960/ 1930–1990	849–894/ 1930–1990	849–894/ 1930–1990	1930–1990	2110–2170
Multiple access scheme	FDMA	TDMA	TDMA	CDMA	CDMA	CDMA
Bandwidth per channel	30 kHz	200 kHz	30 kHz	1.25 MHz	1.25,3.75,7. 5,11.25,15 MHz	5,10,20 MHz
Modulation type	FM	GMSK	$\pi/4$ – DPSK	QPSK and OQPSK	QPSK and BPSK	QPSK and BPSK
Max. output power Base: Mobile:	20 W 4 W	320 W 8 W	20 W 4 W	1.64 kW** 6.3 W	1.64 kW** 2 W	Unspecified 1 W
Users/channel	3	8	3	Up to 63	Up to 253	Up to 250
Data rate	19.2 kb/s*	22.8 kb/s	13 kb/s	19.2 kb/s	1.5 kb/s to 2.0736 Mb/s	100 bps to 2.048 Mb/s

*Using cellular digital packet (CDPD)

**Total *effective isotropic radiated power* (EIRP) for all the carriers within the channel bandwidth.

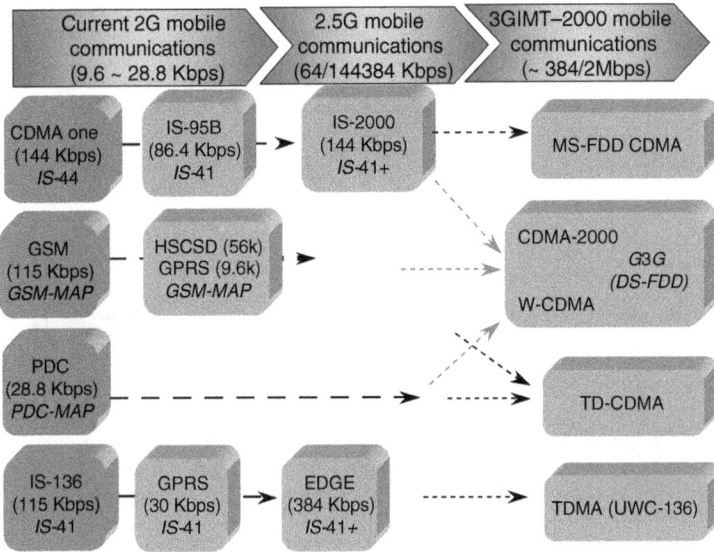

| Current 2G mobile communications (9.6 ~ 28.8 Kbps) | 2.5G mobile communications (64/144384 Kbps) | 3GIMT–2000 mobile communications (~ 384/2Mbps) |

CDMA one (144 Kbps) *IS-44* — IS-95B (86.4 Kbps) *IS-41* → IS-2000 (144 Kbps) *IS-41+* ----→ MS-FDD CDMA

GSM (115 Kbps) *GSM-MAP* — HSCSD (56k) GPRS (9.6k) *GSM-MAP* →

CDMA-2000 *G3G (DS-FDD)* W-CDMA

PDC (28.8 Kbps) *PDC-MAP* ----→ TD-CDMA

IS-136 (115 Kbps) *IS-41* — GPRS (30 Kbps) *IS-41* → EDGE (384 Kbps) *IS-41+* ----→ TDMA (UWC-136)

Figure 4.5 Technology evolution.

- Europe uses the UMTS network specification.
- The United States uses the CDMA2000 and *universal wireless communications* 136 (UWC136) network specification.
- Japan uses the W-CDMA network specification.

Figure 4.5 clearly shows the whole spectrum of today's and future mobile communication systems. This technology evolution with respect to supported bit rates and services is depicted in Fig. 4.6.

3G standards have regional names—UMTS in Europe, Com-A in Japan, CDMA2000 in the United States and some parts of Japan and South Korea, and *International Mobile Telecommunications 2000* (IMT-2000).

The *European Telecommunication Standardization Institute* (ETSI) started the standardization of UMTS in 1991. Within ETSI, the *Technical Committee SMG* (Special Mobile Group) is responsible for UMTS standardization. The *Sub Technical Committee* (STC) SMG2 is in charge of the UMTS radio access system standardization and the SMG3 of the development of possible new UMTS/GSM core network standards.

IMT-2000 is the ITU's[2] initiative for a service that provides radio access to the global telecommunications infrastructure, through both

---

[2]ITU is the United Nations organization responsible for harmonizing global telecommunications for assuring the liaison between different standardization organizations [IMT-2000].

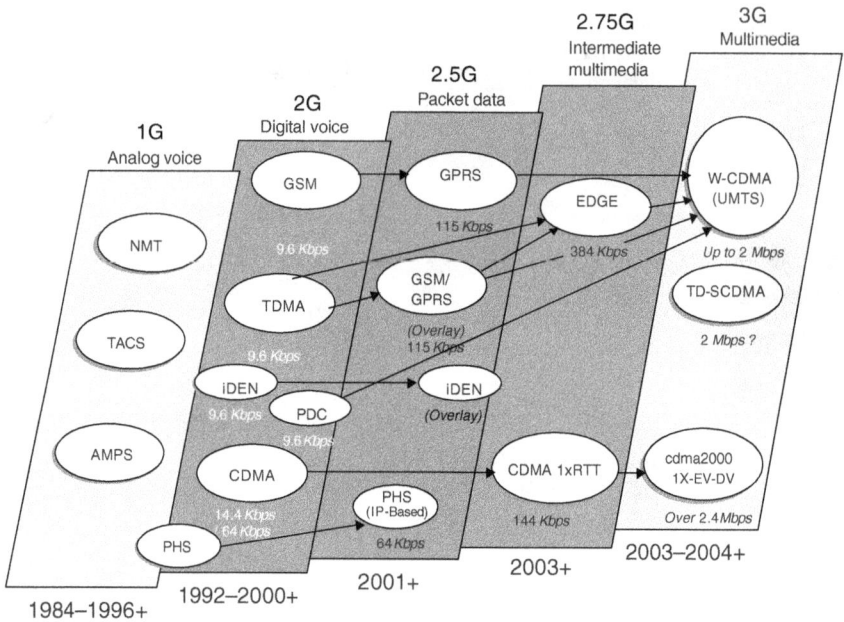

Figure 4.6 Migration to 3G services.

satellite and terrestrial systems, servicing fixed and mobile users in public and private networks. IMT-2000 was previously called the *future public land mobile telecommunication systems* (FPLMTS).

In the following sections we discuss the technology roadmap from 2G to 3G networks and beyond. Our coverage here is necessarily brief, as an entire course would be needed to cover this technology evolution in depth.

4.3 Second Generation (2G) Mobile Systems

4.3.1 Digital enhanced cordless telecommunications (DECT)

The DECT standard [DECT], proposed by the CEPT (the Council of European PTTs), was initially conceived in the mid-1980s as a pan-European standard for domestic *cordless* phones. The objective of the new standard was to use digital radio technology to improve the performance of cordless phones in three important areas—speech quality, security against eavesdropping, and immunity from radio interference between nearby cordless phones.

By the time the DECT standard was finalized in 1992 and published by ETSI (the successor to CEPT), the scope of the standard had broadened beyond domestic cordless phones to include two additional application areas—one being the business cordless telephones (the so-called cordless PBX or wireless PBX) and the other being a cordless access system for subscribers to public telecom networks.

Since 1993, DECT has been regarded as a standard in 26 countries, making it the most widely used digital standard for cordless communications. For this reason, the name of the DECT standard has been revised. In the original form, the letter "E" stood for "European." Now, it denotes "Enhanced." So, today, DECT means *digital enhanced cordless telecommunications*.

Characteristics.   DECT is a digital radio access standard for single- and multicell cordless communications. It is based on a multicarrier *time division multiple access* (TDMA) technology. The standard specifies four layers of connectivity, plus other important functions. The four layers correspond approximately to layers 1 to 3 of the ISO *open systems interconnection* (OSI) model [TANENB96], as follows:

1. *Physical layer.*   Radio parameters such as frequency, timing and power values, bit and slot synchronization, and transmitter and receiver performance

2. *Medium access control layer.*   The establishment and release of connections between portable and fixed parts of the DECT system

3. *Data link control layer.*   Provides very reliable data links to the network layer for signaling, speech transmission, and circuit- and packet-switched data transmission

4. *Network layer.*   The main signaling layer, specifying message exchanges required for the establishment, maintenance and release of calls between portable and fixed elements of the network

5. *Other elements* of the DECT standard cover equipment identities and addressing, security authentication procedures, speech coding and transmission, public access profile, and cryptographic algorithms.

The DECT radio interface.   The DECT radio interface standard, described in ETS 300 175 (Parts 1 to 8), is based on the multi carrier, *time division multiple access, time division duplex* (MC/TDMA/TDD) radio access methodology. Basic DECT frequency allocation uses 10 carrier frequencies (MC) in the 1880 to 1900 MHz range. The time spectrum for DECT is subdivided into time-frames repeating every 10 ms. Each frame consists of 24 time slots each of which may be used for either transmission or reception. For the basic DECT speech service, two time slots are paired with 5 ms separation to provide bearer capacity of typically 32 kbit/s (ADPCM G.726 coded speech) full duplex connections.

To simplify implementations, the 10-ms time-frame is split into two halves (TDD), such that the first 12 time slots are used for downlink (base-to-mobile station) transmissions and the next 12 are used for uplink (mobile-to-base station) transmissions. The TDMA structure allows then up to 12 simultaneous basic DECT (full duplex) voice connections per transceiver. This provides a significant cost benefit when compared with technologies that can support one link per transceiver (e.g., CT2). Due to the advanced radio protocol, DECT is able to offer widely varying bandwidths by combining multiple channels into a single bearer. For data transmission purposes error protected net throughput rates of $n \times 24$ kbit/s can be achieved, up to a maximum of 552 kbit/s with full security as applied by the basic DECT standard.

DECT strengths

- *Seamless handover of calls.* In a DECT business cordless system, as the user moves around from one picocell to another during a call, it is the phone rather than the radio network that initiates handover from cell to cell. A "make-before-break" handover principle ensures that the handover is undetectable to the user.

- *High capacity.* The digital TDMA radio technology used in the DECT standard, with its low radio interference characteristics, allows business cordless systems to handle up to 100,000 users per square kilometer. This allows even the most densely-occupied office buildings and similar locations to be served.

- *High speech quality.* Speech is digitally encoded before transmission, using 32 kbit/s *adaptive differential pulse code modulation* (ADPCM) speech encoding. The resulting speech quality is as good as with an ordinary wired phone.

- *Long battery life.* The radio technology uses discontinuous transmission, occupying only two out of the 16 time slots, which reduces the load on the battery in the cordless phone. Standby and talk times of 45 h and 9 h are commonly available in the latest DECT cordless phones.

- *Data as well as voice.* The DECT standard permits cordless data as well as voice communications, creating the possibility of cordless LANs which could share capacity with cordless telephone systems.

- *Profiles for interworking.* The DECT standard is developed with a number of different interconnection profiles. This allows the DECT system to be linked to other networks, including GSM digital cellular networks, to provide integrated communications mobility. One of the most important profiles is the *generic access profile* (GAP), which ensures that all DECT products from different manufacturers are compatible. This promotes competition and provides users with a wider range of DECT products to choose from.

- *High security.* The DECT standard uses encryption techniques so that radio eavesdropping is virtually impossible.

DECT applications. The three DECT applications that have gained widespread commercial deployment are: home cordless phones, business cordless systems, and as a radio alternative to wired subscriber accesses in public fixed telecom networks known as wireless local loop (WLL).

- In a DECT home cordless phone, a typical DECT system consists of a phone handset and a base unit that contains the radio base station.

- In a DECT business cordless system, the core radio network consists of a number of radio base stations, all connected to a *public branch exchange* (PBX) through a radio exchange. A DECT business cordless system has an architecture that is similar in concept to a cellular mobile phone system, with a network of radio base stations so that users can walk around the premises, making and receiving calls. The cells in a DECT business cordless system are much smaller (picocells) than the ones that are used in a cellular network, which allow much higher user densities. DECT permits the highest user densities for any cordless system, up to $100,000/\text{km}^2$.

- In a DECT WLL system, the radio base station is located somewhere in the neighborhood. Subscribers are equipped with digital cordless (DECT) phones and are provided with a limited degree of mobility in a local area. This solution is commonly termed as *cordless terminal mobility* (CTM).

4.3.2 The global system for mobile communications (GSM)

GSM is the name of the European digital mobile telephone network. The GSM architecture [GSM92] was created in 1980 in France. The first steps were taken in 1982, when CEPT (Conference Europeenne des Administrations des Postes et des Telecommunications) founded the Groupe Special Mobile which initially gave GSM its name. After the foundation of the European standardization institute ETSI, GSM became an ETSI Technical Committee and was renamed to global system for mobile communication, keeping GSM as its acronym. The official start of GSM was in 1992 whereas in late 1993 there were already more than one million GSM subscribers, over 80 percent of them in Germany alone. Since then, more than 200 networks run GSM in over 100 countries, serving around 450 million subscribers around the globe.

GSM is based on *frequency division-time division multiple access* (FD-TDMA) radio access and offers a 9.6 kbit/s rate. It uses the 890 to 915 MHz radio band for the upload traffic and the 935 to 960 MHz radio band for the download traffic.

In 1990, the *digital communication system* (DSC) was developed to offer additional bands:

- DSC1800, which uses the 1710 to 1785 MHz radio band for the upload traffic and the 1805 to 1880 MHz radio band for the download traffic.

- DSC1900, which uses the 1850 to 1910 MHz radio band for the upload traffic and the 1930 to 1990 MHz radio band for the download traffic.

System architecture. A GSM *public land mobile network* (PMLN) is cell-based [LUI01], [SARI01]; that is, the area covered by a cell phone company is virtually divided into hexagons (Fig. 4.7). Within each hexagon there exist one or more *base transceiver stations* (BTS), which serve the *mobile stations* (MS) that currently reside within this cell, to the fixed portion of the network. Cells are further grouped into clusters of size k such that a cluster comprises k cells.

In reality, cells are less likely to be of hexagon shape as well as of the same size. Instead, cells are of different sizes—smaller ones covering areas with high population density (urban areas), and larger ones covering low-density areas (suburban and rural areas).

The GSM equipment is divided into three subsystems:

1. *Base station subsystem* (BSS) to manage the radio interface.

2. *Network and switching subsystem* (NSS) to manage the interconnections with the fixed network.

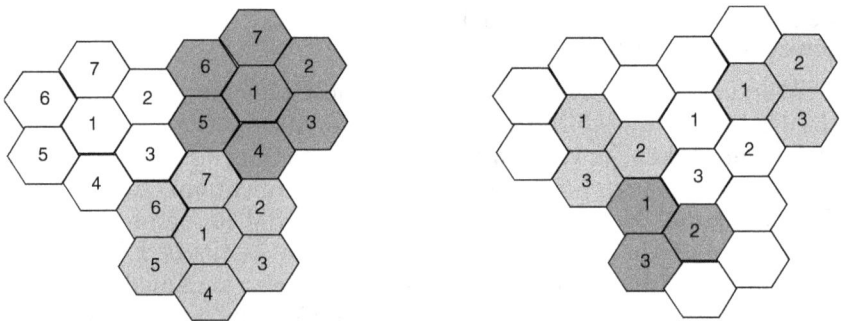

Figure 4.7 Cells covering hexagonal areas.

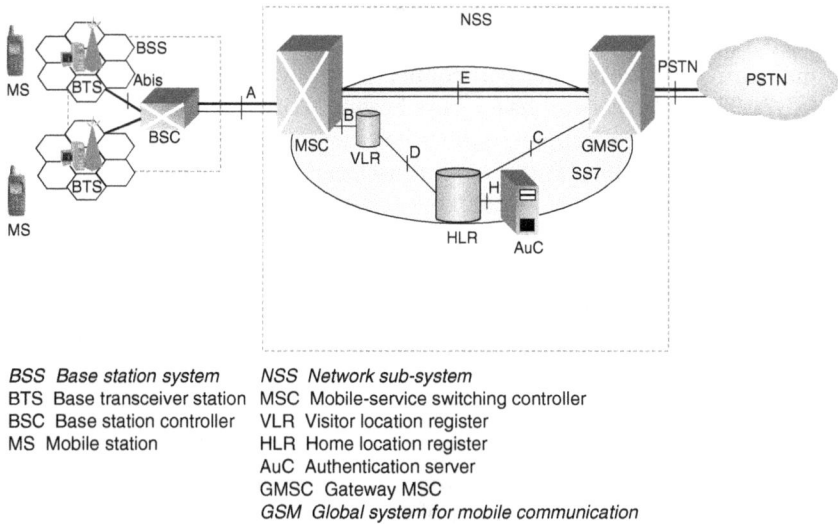

BSS Base station system   NSS Network sub-system
BTS Base transceiver station MSC Mobile-service switching controller
BSC Base station controller   VLR Visitor location register
MS   Mobile station   HLR Home location register
                               AuC Authentication server
                               GMSC Gateway MSC
                               GSM Global system for mobile communication

Figure 4.8 GSM architecture.

3. *Operation support subsystem* (OSS) to supervise the *public land mobile networks* (PLMN).

The GSM architecture with its subsystems is illustrated in Fig. 4.8.

Base station subsystem (BSS).   All radio-related functions are performed in the BSS. The BSS consists of *base station controllers* (BSCs) and the BTSs. Its primary responsibility is to transmit voice and data traffic between the mobile stations. The functionalities of the BTS include modulation, demodulation, and error detection and correction methods to ensure reliable data connections. To maintain simplicity, most of the BTS protocol logic is implemented in BSSs. Each BSS serves several BTSs. The BSS manages the radio channels, the call admission control, and the call transfer between adjacent cells (handovers or handoffs).

GSM mobile stations are wireless terminals capable of receiving and originating short messages as well as voice calls. Mobile stations use radio links to communicate to the BTS they are currently attached to. From the BTS, conventional terrestrial networks take over and route the data to their destination.

Network and switching subsystem (NSS).   NSS manages the connections between mobile stations of the same network, or to other networks, such as public phone networks (PSTN, ISDN, PDN, and the like). NSS is composed of the following system components:

- *The mobile service switching center (MSC).*   It is connected to the *base station controller* (BSC) through the A interface. MSCs create, maintain, and terminate call connections via the *signaling system 7* (SS7). They also control handovers between BSCs and MSCs. In addition, MSCs control supplementary services, as known from ISDNs, like conference calls, blocking certain numbers, and so on. A *gateway MSC* (GMSC) is a special MSC, which manages the interconnections between the PLMN and other networks (PSTN, ISDN, PDN, and the like).

- *The visitor location register (VLR).*   It is connected to the MSC through the B interface. The VLR is a local database that contains temporary information about the mobiles that currently reside within its geographical area. If a mobile leaves this area, data are removed from the VLR. This information is needed by the MSC in order to service visiting subscribers. With each MSC, one VLR is associated.

- *The home location register (HLR).*   It is a database used for permanent storage and management of subscriptions and service profiles. HLR is connected to the MSC through the C interface and to the VLR through the D interface.

Inside the NSS, all the communications are managed by the mobile application part (MAP) protocol, which is an upgraded version of the SS7 telecommunication protocol.

Operation subsystem (OSS).   OSS is responsible for subscription management, network operation, maintenance, mobile equipment management, and billing. The Operation and Maintenance Center (OMC) manages subscribers, billing, controls the state of all network elements, and creates statistical reports about the observed traffic. OMC is based on the *telecommunications management network* (TMN), a hierarchical set of services developed by the ITU-T.

The *authentication center* (AuC) contains all information necessary to protect data transmissions over the radio channels. Here, cryptographic keys and algorithms for authentication are stored.

The *equipment identity register* (EIR) is another database that stores information for all subscribers and mobile equipment. In this database, three lists (white, black, and gray) store identification numbers unique to all mobile terminals. The white list contains allowed terminals, the black contains unallowed terminals (e.g., stolen), and the gray contains known bugs.

Subscriber and equipment identification.   In GSM, there is a strong distinction between subscribers (identified by their SIM) and the hardware they use for making phone calls. In order to identify these during GSM service allocation, several identification numbers exist, which are stored at different locations. The following IDs are stored in the HLR:

1. *International mobile subscriber identity (IMSI).*   Permanent ID assigned to each GSM network subscriber.

2. *International mobile subscriber ISDN number (MSISDN).*   The ISDN number permanently assigned to each GSM subscriber.

3. *Mobile station roaming number (MSRN).*   Temporary ISDN number of a subscriber. This number is assigned by the local VLR each time the subscriber enters its area. The MSRN is then sent to the HLR and the GMSC.

4. *The address of current VLR and MSC.*   Identify the area the subscriber is currently in, if available.

5. *Local mobile subscriber identity (if available).*   A short ID temporarily assigned to an active subscriber by a VLR and sent to the HLR.

The following IDs are stored temporarily at a VLR associated with the MSC that is currently serving an active mobile:

1. IMSI

2. MSISDN

3. MSRN

4. *Location area identity (LAI).*   The ID of the location area (LA) where the subscriber is connected to.

5. *Temporary mobile subscriber identity (TMSI).*   Temporarily assigned to active mobile stations in order to prevent the IMSI from being transmitted too often over radio. The TMSI is periodically changed during a call.

Signaling transmission.   For establishing, controlling, and deleting connections, GSM network modules have to exchange signals with each other. The following interfaces are defined between them:

- MS-BTS: Um
- BTS-BSC: Abis
- BSC-MSC: A
- MSC-VLR: B
- MSC-HLR: C
- VLR-HLR: D
- MSC-MSC: E
- MSC-EIR: F
- VLR-VLR: G

The physical transport of these signals is done via the physical channel in the Um interface and over digital lines of either 2048 kbit/s or 64 kbit/s (ITU-T G.703, G.705, G.732).

Radio channel management

Physical channels   Two bands of 25 MHz width each are reserved for GSM. The band from 890 MHz to 915 MHz is defined for uplink transmissions (from mobile to BTS), and the band from 935 MHz to 960 MHz for downlink transmissions (from BTS to mobile). Frequency multiplexing is done by dividing the available frequency bands into 124 FDM channels, each being separated from its neighbors by 200 kHz.

Each mobile and BTS may send data on one of these FDM channels. To allow for more than 124 mobiles per cell, each FDM channel is then subdivided into 8 time slots of length 0.577 ms. The slots numbered from time slot 0 to 7 form a frame of 4.615 ms length. The recurrence of one particular time slot in each frame makes up one physical channel. 8 time slots are then collected into one TDMA frame. Each mobile is assigned one time slot for sending or receiving (TDMA).

Radio channels in GSM are then based on a TDMA structure that is implemented on multiple frequency subbands (TDMA/FDMA). Base stations are equipped with a certain number of these preassigned frequency/time channels.

In each time slot, data is transported in a burst. There are five different burst types.

1. *Normal burst (NB).*   In these bursts, normal traffic data and control information are transported.

2. *Frequency correction burst (FB).*   This burst is used to synchronize frequencies.

3. *Synchronization burst (SB).*   This is done for time synchronization of mobiles with their BTS.

4. *Dummy burst (DB).*   This burst is sent from each BTS on a special FDM channel (BCCH–see next), in case no other information needs to be sent.

5. *Access burst (AB).*   Burst for random access on the *random access channel* (RACH).

Logical channels.   GSM defines a variety of traffic and signaling control channels of different bit rates. These channels are assigned to logical channels derived from multiframe structuring of the basic eight-slotted TDMA frames just discussed. For this purpose, two multiframe structures have been defined—one consisting of 26 time frames (resulting in a recurrence interval of 120 ms), and the other comprising 51 time frames (or a recurrence interval of 236 ms).

The multiframe structure of 26 time frames is used to define traffic channels (TCH) and their slow and fast associated control channels (SACCH and FACCH) that carry link control information between the mobile and the base stations. The TCH have been defined to provide six different forms of services, that is, full-rate speech or data channels supporting effective bit rates of 13 kb/s (for speech), 2.4, 4.8, and 9.6 kb/s; and the half-rate channels with effective bit rates of 6.5 kb/s (for speech) and, 2.4 kb/s, and 4.8 kb/s for data (note that the gross bit rates on these channels are higher due to required channel coding, 22.8 kb/s for full-rate speech). The full-rate TCHs are implemented on 24 frames of the multiframe, with each TCH occupying one time slot from each frame. The SACCH is implemented on frame 12 (numbered from 0), providing eight SACCH channels, one dedicated to each of the eight TCH channels. Frame 25 in the multiframe is currently idle and reserved to implement the additional eight SACCH required when half-rate speech channels become a reality. The FACCH is obtained on demand by stealing from the TCH, and is used by either end for signaling the transfer characteristics of the physical path, or other purposes such as connection handover control messages. The stealing of a TCH slot for FACCH signaling is indicated by a flag within the TCH slot.

The 51-frame multiframe is used to derive the following signaling and control channels [GSM 05.01]:

- *SDCCH.*   A stand-alone dedicated control channel is used for the transfer of call control signaling to and from the mobile during call setup. Like the TCHs, the SDCCH has its own SACCH and is released once call setup is complete.

- *BCCH.*   Broadcast control channel is used in the BSS to mobile direction to broadcast system information such as the synchronization parameters, available services, and cell ID. This channel is continuously active, with dummy bursts substituted when there is no information to transmit, because its signal strengths are monitored by mobiles for handover determination.

- *SCH.*   The synchronization channel carries information from the BSS for frame synchronization.

- *FCCH.*   The frequency control channel carries information from the BSS for carrier synchronization.

- *CCCH.*   Common control channels are used for transferring signaling information between all mobiles and the BSS for call origination and call paging functions. There are three common control channels:

- *PCH.* The paging channel is used to call (page) a mobile from the system.
- *RACH.* The random access channel is used by the mobiles trying to access the system. The mobiles use the slotted Aloha scheme for this channel for requesting a *dedicated control channel* (DCCH) from the system at call initiation.
- *AGCH.* The access grant channel is used by the system to assign resources to a mobile such as a DCCH channel.

Note that the AGCH and the PCH are never used by a mobile at the same time, and therefore are implemented on the same logical channel. All the control signaling channels, except the SDCCH, are implemented on time slot 0 in different TDMA frames of the 51 multiframes using a dedicated RF carrier frequency assigned on a per-cell basis. The multiframe structure for the SDCCH, together with its associated *slow associated control channel* (SACC), is implemented on one of the physical channels (TDM slots and RF carriers) selected by the system operator.

GSM services.   From the beginning, the planners of GSM wanted ISDN compatibility in terms of the services offered and the control signaling used. However, radio transmission limitations, in terms of bandwidth and cost, do not allow the standard ISDN B-channel bit rate of 64 Kb/s to be practically achieved.

Using the ITU-T definitions, telecommunication services can be divided into *bearer services*, *teleservices*, and *supplementary services*.

Bearer services.   Bearer services are lower-level services that enable the creation of reliable data transport connections. Bearer services are used by higher levels for data transport. Transparent services (T) denote constant bit-rate transport with changing bit error probabilities. Nontransparent services (NT) activate additionally a special protocol called the *radio link protocol* (RLP) between a mobile and the MSC that resends blocks with observed errors. See Table 4.2.

Transparent data transmission.   As the transmission quality of radio signals can change drastically during one connection, the bit-error probability despite *forward error correction* (FEC) can vary between 10^{-2} and 10^{-5}. Transparent bearer services do not try to correct detected errors and rely on FEC only. The sender thus is guaranteed a constant bit rate and may send data at this rate without flow control. Hence, for the sender, the underlying transport system is transparent. The terminal though has to be aware of nonneglectable bit-error probabilities.

Nontransparent data transmission.   Another way of coping with detected bit errors is to resend the data frame. In nontransparent bearer services, the RLP is used. One part of this protocol is located in the

TABLE 4.2   Bearer Services

Service	Structure	BS Nr.	Bit rate	Mode
Data	Asynchronous	21	300	T or NT
		22	1200	T or NT
		23	1200/75	T or NT
		24	2400	T or NT
		25	4800	T or NT
		26	9600	T or NT
Data	Synchronous	31	1200	T
		32	2400	T or NT
		33	4800	T or NT
		34	9600	T or NT
PAD	Asynchronous	41	300	T or NT
		42	1200	T or NT
		43	1200/75	T or NT
		44	2400	T or NT
		45	4800	T or NT
		46	9600	T or NT
Packet	Synchronous	51	2400	NT
		52	4800	NT
		53	9600	NT
Alternating voice/data		61	1300 or 9600	
Voice followed by data		81	1300 or 9600	

mobile, the other in the MSC. In this protocol, data are cut into numbered frames of equal size, where each frame has to be acknowledged by the receiver (one acknowledgment frame can acknowledge several data frames).

If an error is detected inside a frame, the receiver sends a resend command to the sender, either for this particular frame or for all frames beginning from the erroneous frame.

Due to frame resends as a result of bad radio connections, the net bit rate of such a channel may change drastically and the sending terminal must be flow-controlled in order to adapt to the available bit rate. This is done by the *nontransparent protocol* (NTP), where the terminal is connected to. Hence for the sender, the transport system is not transparent anymore.

Teleservices.   Teleservices are well defined services within GSM and use bearer services for transport. They include services like voice, SMS, and *message handling systems* (MHS).

SMS is a bidirectional service for short alphanumeric (up to 160 bytes) messages. Messages are transported in a store-and-forward fashion. For point-to-point SMS, a message can be sent to another subscriber to the service, and an acknowledgment of receipt is provided to the sender.

TABLE 4.3   Teleservices

Category	TS Nr.	Service	Class
Voice	11	Phone	E1
	12	Emergency	E1
Short message services	21	Short message mobile terminated, point-to-point	E3
	22	Short message mobile originated, point-to-point	A
	23	Short message cell broadcast	—
MHS access	31	Access to message handling systems	A
Videotext	41	Videotext Profile 1	A
	42	Videotext Profile 2	A
	43	Videotext Profile 3	A
Teletext	51	Teletext	A
Fax	61	Voice and Fax Group 3 alternating T/NT	E2/A
	62		—

SMS can also be used in a cell-broadcast mode for sending messages such as traffic updates or news updates. In addition, SMS messages can be stored in the SIM card for later retrieval. See Table 4.3.

Supplementary services.   Supplementary services are provided on top of teleservices or bearer services. In the Phase 1 GSM specifications, they include several forms of call forward (such as call forwarding when the mobile subscriber is unreachable by the network), and call barring of outgoing or incoming calls, for example when roaming in another country. Many additional supplementary services are to be (or some already are being) provided in the Phase 2 GSM specifications, such as caller identification, call waiting, and multiparty conversations. See Table 4.4.

TABLE 4.4   Supplementary Services

Category	Short name	Service	Class
Call offering	CFU	Call forwarding unconditional	E1
	CFB	Call forwarding on mobile subscriber busy	E1
	CFNRy	Call forwarding on no reply	E1
	CFNRc	Call forwarding on mobile subscriber not reachable	E1
Call restriction	BAOC	Barring of all outgoing calls	E1
	BOIC	Barring of outgoing international calls	E1
	BAIC	Barring of all incoming calls	E1
	BOIC-exHC	Barring of outgoing international calls except those to home PLMN	A
	BIC-Roam	Barring of incoming calls when roaming outside the home PLMN	A

Additionally, GSM defines the above services to be either essential (E) or additional (A). For E-services, three phases have been defined. E1-services must be implemented from 1991 onward, E2 from 1994, and E3 from 1996 onward. A-services are not compulsory but can be offered by the cell phone company.

GSM handover.   One of the most important features of a cell-based mobile phone system is its ability to hand over an active call from one BTS to another. This is done in five steps:

1. The quality of the mobile's connection to its serving BTS and to its neighbor BTSs is constantly measured (link-quality monitoring). Measurement data are sent to the BSC. Measuring the connection quality is done twofold—First, the field strength of each downlink burst of the current BTS is measured by the mobile. An average of 100 such bursts is then computed and stored as the current *received signal level* (RXLEV). Additionally, the mobile measures the field strength of the logical channel BCCH of each of the neighbor BTSs. This is done in the intervals between uplink and downlink communication. Secondly, the mobile measures the channel quality of the traffic channel (logical channel TCH), which contains a set of fixed training bits. The parameter RXQUAL denotes the mean bit error in this sequence prior to error correction.

   The current RXLEV, the best six neighbor RXLEVs and RXQUAL are then attached into a measurement report message and sent to the BSC.

2. Measurements must be averaged over a minimum period of time. This inhibits handovers due to short-term quality drops. If BTS-to-mobile link quality drops below a certain margin, BTS and mobile will increase their field strength in steps of 2 dBm.

3. If the maximum field strength is reached and still the measured connection quality is below a certain level, though there is a much better connection to one of the neighbor BTSs, the BSC sends a handover request to its MSC.

4. The MSC decides whether the handover should be carried out.

5. If a positive decision is made, the MSC handles the handover and updates its VLR.

There exist different handover scenarios.

- *Intracell handover.*   The channel used is changed in one particular cell. The mobile is assigned either another FDM channel or another time slot.

- *Intercell/intra-BSC handover.* Here, the mobile changes to a new BTS (is assigned a new FDM channel) but of the same BSC.

- *Inter-BSC/intra-MSC handover.* The mobile is assigned to a new BTS and BSC, the MSC is the same.

- *Inter-MSC handover.* The mobile is assigned a new BTS, BSC, and MSC.

User data transmission. For carrying user data, the path is divided into two parts—the mobile-to-BSS communication part and the BSS-to-MSC communication part. The former uses GSM-specific protocols, whereas the latter is carried over the A-interface in an ISDN-compatible manner. The Abis-interface between BTS and BSC in general is transparent to user data.

Transparent data transmission. As the transmission quality of radio signals can change drastically despite FEC, the bit-error probability can vary between 10^{-2} and 10^{-5}. Transparent bearer services do not try to correct detected errors and rely heavily on FEC. The sender is thus guaranteed a constant bit rate with no flow control.

Nontransparent data transmission. Another way of coping with bit errors is to resend (retransmit) the data frame. In nontransparent bearer services, the RLP is used for monitoring the RL quality. In RLP, there are two different frame types—*information frames* that carry the user data and *control frames* that carry information for controlling the connection and sending acknowledgments. According to RLP, data are segmented into a sequence of equal size numbered frames. The receiver has to acknowledge the reception of each frame (one acknowledgment can in fact acknowledge several data frames).

If an error is detected inside an information (data) frame, the receiver in response sends a resend command to the sender either for this particular frame or for all frames beginning from the erroneous frame. As a consequence of resending frames, the net channel bit rate may change drastically and the sending TE must use a flow control practice in order to adapt to the available bit rate. This is achieved with the NTP, where the TE is connected to (generally over a V.24 interface). Hence, for the sender, the transport system is not transparent anymore.

Voice transmission. GSM and ISDN both use different codecs and transmission rates. GSM generates data at 13 kbit/s, using the *regular pulse excitation-long term prediction* (RPE-LTP) codec. ISDN-B channels work at 64 kbit/s and use ITU-T A-law coding, a nonuniformly spaced logarithmic codec. At some point in the transmission path, the two different streams have to be converted into each other. This is done in the *transcoding and rate adaption unit* (TRAU).

On possible location of the TRAU is inside the BTS. In this case, the data leaving the BTS will flow at the rate of 64 kbit/s into the adjacent BSC. Another location can be between a BSC and an MSC. Here, the TRAU can be located at the BSC or MSC. If it is located at the BSC, the BSC can multiplex four 13 kbit/s channels onto one ISDN-B-channel. As the TRAU depends on the radio link for synchronization, the remaining 3 kbit/s ($16 = 13 + 3$) are used to carry signaling information (inband signaling). This must also be done, if the TRAU is physically placed at the MSC.

Once data are transported inside an ISDN-B-channel, they travel using the ITU-T ISDN standard protocols—G.703, G.705, and G.732.

The GSM protocol layers. The following sections describe briefly the protocols at the Um, Abis, and A interfaces (Fig. 4.9).

Layer 2 protocols

- *Link access procedure for the Dm-channel (LAPDm).* It is a GSM-specific layer-2 protocol to provide secure Dm-channels between mobile and BTS for layer 3 protocols. LAPDm is similar to HDLC and works in two modes. Unacknowledged operation means sending UI-frames (unacknowledged information frames) without acknowledgment. Flow control and L2-error correction are not carried out. In acknowledged operation, data are transported in I-frames (information-frames) and

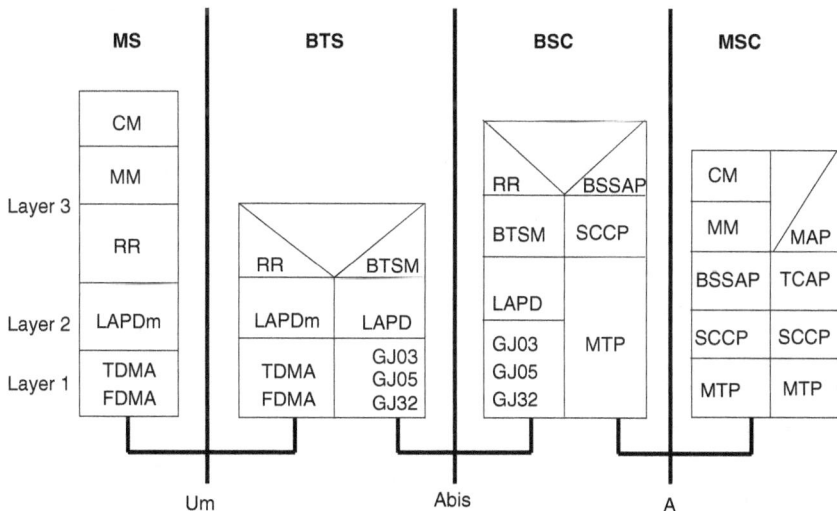

Figure 4.9 GSM protocol layers.

must be acknowledged. Error correction by resending and flow control is carried out.

- *Link access procedure for the D-channel (LAPD).* It is a layer-2 protocol to provide secure D-channels for ISDN.

- *Message transfer part (MTP).* It is the standard ISDN message transport part for SS7. Generally, it includes the lower three layers of the ISDN network, i.e., it routes and transports signaling messages. Mobility and connection management need identifiable connections for signals, the *signaling connection control part* (SCCP) is inserted at layer 3.

Layer 3 protocols

- *Radio resource management (RR).* It is a layer-3 protocol for creating, maintaining, and deleting of radio link channels. It is also responsible for measuring the channel quality, controlling the radio field strength and synchronization, handover, and data ciphering. An RR message contains a protocol discriminator for protocol identification, a transaction ID, and a message type. The data itself is carried in a so-called information element (IE) of fixed or variable length.

- *Mobility management (MM).* It is a layer-3 protocol for enabling the network to locate and keep track of locations of users at varying degrees of granularity, depending on the task at hand. In general, mobility management can be classified into in-session and idle mobility. In-session mobility management has the urgent task of seamless handoffs for ongoing sessions to new radio resources belonging to new access points or new systems. Idle mobility management typically requires less granular location information, such as the location area, to keep track of the whereabouts of users. The aim of the latter is to find the user in time in order to be able to deliver services to the user within a given delay budget and overhead signaling volume.

  MM procedures need a preestablished RR connection, i.e., a logical channel and a LAPDm connection. Signaling is carried out between the mobile and the MSC and it is thus transparent to the BSS.

  Lately, with wireless geolocation systems and services, determining the coordinates of mobile terminals has become a reality. Geolocation information on mobiles can not only serve mobility management purposes, but also facilitate a new set of services that can be called *location-sensitive services*. The use of geolocation information for mobility management is exemplified by geolocation-aided handoff, where ping-pong handoff requests (due to stationary mobiles in cellular overlap regions) can be avoided; system boundary ambiguities can be resolved to alleviate unfair roaming charges; zone-based billing can be enabled with greater accuracy, radio resources can be reserved ahead of time for highly mobile users, and so on. However, wireless geolocation in itself generates additional signaling complexity and overhead signaling volume, which need to be optimized. The combination of mobility management and location-specific service provisioning can be termed as *location management*. An optimal location management technique would aid in the provisioning of guaranteed services and improving quality of service while minimizing the cost of overhead signaling associated with geolocation and dissemination of location information.

- *Call management (CM).*   It is a layer-3 protocol containing three subprotocols. *Call control* (CC) creates, maintains, and deletes calls. Several parallel calls can be established, thus for each call, one CC instance is created in the mobile, and one in the MSC. CC instances communicate with each other via dedicated MM instances they own. The SMS is divided into the SMS control layer (SMS-CL) and the SMS relay layer (SMS-RL). They need previously established MM, RR, and LAPDm connections. As the user data are packet oriented, they are marked SAPI = 3 in the LAPDm layer. Finally, *supplementary services* (SS) provide an entry point to access the GSM supplementary services. Applications from upper layers may enter the CM via the three *service access points* (SAP)—MNCC-SAP, MNSS-SAP, and MNSMS-SAP. However, they can also bypass the CM by directly entering the MMREG-SAP of MM.

- *Signaling connection control part (SCCP).*   It is a layer-3 SS7 protocol for establishing and maintaining identifiable control connections. At the A-interface, it offers connection-oriented and connectionless transport services.

- *Base station system application part (BSSAP).*   It is a layer-3 protocol for signaling at the A-interface, using services offered by the SCCP. It is further divided into three subparts. The *direct transfer application part* (DTAP) offers services for signaling between the mobile and the MSC (CM, MM). Here, signals are transported transparently through the in between BSS. DTAP signals only use connection oriented SCCP services. The *base station system management application part* (BSSMAP) transports signals between an MSC and a BSC. These signals may concern single mobile, physical channels of the radio link as well as global commands for the BSC resource management. BSSMAP procedures use connection-oriented and connectionless SCCP services. Finally, the *base station system operation and maintenance application part* (BSSOMAP) transports network management messages from the OMC over the MSC to a BSC.

- *Mobile application part (MAP).*   It is the GSM-specific enhancement of SS7. It manages roaming functions like location registration/ updating, IMSI attach/detach, handover, subscriber management, IMEI management, authentication and identification, and SMS.

4.4   2.5G Mobile Wireless Systems

4.4.1   GPRS: The gateway to 3G

The GSM MoU Association, representing 239 GSM network operators, telecommunications regulators, and administrations from over 109 countries/areas of the world, placed a high priority on developing its vision of future mobile multimedia services based on an evolution of the GSM platform (GSM Phase 2+). In order to satisfy the increasing user requirements, one major concern of GSM Phase 2+ development, led by ETSI, was to specify a GPRS that accommodates data connections with high bandwidth efficiency. GPRS was first discussed in 1992 and the first release of the standard was produced by the end of 1997 [3GSM01], [GOOD97]. This release covered all the major functionalities of GPRS, including point-to-point transfer of user data, Internet, and X.25 interworking, filtering functionality for security reasons, volume-based charging tools, anonymous access that can be used for prepaid cards in the road traffic information systems, for example, and roaming between PLMN. A follow-up release 1 year later included point-to-multipoint (PTM) transfer (PTM-group call, and PTM-multicast), supplementary services, and additional interworking functionality (e.g., ISDN and modem interworking).

GPRS is a packet-based data bearer service for wireless communication services that is delivered as a network overlay for GSM, CDMA, and TDMA (ANSI-136) networks. The main intention of the GPRS specification was to increase the range of existing GSM data services that offer data rates up to 9.6 kbit/s only. In order to enable support of new data applications with a convenient quality of service, the GPRS concept foresees bit rates of nearly 170 kbit/s that can be flexibly allocated according to actual user demands.

GPRS applies a packet radio principle to transfer user data packets in an efficient way between GSM mobile stations and external packet data networks. Packet-switched technology is introduced to optimize bursty data transfer and occasional transmission of large volumes of data. Packet switching is where data are split into packets that are transmitted separately and then reassembled at the receiving end. In fact, GPRS supports the world's leading packet-based Internet communication protocols, Internet protocol (IP) and X.25, thus enabling any existing IP or X.25 application to operate over a GSM cellular connection.

Combining mobile access with IP-based services, GPRS makes a highly efficient use of radio spectrum and enables high data speeds for packet data transmissions.

With GPRS, it becomes possible to remain constantly connected ("always on" connectivity) in a cost-effective way. GPRS users are able to log on to an *access point name* (APN) and have access to many services or an office network (without the need to dial-up) and remain continuously connected until they log off, only paying when data are actually transmitted. A physical end-to-end connection is not required because network resources and bandwidth are used only when data are actually transferred. The same radio resource is shared by all mobile stations in a cell, providing effective use of the scarce resources.

For users, the most important advantage of GPRS is the possibility for charging, which is based on traffic volume. There is no need to pay for unused transmission capacity. During the idle periods, the spectrum is effectively given to other users. This makes extremely efficient use of available radio bandwidth. Therefore, GPRS packet-based services should cost users less than circuit-switched services since communication channels are being shared and are on a "as-packets-are-needed" basis rather than dedicated to only one user at a time. Further on, faster data rate means that middleware currently needed to adapt applications from fixed line rates to the slower speed of wireless systems will no longer be needed; thus making availability of the application to mobiles much easier.

From the point of view of network operators, the scarce system resources must be treated as efficiently as possible. Especially in data services, the burstiness of transmission enables sharing of the radio interface and network resources by several users, without decreasing the efficiency of each individual transmission too much. That enables increased income per unit of available spectrum, even if each user experiences decreased operational expenses. The service is targeted mainly at applications with traffic characteristics of frequent transmission of small volumes and infrequent transmission of small or medium volumes. This enables the system to attract new services and applications. The transmission of large volumes should still remain via circuit-switched channels to prevent the blocking of the packet radio spectrum.

GPRS uses the same radio channel as GSM voice calls (a 200-kHz wide channel). This radio channel carries a raw digital radio stream of 271 kbit/s that, for voice calls, is divided into eight separate data streams, each carrying approximately 34 kbit/s. After protocol and error correction overhead, 13 kbit/s is left for each voice connection or about 14 kbit/s for data connection. Circuit-switched data today use one voice channel. GPRS can combine up to eight of these channels, and since each of these can deliver up to 14 kbit/s of data throughput, the net result is that users are able to enjoy rates of over 100 kbit/s. The higher data rates will allow users to take part in video conferences and interact with multimedia Web sites and similar applications using mobile handheld devices as well as notebook computers. In addition, in GPRS, the number of classes is extended to cover the half-duplex/asymmetric functionality more thoroughly. The standard allows half-duplex/asymmetric classes that allow six or eight time slots in the downlink while the uplink operates with a lower number of time slots. This is possible since the GPRS radio protocols allow flexible scheduling between uplink and downlink transmission.

A concise list with the key drivers for GPRS network operators as well as for end users is presented as follows.

For operators:

- To *increase revenues* by moving into the mobile data market, especially since the voice market has had profit margins squeezed with the commoditization of voice services.

- To *gain new subscribers* who require mobile data services. There are more mobile phones in general use than there are PCs in people's homes. This means that the potential market for GPRS is high and that new Internet users are more likely to upgrade to a GPRS handset rather than making a larger investment in a PC.

- To *retain current subscribers* by offering new services.

- To *reduce costs* by the efficient use of network resources. With packet switching, radio resources are used only when users are actually sending or receiving data. This efficient use of scarce radio resources means that a large number of GPRS users can potentially share the same bandwidth and be served from a single cell. GPRS spectrum efficiency means that there is less need to build in idle capacity that is only used in peak hours. Packet switching enables new charging and tariff schemes that are more appropriate for mobile services and on the other hand it saves operators money as it utilizes network resources much more efficiently compared to circuit-switched architectures (e.g., GSM).

- To *ease of adapting applications* for mobile users because high data speeds mean that middleware is no longer required to convert fixed applications for mobile use.

- Offer *innovative tariffs* based on new dimensions such as the number of kilobytes or megabytes.

- *Return on investment.* Investment in GPRS will be twofold since the new network infrastructure pieces will be used as part of the UMTS network requirements as well as GPRS. GPRS provides an upgrade path and test bed for UMTS.

For end users:

- New data services are enabled.

- Higher data speeds. Higher levels of bandwidth means higher speeds for data transactions.

- Cost-effectiveness. Only charged when data are transmitted and not for the duration of the connection.

- "Always-on" connectivity. GPRS enables instant connections and the ability to remain logged-on at all times (Internet or corporate virtual private networks (VPN). For example, a user with a laptop computer could be working on a document and at the same time receiving new e-mails or browsing the Internet.

Network architecture

Network and switching subsystem (NSS).   The GPRS network architecture is built on top of the existing GSM network infrastructure, enriched with new network elements that allow migration to packet-switched functionalities. The GPRS architecture is illustrated in Fig. 4.10.

The key network elements in a GPRS network are:

The base station subsystem.   The operation of low-level radio interface protocols is taken care of by the BSS. The GSM-based BSS is enhanced in

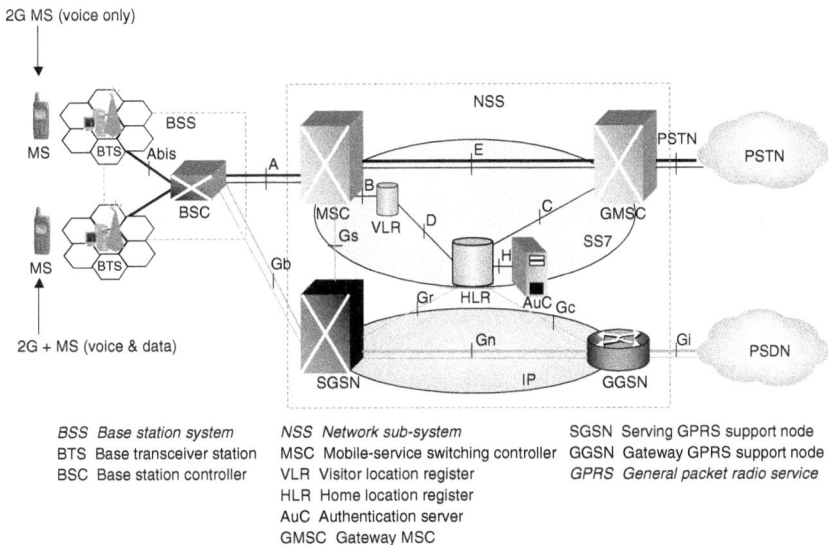

BSS	Base station system	NSS	Network sub-system	SGSN	Serving GPRS support node
BTS	Base transceiver station	MSC	Mobile-service switching controller	GGSN	Gateway GPRS support node
BSC	Base station controller	VLR	Visitor location register	*GPRS*	*General packet radio service*
		HLR	Home location register		
		AuC	Authentication server		
		GMSC	Gateway MSC		

Figure 4.10   Logical architecture for GPRS PLMNs.

GPRS to recognize and send user *data* to the serving SGSN. Similar to GSM, the radio equipment of a BSS may support one or more cells.

The BSS consists of one base station controller and one or more base transceiver stations. A BSC is a network component in the PLMN with the functions for control of one or more BTS. A BTS is a network component that serves one cell. The protocol architecture between the BSS and the SGSN is based on Frame Relay [TANENB96], where data from several mobile stations are multiplexed using virtual circuits.

In case the control of the radio interface is not taken care of by the base transceiver station, a remote *packet control unit* (PCU) may be optionally implemented. In that case the radio control functions are located remotely, in the BSC or SGSN site.

The mobile-services switching center (MSC). The mobile-services switching center performs all the switching and signaling functions for mobile stations located in a geographical area; this area is designated as the MSC area. The MSC handles the allocation of radio resources as well as the location registration and handover procedures.

The home location register (HLR). Like in GSM, the HLR in GPRS is a database in charge of the mobility management of mobile subscribers. A PLMN may contain one or several HLRs—this depends on the number of mobile subscribers, on the capacity of the equipment, and on the organization of the network. Two kinds of information are stored in HLRs.

- Subscription information

- Some location information enabling the charging and routing of calls toward the MSC where the MS is located (e.g., the mobile roaming number, the VLR address, the MSC address, and the local MS identity)

The visitor location register (VLR). When an MS enters a new location area and starts a registration procedure, the MSC in charge of that area notifies the local VLR about the identity of the location area where the MS is situated. If this MS is not yet registered in the VLR, the VLR proceeds and notifies the legitimate HLR of the MH about its current location for call handling purposes.

The authentication center (AuC). The AuC is associated with an HLR over the H-interface. It stores an identity key for each mobile subscriber registered with the associated HLR. This key is used to

- Generate data which are used to authenticate the *international mobile subscriber identity* (IMSI)

- Cipher communication over the radio path between the mobile station and the network

The equipment identity register (EIR). This functional entity comprises one or several databases which store(s) the IMEIs used in the GSM system. The mobile equipment may be classified as "white listed," "grey listed," and "black listed" and therefore may be stored in three separate lists.

Charging gateway. The charging gateway makes a log entry whenever there is a network activity such as data being transferred, the charging terms (peak/off-peak), an alteration in the quality of service, or if a GPRS session ends (known as a packet data protocol—PDP context). The main functions of the charging gateway are the collection of GPRS data records from the GPRS nodes, intermediate data record storage, buffering, and transfer of data records to the mediation/billing systems. The charging gateway functionality can be either a separate centralized network element or a distributed resident in SGSN and GGSN support nodes (see next).

GPRS support nodes (GSN)—SGSN and GGSN. The main routing functionality in GPRS is carried out by the so-called support nodes. There exist two nodes—a *gateway GPRS support node* (GGSN) and a *serving GPRS support node* (SGSN).

The SGSN is the node within the GSM infrastructure that sends and receives packet data to and from the mobile stations over the Gb interface in Fig. 4.10. SGSN participates in routing, user verification, collection of billing data, mobility management functions (tracking a mobile location), detects new GPRS mobile stations in a given service area and, sends queries to HLRs to obtain profile data of GPRS subscribers.

The GGSN is the node that interfaces to external public data networks (PDNs) such as the Internet and X.25. GGSNs maintain routing information for attached GPRS users. The routing information is necessary to tunnel *protocol data units* (PDUs) to the mobile's current point of attachment, i.e., the SGSN. Also, the packets transmitted by a mobile station via SGSN to GGSN are routed to the external data network. The GGSN is thus the first point of *packet data network* (PDN) interconnection with a GSM PLMN supporting GPRS (that is, the Gi reference point is supported by the GGSN). Other functions include network and subscriber screening and address mapping.

The SGSN and GGSN functionality may be combined in the same physical node or may reside in different physical nodes. SGSN and GGSN contain IP routing functionality, and they may be interconnected with IP routers. When SGSN and GGSN are in different PLMNs, they are interconnected via the Gp interface. The Gp interface provides the functionality of the Gn interface plus security functionality required for inter-PLMN communication. The security functionality is based on mutual agreements between operators.

The SGSN may send location information to the MSC/VLR via the optional Gs interface. The SGSN may receive paging requests from the MSC/VLR via the Gs interface.

GPRS network interfaces.   In the GSM-GPRS network architecture the following interfaces are defined:

1. Gb—Interface between an SGSN and a BSS.

2. Gc—Interface between a GGSN and an HLR.

3. Gd—Interface between an SMS-GMSC and an SGSN, and between an SMS-IWMSC (interworking MSC) and an SGSN.

4. Gf—Interface between an SGSN and an EIR.

5. Gi—Reference point between GPRS and an external packet data network.

6. Gn—Interface between two GSNs within the same PLMN.

7. Gp—Interface between two GSNs in different PLMNs. The Gp interface allows support of GPRS network services across areas served by the cooperating GPRS PLMNs.

8. Gr—Interface between an SGSN and an HLR.

9. Gs—Interface between an SGSN and an MSC/VLR.

10. Um—Interface between the mobile station and the GPRS fixed network part. The Um interface is the GPRS network interface for providing packet data services over the radio to the MS.

GPRS radio interface.   GPRS follows a 52-multiframe structure. In a 52-multiframe, every 13th TDMA frame is idle. Mobiles utilize the idle frames for the identification of the base station identity codes, the continuous timing advance update procedure, and the interference measurements for power control purposes. The remaining frames are used for the GPRS logical channels.

GPRS defines the following set of logical radio channels.

- *The packet common control channel (PCCCH).*   It is a set of logical channels used for common control signaling. PCCCH consists of:
  - *Packet random access channel (PRACH).*   It is used by the mobile station to request access for uplink transfer.
  - *Packet paging channel (PPCH).*   It is used by the network to inform the mobile station about downlink transfer.
  - *Packet access grant channel (PAGCH).*   It is used by the network to reserve a packet traffic channel for the mobile station.
  - *Packet notification channel (PNCH).*   It is used by the network to notify a group of mobile stations prior to point to multipoint-multicast (PTM-M) packet transfer.

- *Packet broadcast control channel (PBCCH).* It is used by the network to send system information messages for GPRS mobile stations. Packet traffic channels are dedicated temporarily for mobile stations.

- *Packet data traffic channel (PDTCH).* It is used for data transfer. Data may be point-to-point or point-to-multipoint messages or GPRS mobility management messages.

- Low-level signaling related to a given mobile station is transmitted on a *packet associated control channel* (PACCH). Such signaling includes acknowledgments of data, resource allocations, and exchange of power-control information.

Transmission and signaling planes

Transmission plane. The transmission plane consists of a layered protocol structure that provides information on transfer control procedures (e.g., flow control, error detection, error correction, and error recovery) as well as the actual user data information transfer. The protocol hierarchy of the GPRS transmission plane is described in Fig. 4.11.

Subnetwork dependent convergence protocol (SNDCP). SNDCP, specified in GSM 04.65, segments MS-to-SGSN network protocol data units into one or several logical link control (LLC) frames. In other words, SNDCP maps network-level characteristics onto the characteristics of the underlying network. SNDCP may support, in addition to

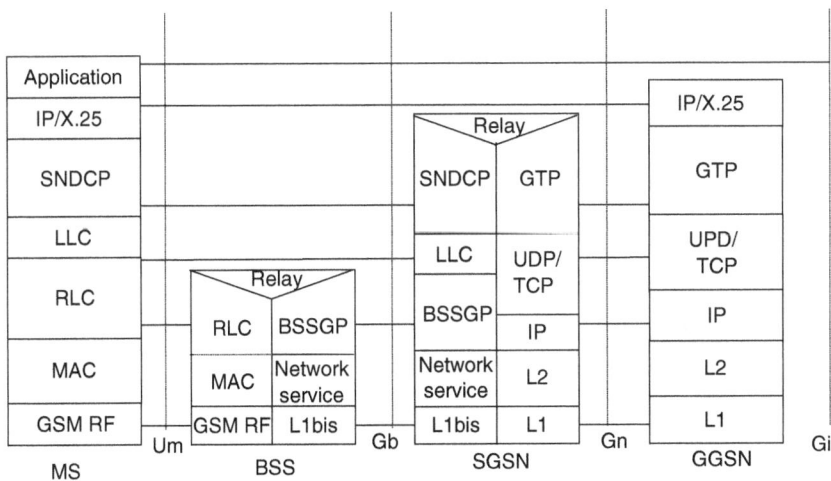

Figure 4.11   Transmission plane.

the segmentation function, multiplexing of user data, user data compression,[3] TCP/IP header compression, and transmission based on the requested quality of service (QoS).

Logical link control (LLC). The LLC protocol, specified in GSM 04.64, provides a highly reliable logical link between MS and SGSN. The functionality of LLC includes maintenance of the communication context between MS and SGSN, transmission of acknowledged and unacknowledged frames, and detection and retransmission of corrupted frames. The LLC frame includes a frame header with numbering and temporary addressing fields, a variable length information field, and a frame check sequence. The maximum length of the LLC frame is almost 1600 octets. However, the mobile station and the network may negotiate a shorter maximum frame length for the logical link.

LLC frames are transmitted in one or several radio blocks. LLC is thus independent of the underlying radio interface protocols in order to allow introduction of alternative GPRS radio solutions with minimum changes to the NSS. The logical link is maintained when the mobile station moves between cells of the same SGSN. A new logical link must be established, however, when the mobile station moves into a cell of another SGSN.

GPRS ciphering is operated at the LLC layer. The ciphering key is exchanged during the "attach" procedure.

TCP/IP. Tunneling of data and signaling messages between GPRS support nodes is carried over a GPRS backbone network. The protocol architecture of the backbone network is based on the Internet protocol. For network protocols requiring reliable transfer over the GPRS backbone (e.g., X25), *transmission control protocol* (TCP) is used with IP. Otherwise, the *user datagram protocol* (UDP) is used with IP (e.g., for Internet communication).

GPRS tunneling protocol (GTP). GTP is a specialized protocol that operates over the top of standard TCP/IP protocol to tunnel protocol data units (IP or X.25 packets) through the IP backbone. ETSI has defined the functions of GTP as:

- *Call data record* (CDR) transfer mechanism between GPRS nodes—SGSN and GGSN—generating CDRs and the *charging gateway functionality* (CGF)

- Redirection of CDR transfer to another CGF

- Ability to detect communication failures between the CDR handling GPRS network elements by echo messaging

---

[3]The V.42bis algorithm is used for compression of user data.

- Ability of a CDR-handling node to advertise the peer CDR handling GPRS network elements about its CDR transfer capability (e.g., after a period of service downtime)

- Ability to prevent duplicate CDRs that might arise during redundancy operations. If so configured, the CDR duplication prevention function may also be carried out by marking potentially duplicated CDR packets and delegating the final duplicate deletion task to CGF or the billing system (instead of handling the possible duplicates solely by GTP messaging)

- The aim of the duplication prevention support of GTP is to reduce the number of duplicated CDRs sent to the billing system and to support the billing system in keeping the efforts for duplicate CDR checking as small as possible

Relay. In the BSS, this function relays LLC PDUs between the Um and Gb interfaces. In SGSN, this function relays PDP PDUs between the Gb and Gn interfaces.

Base station system GPRS protocol (BSSGP). BSSGP provides transfer procedures for data and paging messages, and mechanisms for link management. The BSSG protocol is used on top of Frame Relay.

Network service (NS). This layer transports BSSGP PDUs. NS is based on Frame Relay connection between BSS and SGSN, and may be multihop (in the sense, traversing a network of Frame Relay switching nodes).

RLC/MAC. The radio link control (RLC) function provides a radio-solution-dependent reliable link whereas the MAC function controls the access signaling (request and grant) procedures for the radio channel, and the mapping of LLC frames onto the GSM physical channel. The RLC/MAC is defined in GSM 04.60.

The function of RLC between a mobile station and a base station subsystem is to detect the corrupted RLC blocks and request a selective retransmission of the corrupted blocks. The retransmission request is an acknowledgment consisting of a bitmap that indicates each of the received RLC blocks either as corrupted or successfully received.

When the mobile station has a packet to transmit, a packet channel request is first transmitted. The access may proceed in one or two phases. In the one-phase approach, the packet channel request message contains all the information needed for the establishment of the channel, e.g., multislot-related information and quality of the requested service. The BSS acknowledges the request by sending a packet immediate assignment message that contains information about the physical channels and a temporary flow identity reserved for the mobile station. At this point the mobile station may start data transmission on the allocated packet data channels.

In the two-phase access, the packet channel request leads to the reservation of a packet-associated control channel. The base station subsystem allocates the control channel where the mobile station may transmit a packet resource request containing all the details of the requested service. A packet resource assignment message containing information about the physical channels, which is reserved for the mobile station, is used as an acknowledgment for the packet resource request.

In addition, to enable optimizations in power consumption, mobile-specific discontinuous reception periodicity may be applied for the paging. Discontinuous reception may also be applied to the reception of the packet resource assignment. The packet resource assignment includes all the details of the reserved physical channel on the downlink, allowing the mobile station to start receiving data in the frame indicated in the packet resource assignment message.

GSM RF.   As defined in GSM 05 series.

Signaling plane.   The signaling plane is responsible for the control as well as support of the transmission plane functions. These are:

- Controlling the GPRS network access connections, such as attaching to and detaching from the GPRS network

- Controlling the attributes of an established network access connection, such as activation of a PDP address

- Controlling the routing path of an established network connection; this is needed to support user mobility

- Controlling the assignment of network resources to meet changing user demands

- Providing supplementary services

High-level functions required for data transmission in a GPRS network.
Although a GPRS user experiences a continuous connection, a network connection opens and closes for each transaction. In the context of the protocol layers described earlier, data transmission in a GPRS network undertakes (by virtue of packet switching) several steps. These include

- Network access control functions
- Packet routing and transfer functions
- Mobility management functions
- Logical link management functions
- Radio resource management functions
- Network management functions

Network access control functions.   Network access is the means by which a user is connected to a telecommunication network in order to use the services and/or facilities of that network. An access protocol is a defined set of procedures that enable the user to employ the services and/or facilities of the network.

User network access may occur from either the mobile side or the fixed side of the GPRS network. The fixed network interface may support multiple access protocols to external data networks, for example X.25 or IP. Individual PLMN administrations may require specific access-control procedures to limit the set of users permitted to access the network, or to restrict the capabilities of individual users, such as by limiting the type of service available to an individual subscriber.

Common to GSM cellular networks, several administrative functions are performed to validate a user, including

- *User registration.*   It associates the mobile ID with the user's PDP and PLMN address. Within the MS's home area, traditional HLRs are enhanced to reference GPRS subscriber's data, whereas outside the home area, dynamically allocated records are referenced in VLRs.

- *Authentication.*   It ensures the validity of the GPRS mobile station and its associated services.

- *Call admission control (CAC).*   It determines the required network resources for the quality of service that is requested. If these resources are available, they need to be reserved.

In addition to the standard point-to-point (PTP) data transfer, GPRS may support anonymous access to the network. The service allows an MS to exchange data packets with a predefined host that can be addressed by the supported interworking protocols. Only a limited number of destination PDP addresses can be used within this service. IMSI or IMEI are not used when accessing the network, thus guaranteeing a high level of anonymity. However, no authentication and ciphering functionality is foreseen for anonymous access.

Routing and data transfer functions.   Routing is the process of determining and using a route for the transmission of a message within and between PLMN(s). Once a mobile station begins its data transmission, routing is performed by GSNs on a hop-by-hop basis through the mobile network using the destination address indicated in the header of the message. Routing tables are maintained in GSNs using the GTP layer. The GTP layer performs the address translation and mapping functions to convert the external PDN addresses to an address that is usable for routing within

PLMNs. The data go through several transformations as they travel through the network. Depending on the destination PDN, data can be:

- *Forwarded*, from one node to the other in the route, using relay functions.

- *Tunneled*, to transfer data from one PLMN to another.

- *Compressed*, to use the radio path in an efficient manner. (Note that compression algorithms may be used for manufacturers to differentiate themselves.)

- *Encrypted*, to protect the mobile station from eavesdropping (encryption algorithms can also be used as a differentiating factor).

Mobility management functions.   Similar to GSM, mobility management functions in GPRS are used to track a mobile station's location and support seamless continuation of ongoing sessions. The mobile station's profiles are preserved in VLRs, which are accessible to SGSNs via the local MSC. SGSNs communicate with each other and update the user location. A logical link is established and maintained between the mobile station and the SGSN at each PLMN. At the end of the transmission or when a mobile station moves out of the area of a specific SGSN, the logical link is released and the resources associated with it can be reallocated.

There exist three mobility management profiles in GPRS—*idle*, *standby*, and *ready*. In the idle state the mobile station may perform PLMN selection, GPRS cell selection, and reselection. However, the mobility management and routing contexts are not active in the mobile station and the SGSN. The mobile station may receive only point to multipoint-multicast (PTM-M) data.

Standby is a state where mobile stations are normally prepared for data transfer, but are not actively transferring at the very moment. In the standby state, the mobility management context between the mobile station and the SGSN is active. The mobile station informs SGSN every time it changes from one routing area to another. The routing area is a set of cells (from one cell up to the size of a location area) defined by the operator. The mobile station may receive pagings for circuit-switched services, as well as pagings for GPRS PTP and PTM data.

The PTP service provides a transmission of one or more packets between two users initiated by a service requester and received by a receiver. There are two PTP services.

- PTP connectionless network service (PTP-CLNS)
- PTP connection orientated network service (PTP-CONS)

TABLE 4.5   Relationship of Service Request and Service Requester/Receiver

Service requester/receiver AP = access point	PTP-CONS and PTP-CLNS	Types of service request		
		PTM-M	PTM-G	IP-M
From fixed AP to mobile AP	Supported	Supported	Supported	Supported
From mobile AP to mobile AP (see note 1)	Supported	Supported	Supported	Supported
From mobile AP to fixed AP	Supported	Not applicable	Supported (limited See note 2)	Supported

NOTE: 1: It shall be possible to transfer data between two mobiles of the same operator without the use of non-GSM external data networks.
NOTE: 2: All PTM-G features may not be supported for the fixed AP, e.g., paging.

The PTM service provides a transmission of packets between a service requester and a receiver group. There are three PTM services.

- PTM multicast (PTM-M)
- PTM group call (PTM-G)
- IP multicast (IP-M)

For PTM-M and PTM-G, data transmission is restricted to the members of a receiver group currently located within a geographical area. The service requester specifies both the receiver group and the geographical area. The geographical area addressing mechanism is not applicable to IP-M.

An invocation of the service request by a service requester is possible from the fixed and mobile access points. Table 4.5 presents the relationship between service requests and the service requester/receiver.

When the mobile station has data to send or receive (except PTM-M) it must enter into the ready state. In the ready state, reception of data is possible without the paging procedure since the network is aware of the location of the mobile station. The mobile station informs SGSN every time it switches between cells. The ready state is guarded by a timer. This timer is reset after the reception or transmission of a packet. When the timer elapses, the mobile station enters into the standby state. A change from the standby to the ready state may be initiated by the network, using a paging procedure. This is used when there are data to be sent to the mobile station.

Logical link management functions.   Logical link management functions are concerned with the maintenance of a communication channel between an individual MS and the PLMN across the radio interface

These functions involve the coordination of link-state information between the MS and the PLMN as well as the supervision of the data transfer activity over the logical link.

Radio resource management functions. Radio resource management functions are concerned with the allocation and maintenance of radio communication paths.

Example of GPRS transaction. We next illustrate how the GPRS signaling works for the establishment of data transfer within the GPRS context, using a simple example.

Let us assume that a GPRS subscriber wants to send and receive data During the call setup, an attach procedure is initiated by the mobile station. During the attach procedure (GPRS attach, IMSI attach, or combined GPRS/IMSI attach), the mobile station enters into the ready state. In the attach procedure, the identity of the mobile station, the ciphering key sequence number, and the classmark of the mobile station are delivered to the network. The identity of the mobile station is often a *temporary logical link identity* (TLLI) that is combined with the routing area identity, identifying thus the routing area where the TLLI has been allocated. In case the mobile station has never had a TLLI before, IMSI may be used instead. If the identity is known by another SGSN, anidentification from the other SGSN is queried. If so, the SGSN requests authentication of the mobile station, the ciphering is initialized, and the IMEI of the mobile station is checked. If necessary, the location information in the HLR is updated, the context in the old SGSN is deleted, and the subscription information in the SGSN is updated. In addition, the location information is delivered to the new MSC/VLR and the information in the old MSC/VLR is deleted. Once the attach procedure is completed, the mobile station may transfer short messages (SMS), receive PTM-M messages, or activate a PDP context for some packet data protocol.

The activation of the PDP context is a natural continuation for the attach procedure. PDP context activation may be requested either by the network or by the mobile station with a desired QoS. In PDP context activation, the routing context in GGSN is activated. Routing between SGSN and GGSN is enabled by the activation of the tunneling identifier in the SGSN and GGSN. When the mobile station is successfully attached to the network and the PDP context is activated, the mobile station is ready for receiving and transmitting data.

When transmitting data, the terminal equipment passes the user data for the GPRS mobile terminal. The IP datagram is compressed and encapsulated in an SNDC protocol unit. This information is sent over the radio interface, using the LLC, RLC/MAC and physical protocols, to the SGSN that is currently serving the mobile station. When the

SGSN receives the packet, it tunnels the packet through the GPRS backbone network to the GGSN, using the GPRS tunnel protocol. The GGSN removes the tunneling and forwards the IP datagram to the Internet. Standard Internet routing procedures will then route the packet to its intended recipient.

On the other hand, a host connected to the Internet does not need any special skills to communicate with a GPRS mobile station. The host transmits IP datagram through the Internet using the IP address for the GPRS mobile station. Internet routing protocols are used to route the packet to the subnetwork from which the IP address of the mobile station was acquired. When the IP datagram arrives at GGSN, the IP address of the receiver is extracted and mapped to the current location of the mobile station. The IP datagram is tunneled through the backbone network to the SGSN that is currently serving the mobile station. The SGSN removes the tunneling and the original IP datagram is encapsulated in the SNDC protocol data unit. It is then transmitted to the mobile station over the radio interface, using the LLC, RLC/MAC, and physical protocols.

Applications for GPRS.   The amalgam of GPRS technologies based on packet and circuit-switched platforms intends to lead to a proliferation of unique mobile e-services.

- *Web browsing.*   The general information services such as train timetables, interactive data access, toll roads charging systems, and Internet browsing using the World Wide Web, are provided to the mobile phone via a wireless application protocol (WAP) server. WAP is a light browser for cellular phones in which the information is only textual.

  WAP is established in the market place and its value to the end user is greatly enhanced by the always-on connectivity of GPRS, which reduces the average setup time from 20 to 30 s to around 5 s. The WAP experience is further enhanced with the introduction of WAP push where subscribers can specify the type of information that is important to them—stock price changes for example—and a service notification is then automatically sent to them as it becomes available. This is enhanced with a new technology that allows users to make voice calls during their WAP browsing session. For instance, users could be sent details of special flight deals and if one particular offer was attractive they could simply click on the phone number displayed on the screen and be put through to an operator to make a booking.

  The WAP forum has completed the standardization of WAP 2.0, the implementation of which significantly improves the user experience. WAP 2.0 is based on XHTML rather than the wireless markup language (WML) standard used in WAP 1.0. This is important as XHTML enables users to browse the Web directly rather than only giving access to the Internet pages written in WML. XHTML provides not only a bridge between the fixed and mobile Internet, it also gives tremendous flexibility in displaying information, offering a wide variety of text formats, fonts, and text positioning.

- *Chat.* GPRS, considered as an extension of the Internet, allows mobile users to use existing Internet chat groups rather than needing to set up their own groups dedicated to mobile users. GPRS does not support point-to-multipoint services in its first phase, hindering the distribution of a single message to a group of people.

- *Internet email.* Most Internet email users do not get notified of new email on their mobile phone. When they are out of the office, they have to dial in speculatively and periodically to check their mailbox contents. However, by linking Internet email with an alert mechanism such as SMS or GPRS, users can be notified when a new email is received.

- *Information services as text or graphics.* Information content includes services such as share prices, sports scores, weather reports, news headlines, flight information, news headlines, traffic reports, maps, graphs, and lottery results. GPRS is likely be used for qualitative information services when end users have GPRS-capable devices, but SMS (which is limited to 160 characters) is likely to continue to be used for delivering most quantitative information services.

- *Still images.* Photographs (either scanned or from a digital camera), pictures, postcards, greeting cards, presentations, and the like.

- *Moving images.* Video messages, movie previews, security camera images, patient images (e.g., from a crash site to a hospital), video conferencing, and the like.

- *Document sharing and remote collaborative working.* This lets different people at different places work on the same document in a problem solving exercise, such as medical treatment, journalism, advertising copy, and fire-fighting.

- *Audio reports.* For broadcasting or analyzing. For example, high-quality voice clips for television and radio or sounds for police to use as evidence, occupy large file sizes that need high-speed mobile data rates such as GPRS.

- *Job dispatch.* This includes advising mobile engineers where their next job is, providing full briefing details and images, as well as enabling the engineer to report back to the head office corporate email, allowing mobile employees to access their internal email system from their LAN in the office.

- *LAN applications.* Mobile employees can access any applications normally available on their PC at the office.

- *Vehicle positioning.* An application that integrates a satellite positioning system that tells people where they are with nonvoice mobile services that let people tell others where they are.

- *File transfer.* Downloading sizable data across the mobile network such as a presentation for a sales person or a manual for a service engineer or a software application such as Adobe Acrobat Reader to read documents.

To put these "vertical" services into context, we highlight some scenarios of integrated GPRS services ranging from the corporate to consumer application segments.

To start with the corporate segment, GPRS is ideally suited to industrial and telemetry applications that might involve asset monitoring, remote control, and home/office security. The automotive industry has taken the initiative here, incorporating real-time connectivity to service centers from dedicated mobile terminals installed in motor vehicles. GPRS terminals in vehicles connect to operation centers in real time as the vehicle becomes a simple IP address on the Internet, speeding up emergency help and enabling vehicles to be monitored continuously for reasons of tracking, safety, protection, and assistance. Full-time connection will support on-demand or automatic access to pay per-use services such as traffic information, stock market listings, breaking news, and weather reports. In addition, a dashboard display with a personalized driver portal will offer travelers Web surfing, voice news, email, updated online road guides, tourist information, m-commerce, and many other services.

All these services are georeferenced and refer to the vehicle's actual location at the time. This could be useful if a car is stolen. In this case, hidden onboard cameras and biometric fingerprint sensors in the steering wheel could transmit the identity of the thief in real time.

Location-based services are being talked up a great deal by operators and suppliers alike. These will play a pivotal role, providing visual information to GPRS users who need it to find out more about a particular area and task. While location standards are still incomplete, information based purely on the ubiquitous cell ID is still a powerful tool when combined with WAP/GPRS. A classic example would be the movie theater. In this case, customers use their personalized smartphones with a 32K SIM and protected PKI PINs to search for a movie, see a review, preview a portion, find the nearest theater, book and pay for their ticket, be guided to the theater location, and, if they wish, also pay for parking and food items. If the theater is in a shopping center, stores with picocells might send teaser SMS ads to these users as they walk by, inviting them to come in and enjoy a discount on any purchases in that store for that hour, or even provide a sponsored 3 min call within a specific promotional area. Technology might even allow them to use a bar-code scanner in their smartphone to compare prices of items with other stores.

Contents could also include rich multimedia services like full-motion news and financial stories, sports highlights, short entertainment clips and music videos, weather and traffic reports, security camera monitoring, health applications, stereo music broadcast, and visual navigation services that augment current text-based services. Doctors, for example, could remotely view and control ECG and ultrasound facilities with their GPRS smartphones, while delegates to conferences can download updated rich-multimedia PowerPoint presentations via Bluetooth/GPRS units.

GPRS can also facilitate the purchase and delivery of personalized MP3 stereo audio or videos for local smartphone storage. Users will even be able to use MP3 stereo audio for their ringtones. Here the *secure digital music initiative* (SDMI) is working with vendors to ensure that encrypted unique digital "watermarks" on downloaded audio are not disseminated without additional payment. Along with audio transfer, the power and relative simplicity of still-image (including stored Web pages) transfer over GPRS could herald a new, massmarket "killer application" category replacing the already dominant picture and logo messaging offerings. It is likely to appeal to a diverse group of users, ranging from real estate agents, to journalists, to the disposable-camera generation.

4.4.2   Enhanced data rates for GSM

The EDGE (Enhanced Data rates for GSM) is a 2.5G network. It is basically considered to be the second step toward 3G for GSM/GPRS networks. EDGE can be deployed within the existing GSM900 and GSM1800 frequencies. Also called *enhanced-GPRS* (EGPRS), EDGE delivers higher data speeds than GPRS, although the precise level of improvement is contentious. Specifically, EDGE could offer data rates up to 384 kbit/s in wide areas and 554 kbit/s in local areas [SKOLD97]. These are theoretical values and unrealistic at best. A more realistic figure is that EDGE can deliver around three times the best average speed of GPRS—that is, 30 to 40 kbit/s—which is equal to 100 kbit/s plus, in good conditions.

EDGE introduces higher-level modulation and new coding schemes for packet-switched and circuit-switched data communication. In addition to GMSK modulation, EDGE introduces *eight-symbol phase shift-keying* (8PSK) modulation. The symbol rates for GMSK and 8PSK are the same, that is, approximately 271 kilosymbols per second. The introduction of EGPRS increases maximum bit rates to approximately three times that of standard GPRS. New techniques introduced with EDGE optimize the data throughput for each radio link. One such technique, called *link quality control* (LQC), combines *link adaptation* (LA) and incremental redundancy. The LA functionality adapts coding and modulation relative to signal quality. In poor radio conditions robust coding and GMSK modulation are selected, whereas in good radio conditions less-robust coding and 8PSK modulation are employed. EGPRS also features backward error-correction functionality, which means that it can request the retransmission of erroneously received blocks.

This mechanism is called *automatic repeat request* (ARQ). EGPRS uses an enhanced variant of ARQ called *incremental redundancy* (IR). With IR, all information is coded with a convolution code at a rate of 1/3. The code is punctured to a certain rate and transmitted over the air. If the decoding fails, a retransmission is formed using a different puncturing scheme. Because the retransmission is combined with the previously-transmitted block, the process yields a lower bit rate, which facilitates decoding. At present, nine coding schemes have been defined for EGPRS (see Table 4.6).

The "Family" column in Table 4.6 indicates the coding schemes that can be used for retransmission. For example, if the initial transmission with MCS-9 fails and the quality of the radio channel deteriorates, the retransmission can use more robust coding schemes from the same family. Had the retransmission used a different coding scheme than the original transmission, it would have to be resegmented into new RLC blocks. This is what limits the selection of coding schemes. Blocks that are initially transmitted with MCS-8 can be retransmitted using MCS-6 or MCS-3, by adding padding bits to the data field.

It should also be noted that even when there is nothing with which to combine a retransmission—that is, if the initial transmission is lost altogether—it is possible to decode a retransmission. There is flexibility in the implementation of LQC for EGPRS in the system—LA will be mandatory, while IR is optional.

TABLE 4.6 Modulation and Coding Schemes for EGPRS

Scheme	Modulation	Maximum rate (kbit/s)	Code rate	Header code rate	Family
MCS-9	8PS K	59.2	1.0	0.36	A
MCS-8	8PS K	54.4	0.92	0.36	A
MCS-7	8PS K	44.8	0.76	0.36	B
MCS-6	8PS K	29.6	0.49	1–3	A
MCS-5	8PS K	22.4	0.37	1–3	B
MCS-4	GMS K	17.6	1.0	0.53	C
MCS-3	GMS K	14.8	0.80	0.53	A
MCS-2	GMS K	11.2	0.66	0.53	B
MCS-1	GMS K	8.8	0.53	0.53	C

With technological compression having put paid to that concept, the question that arises is "If GSM operators do plan to deploy EDGE, how will it fit into their overall strategy, particularly with respect to the ongoing implementation of 3G?" In response, EDGE proponents advocate that EDGE should be deployed as a complement to 3G air interface (W-CDMA). Operators already have the GSM infrastructure nationwide, so going for EDGE is a minor investment. This becomes evident by the fact that EDGE is a "simple" upgrade rather than a new network and would inherently be covered by the ongoing frame contracts which all operators have in place with their vendors.

Both the efficiency and speed of service delivery on the GSM bands can be increased considerably with an easy and cost-efficient upgrade of the existing network. Its appeal appears to be widening by the fact that EDGE can work on the existing GSM spectrum as well as that it requires "minimal tweaking" to GPRS networks. This also allows operators to provide 3G services immediately to subscribers without W-CDMA terminals. In regions where 3G licenses are yet to be awarded or where operators are not committed to a specific 3G rollout timetable, or even failed to get a 3G license, EDGE will provide a means of ensuring seamless services complimenting the W-CDMA coverage. Figure 4.12 depicts such complimentary radio coverage scenarios from a higher perspective.

An incumbent cellular operator with a UMTS/IMT-2000 license would benefit from EDGE to win a higher market share for various reasons. Primarily, this is so because EDGE

- Uses existing resources, competencies, and processes
- Builds coverage more quickly and economically

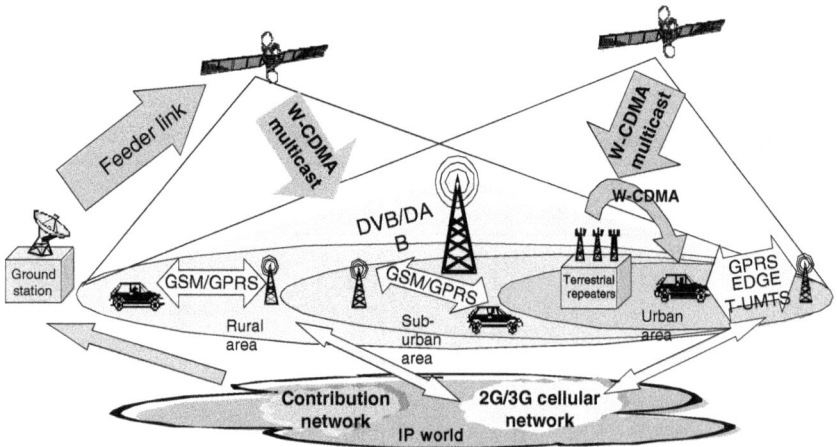

Figure 4.12 Complimentary 3G radio coverage scenarios.

- Leverages earlier handset volumes, thus helping operators to gain early market entry
- Offers low-cost capacity for voice and 3G data mass market
- Offers an attractive business case
- Delivers the extra capacity that high packet usage demands
- Allows operators to focus UMTS deployment on key areas using EDGE as a complementary service
- De-risks UMTS investments
- Allows a very fast global footprint

In the United States—where, at the time of this writing, allocation of the 3G spectrum is uncertain—AT&T Wireless, Cingular Wireless, and T-Mobile US have all started deploying EDGE networks. In the Asia-Pacific, People's Telephone, Sunday, Hong Kong CSL, Optus, Telstra, and TAC have all either announced plans to deploy EDGE or are in the process of trailing it. Not burdened by high 3G license fees and perhaps having a more pragmatic attitude to the W-CDMA roadmap, the Asia-Pacific players do not seem to be as concerned about the "image" of EDGE as their European counterparts.

While vendor attitudes are changing, one can only assume that mobile operators in Europe still feel reticent as not one—save for Telecom Italia Mobile—has publicly declared its association with EDGE. And perhaps this is understandable. To talk openly about EDGE, having already committed around US$100 billion in 3G license fees may be interpreted as an open admission by European operators that they severely miscalculated their 3G strategy. Many 3G licenses were awarded on the premise that there would be near-nationwide UMTS coverage by a certain date and regulators have proven difficult to budge on that. In Sweden, the regulator declined in 2001 to extend the 99.8 percent UMTS population coverage deadline beyond 2003, prompting Orange Sverige to pack up its Swedish 3G ambitions. What is more, the gaping cost of UMTS has to be covered before UMTS services are in place to recoup them, at the same time as regulators are likely to squeeze voice charges. GPRS, together with EDGE, are supposed to heal the breach.

4.4.3   Evolving GSM to 3G

The promise of 3G technologies is a combination of high-speed wireless access with IP-based services, which develops an intuitive feel of producing a greatly enhanced user experience bringing the world to our fingertips. It is a world in which we will be able to check emails, book holidays, organize share portfolios, hold videoconferences, or download video clips of the latest film instantly and simply through our mobiles.

Mobile networks are, however, not yet equipped to deal with high-speed mobile Internet, video, and other data-oriented services. This is perhaps why the migration to 3G is often presented as a revolution, requiring new network structures and the establishment of a new communications system that should evolve from the existing GSM/GPRS infrastructure. Recognizing this need, during a plenary meeting back in 1997, GSM MoU Association members agreed that GPRS would be a key link between GSM and the 3G, as many of the characteristics of GPRS (e.g., packet-based transport, new control) are quite relevant to provide the fastest, easiest, and most cost-effective path to 3G services.

For GSM/GPRS operators with the right planning, however, this could be an evolutionary rather than a revolutionary process. GSM/GPRS are feature-rich and have the lowest cost as a result of open standards and global economies of scale, and provide a clear, smooth future-proofed roadmap to 3G services. GPRS builds on an installed base of over 100 million GSM users, leveraging previous investments and industrywide expertise in developing GPRS-enabled handsets. Combined with WAP, the "always on" capability of GPRS makes access to mobile Internet and future multimedia services virtually instantaneous to users. Users can have a virtual "connection" at all times—network resources are only used when data are actually being transmitted or received.

Table 4.7 illustrates this evolution.

GPRS is already having a major impact on the global mobile communications scene providing the core network packet data handling capabilities that 3G needs. By assuring backward compatibility with GSM/GPRS, a 3G system also assures its economical viability.

The evolution from GSM to 3G is important for another reason—to facilitate seamless coverage using multimode GSM/UMTS terminals that precludes one of the main objectives of the UMTS standards, which is *global roaming*, as well as a spectrum- efficient utilization of GSM900, GSM1800, GSM1900, and 3G frequency bands, especially in the beginning of the UMTS life time.

To the efficiency spectrum utilization, one must also add the cohabitation and internetworking with second-generation systems. Figure 4.13 shows the allocation of UMTS/IMT-2000 spectrum and its use in the United States, Europe, and Japan. As illustrated, the spectrum is allocated differently in different parts of the world. Within Europe, the European Radiocommunications Committee (ERC) of the CEPT (European conference on Postal and Telecommunication administrations) is responsible for the actual allocation of radio frequencies. The ERC specifies that 155 MHz of the spectrum shall be reserved for the terrestrial component of 3G systems. The 155 MHz is split into *paired bands* 2×60 MHz (1920 to 1980 MHz and 2110 to 2170 MHz bands) for frequency division duplex (FDD) and *unpaired bands* separate into

TABLE 4.7   Evolution of GSM Toward UMTS [HUBER00]

	GSM Phase 2	GSM Phase 2+ (GPRS)	UMTS
Multiple access	FDMA/TDMA	FDMA/TDMA	FDMA/CDMA
Max bit rate	9.6 kb/s	64 kb/s, 115 kb/s	384 kb/s, 2 Mb/s
Speech quality	Full rate	Enhance full rate, transcoder-free operation	Adaptive multirate
Capacity	900 MHz	Tri-band (900,1800, and 1900 MHz)	2000 MHz with spectrum efficiency
Roaming	International	Global	Seamless global roaming in multiradio environments and application areas
Security	Authentication, encryption	Fraud information gathering, SS7 security, lawful interception	Enhanced authentication and user identity confidentiality, network domain security
Bearers	Circuit-switched bearers	64 kb/s circuit bearer, packet bearers be general packet radio service	Circuit—and packet-switched bearers real-time packet bearer
Services	Speech and low-speed circuit-switched data, supplementary services, short message services	Services customization, service portability, value–added services, mobile Internet access, and web–like information services	Full Internet capability, speech data, multimedia, virtual home environment

20 MHz and 15 MHz (1900–1920 MHz and 2010–2025 MHz bands) for time division duplex (TDD). FDD systems use different frequency bands for uplink (from the subscriber to the network) and downlink (from the network to the subscriber) separated by the duplex distance, while TDD systems utilize the same frequency for both uplink and downlink.

Whereas European and Japanese third-generation frequency allocations are very similar, the IMT-2000/FPLMTS spectrum in the United States is already used for PCS (see Table 4.8). The allocation of the spectrum in the United States differs from that in Europe and Japan, since parts of the 2-GHz frequency band have already been allocated for use by personal communication services systems.

Although the spectrum has been reserved in certain parts of the world for IMT-2000 services, this does not mean that similar services cannot be provided in other bands. For instance, EDGE (which is a migration path for GSM and TDMA/136) and multicarrier cdma2000 (which is a

TABLE 4.8 Wireless Personal Communication Systems (PCS) Technologies

| | High-power systems | | | | Low-power systems | | | |
| | Digital cellular (high tier PCS) | | | | Low tier PCS | | Digital cordless | |
System	IS-54	IS-95 (DS)	GSM	DCS-1800	WACS/PACS	Handi-Phone	DECT	CT-2
Multiple access	TDMA/FDMA	CDMA/FDMA	TDMA/FDMA	TDMA/FDMA	TDMA/FDMA	TDMA/FDMA	TDMA/FDMA	FDMA
Freq. Band (MHz) Uplink	869–894	869–894	935–960	1710–1785	Emerg. Tech. (U.S.)	1895–1907 (Japan)	1880–1900 (Europe)	864–868 (Europe and Asia)
(MHz) Downlink	824–849 (U.S.)	824–849 (U.S.)	890–915 (Europe)	1805–1880 (UK)				
RF ch. Spacing Downlink (KHz)	30	1250	200	200	300	300	1728	100
Uplink (KHz)	30	1250	200	200	300			
Modulation	π/4 DQPSK	BPSK/QPSK	GMSK	GMSK	π/4QPSK	π/4QPSK	GFSK	GFSK
Portable txmit Power, max./avg.	600 mW/ 200 mW	600 mW	1 W/ 125 mW	1 W/ 125 mW	200 mW/ 25mW	80 mW/ 10 mW	250 mW/ 10 mW	10 mW/ 5 mW
Speech coding	VSELP	QSELP	RPE-LTP	RPE-LTP	ADPCM	ADPCM	ADPCM	ADPCM
Speech rate (kb/s)	7.95	8 (var.)	13	13	32/16/8	32	32	32
Speech ch./RF ch.	3	—	8	8	8/16/32	4	12	1
Ch. Bit rate (kb/s) Uplink (kb/s)	48.6		270.833	270.833	384	384	1152	72
Downlink (kb/s)	48.6		270.833	270.833	384	384		
Ch. Coding	1/2 rate conv.	1/2 rate fwd 1/3 rate rev.	1/2 rate conv.	1/2 rate conv.	CRC	CRC	CRC (control)	None
Frame (ms)	40	20	4.615	4.615	2.5	5	10	2

*Spectrum is 1.85 to 2.2 GHz allocated by the FCC for emerging technologies; DS is direct sequence.

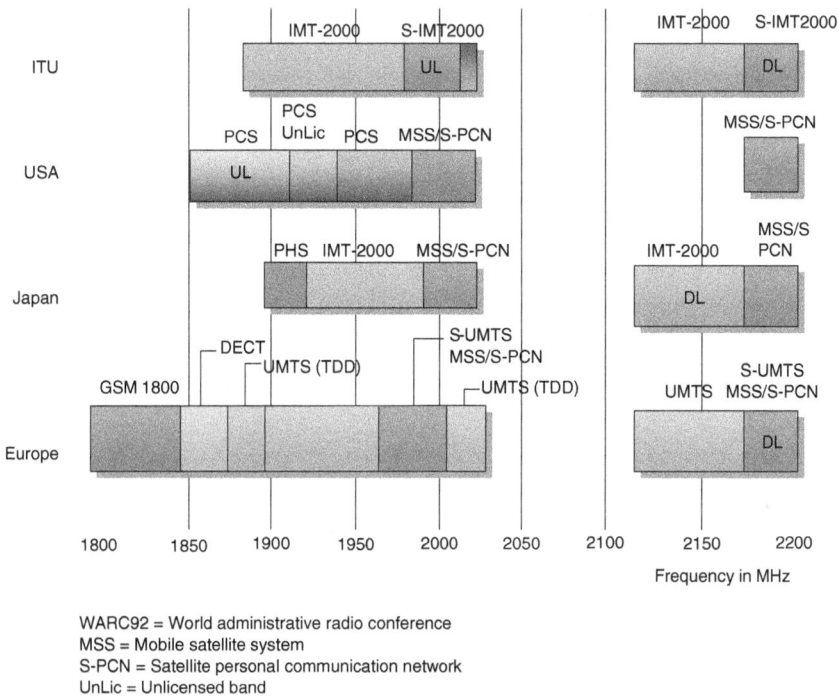

WARC92 = World administrative radio conference
MSS = Mobile satellite system
S-PCN = Satellite personal communication network
UnLic = Unlicensed band

Figure 4.13   Worldwide spectrum allocation.

migration path for IS-95) support the majority of IMT-2000 services. Accordingly, we can expect to see the following developments in the market:

- Operators will be allocated a new spectrum, either paired (FDD) or unpaired (TDD) bands
- Operators will migrate the existing second-generation spectrum, adding support for third-generation services.

Availability of the spectrum as well as licensing are the main factors that impact the decision for a UMTS deployment scenario. A GSM operator without UMTS spectrum needs most likely a different solution to provide third-generation services within the limited GSM spectrum than an operator with both GSM and UMTS spectrum licenses. Existing cellular (GSM900, GSM1800, or PCS1900) operators without UMTS frequency license will most likely start UMTS with local coverage relying on EDGE to obtain better spectrum efficiency and to introduce higher bit-rate services.

An existing GSM operator with UMTS spectrum license can use only wideband TDMA to introduce high bit-rate services. On the other hand, new UMTS operators can use TDMA-based UMTS, both with narrowband and wideband carriers, to provide a full set of UMTS services for the coverage of a wide area and hot spots coverage.

Besides the previously mentioned considerations, transition to 3G is factored by a plethora of other factors including market needs, regulatory conditions, and technical possibilities. One of the principal success factors in the upcoming third-generation wireless communication systems is the existence of increased spectrum efficiency and data rates. With these in mind, 3G design is primarily focused on new air interfaces that are primarily based on the code division multiple access (CDMA) technology. "Overview of the CDMA Technology" in Sec. 4.5.4 deals with the theory of CDMA in detail.

4.5   3G Mobile Wireless Systems: The Era of Universal Mobile Telecommunications System (UMTS)

Whereas the 2G systems were built mainly to provide speech services in macrocells, third-generation systems are designed for multimedia communications on the move—with them person-to-person communication can be enhanced with high-quality images and video, and access to information and services on public and private networks is enhanced by the higher data rates and new flexible communication capabilities of 3G systems [RAPELI95]. The UMTS Forum defines UMTS as "a mobile communications system that can offer significant user benefits including high-quality wireless multimedia services to a convergent network of fixed, cellular, and satellite components. It will deliver information directly to users and provide them with access to new and innovative services and applications. It will offer mobile personalized communications to the mass market regardless of location, network or terminal used." For the first time, users will have access to extremely sophisticated services (see Fig. 4.14). Some of these will resemble those delivered by broadband fixed networks, while others will be entirely new and specific to 3G mobiles.

With these requirements in mind, UMTS standards aim at advancing and unifying the diverse mobile communications systems of today into a common flexible radio infrastructure with the goal to provide communications at any time, any place, in any form, toward the future world of global personal mobile telecommunications [IMT-ITU]. The integration of fixed structures with wireless (both terrestrial and satellite) ones along with the use of international common frequency bands

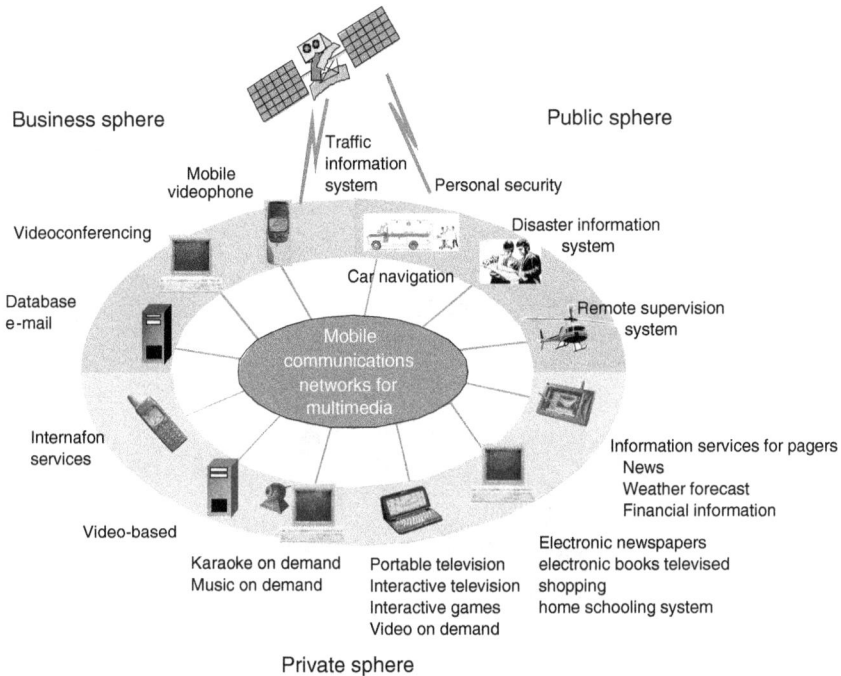

Figure 4.14 Main user types of UMTS. (*Source: www.imt-2000-online.com*)

and roaming will enable the birth of a truly global voice, data, and multimedia communications network.

To bring these about, UMTS will require at least:

1. A mobile technology that supports a very broad mix of communication services and applications

2. High-speed mobile multimedia services
   - Wide area mobility
     - Up to 144 kbit/s for fast moving mobiles
     - Up to 384 kbit/s for pedestrians
   - Low-mobility indoor up to 2 Mbit/s

3. Provision of a unified presentation of services to the user in wireless and wired environments
   - Advanced service creation, customization, and portability by the virtual home environment

4. Standardization that allows seamless roaming and interworking capability between indoor, outdoor, and far outdoor environments.

IMT-2000 vision
The emerging network of the 21st century

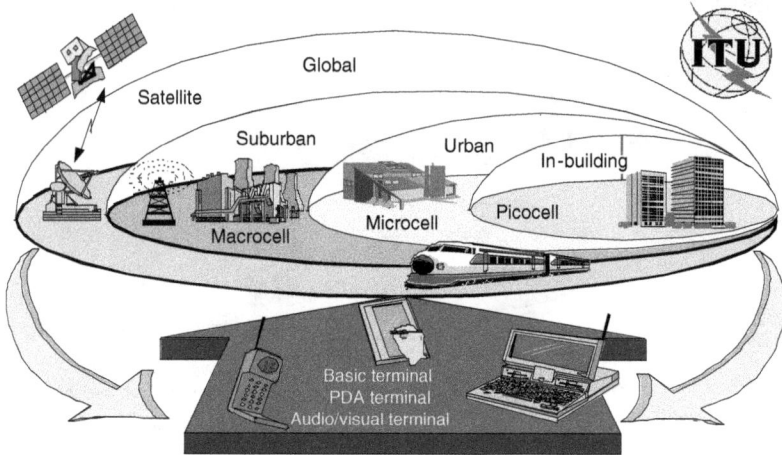

Figure 4.15   IMT-2000 [IMT-2000].

Figure 4.15 illustrates the general idea of a unified, seamless operation in different environments in UMTS/IMT-2000.

As illustrated, pico- and microcells are used in indoors/urban areas while macrocells in outdoor and rural areas. Since human presence is not spread over the entire surface of the globe (yet) as well as due to the surface propagation properties at 2 GHz—the UMTS operation band—macrocells cannot provide full rural coverage. Satellite networks [JAMALI01, WOOD01] are thus needed to complete the *UMTS terrestrial radio access* (UTRA) network to provide a seamless rural outdoor coverage using the IMT satellite frequencies (adjacent to terrestrial ones – see Fig. 4.13) with a minimum amount of modifications in services and user terminals.

4.5.1   Third-generation standards

Third-generation systems are intended to be a catalyst for increased variety and competition in mobile services, thus giving worthwhile return for a strong support by worldwide research funding of projects creating new technologies and telecommunications applications [RAPELI95]. In Europe, research work on the development of future generations of mobile communications concepts, systems, and networks [HJELM00] was initiated in the European Union R&D projects RACE

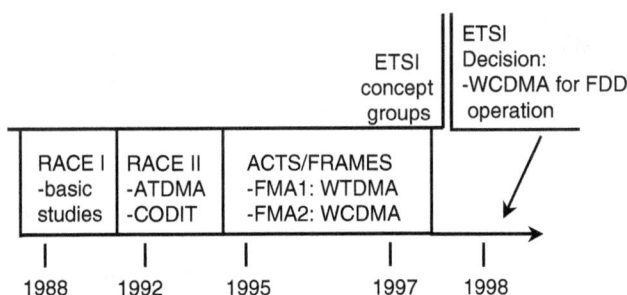

Figure 4.16 European research programmes for third-generation systems and the ETSI decision.

I and II (*research of advanced communication technologies in Europe*) and *advanced communication technologies and services* (ACTS–see Fig. 4.16), and also within large European wireless communications companies, at the start of the 1990s [OJANPE96]. The RACE I programme started the basic third-generation research work in 1988 and lasted until June 1992. This programme was followed by the RACE II program. During RACE II, the *code division testbed* (CODIT) [PIRHO99] and the *advanced TDMA mobile access* (ATDMA) projects developed air-interface specifications and testbeds for UMTS radio access [ANDER95, BAIER94, URIE95]. In addition, wideband air-interface proposals were studied in a number of industrial projects in Europe (see, for example, Ref. [PAJU97]).

The European research program ACTS was launched at the end of 1995 in order to support mobile communications research and development. Within ACTS the *future radio wideband multiple access system* (FRAMES) project [NKULA98] was set up with the objective of defining a proposal for a UMTS radio access system. The main industrial partners in FRAMES were Nokia, Siemens, Ericsson, France Telecom, and CSEM/Pro Telecom, with participation also from several European universities. Based on an initial proposal evaluation phase in FRAMES, a harmonized multiple access platform was defined, consisting of two modes—FMA (*FRAMES multiple access*) 1—a wideband TDMA [KLEIN97], and FMA2—a wideband CDMA [OVESJO97, OJANPE97]. The FRAMES wideband CDMA and wideband TDMA proposals were submitted to ETSI as candidates for UMTS air interface and ITU IMT-2000 submission.

Furthermore, a Study Committee, which was established in April 1993 within the Association of Radio Industries and Business (ARIB) with the goal to coordinate the Japanese research and development activities for IMT-2000 FPLMTS systems, established a Radio Transmission Special

Group for radio transmission technical studies and production of draft specification for FPLMTS [ARIB]. The Special Group, which consisted of a CDMA and TDMA group [SASAKI97], produced 13 wideband CDMA/FDD radio interface proposals [ADACHI95, ONOE97], as well as a W-TDMA and OFDM scheme. From these, the CDMA group submitted into ETSI three of the original W-CDMA schemes, whereas the TDMA group submitted an OFDM-based scheme called *band division multiple access* (BDMA).

By 1996, Japan had decided to propose a W-CDMA technology, but in Europe and the United States, there were other proposals. Europe had to choose between W-CDMA and TD-CDMA technologies (and also W-TDMA), whereas the U.S. groups favored a particular CDMA approach (which eventually became cdma2000) and a TDMA option, UWC-136. See Fig. 4.17.

The proposals for UMTS air interface received by the milestone were grouped into five concept groups in ETSI in June 1997. For the sake of completeness, we present an overview of these candidate schemes. You can find a more in-depth description in Ref. [HOLMA00].

Wideband CDMA (W-CDMA).   A synopsis of W-CDMA is provided here. The W-CDMA technology is covered in more detail in "Summary of main parameters in W-CDMA" in Sec. 4.5.4.

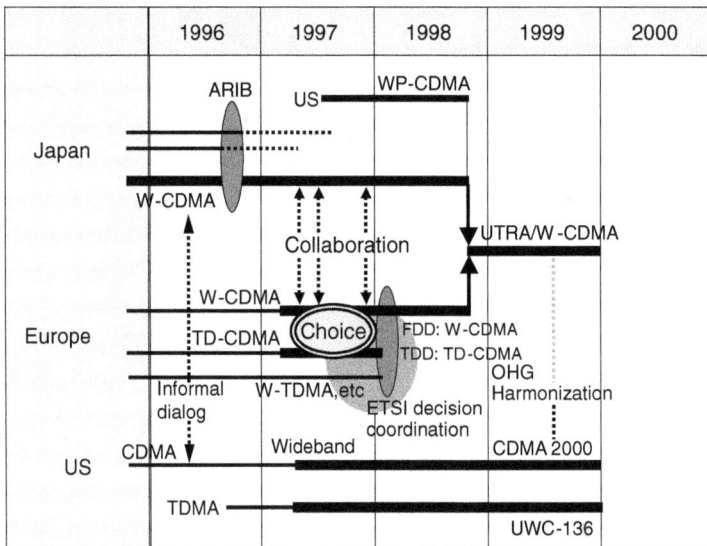

Figure 4.17   History of 3G radio access technology (3GSM World Focus 2001).

The W-CDMA concept achieved the greatest support, one of the technical motivating issues being the flexibility of the physical layer for accommodating different service types (data rates) simultaneously on a single carrier. This was considered to be an advantage, especially with respect to low and medium bit rates. The enhancements covered include transmit diversity, adaptive antenna operation, and support for advanced receiver structures.

The main parameters of the current scheme are based on the uplink of the FMA2 scheme and on the downlink of the ARIB's wideband CDMA. Also, contributions from other proposals and parties have been incorporated to further enhance the concept.

Wideband CDMA has several attractive properties over second-generation narrowband CDMA. A radio channel is reflected from obstacles such as houses and hills and the receiver gets several copies of the signal over different radio paths (multipath phenomenon). Wideband CDMA can resolve these signals and is able to combine them to improve system performance. Furthermore, wideband CDMA is able to average interference from other users more effectively. This is especially important for high-rate data users. In addition, wideband CDMA signal structure is well-suited for multiplexing of different services with different quality of service requirements. It can also vary the bit rate and service parameters on frame-by-frame basis. The multirate concept utilizes variable spreading for low and medium bit rates, and multicode for the highest bit rates. W-CDMA supports several carrier bandwidths—5, 10, and 20 MHz. However, the main bandwidth alternative is 5 MHz providing bit rates up to 384 kbit/s with appropriate coding for BER 10^{-3} and 10^{-6}.

Furthermore, to improve performance against the fading channel, fast power control is used both in the uplink as well as in the downlink.

Coherent detection is used in the uplink. This contributes to a 3-dB increase in performance compared to noncoherent detection. Coherent detection is performed with the help of user-dedicated pilot symbols, which further eases the implementation of adaptive antennas.

W-CDMA supports asynchronous cell sites as an asynchronous network does not require any external timing reference such as a GPS. GPS would increase costs since it requires a separate receiver and, for indoor solutions, an outdoor antenna installation to catch the reference signal. Thus, wideband CDMA with an asynchronous network is a cheaper and more reliable solution compared to a synchronous network.

In a high-capacity network there will be several carrier frequencies and hierarchical cell structures. W-CDMA supports seamless interfrequency handover between two different carrier frequencies. Seamless interfrequency handover has two implementation alternatives—compressed mode/slotted mode and dual receiver [OJANPE97].

Among the drawbacks of W-CDMA, it is recognized that in an unlicensed system in the TDD band—with the continuous transmit and receive operation—pure W-CDMA technology does not facilitate interference avoidance techniques in cordlesslike operating environments.

Wideband TDMA (W-TDMA).   W-TDMA is basically the FRAMES FM1 without spreading. The W-TDMA concept aims at a high system capacity increase with the aid of interference averaging over the operator bandwidth, with fractional loading and frequency-hopping.

The basic TDMA operation supports two basic burst types of length 1/16th and 1/64th, for high and low data rates, respectively. Advanced resource allocation algorithms, such as dynamic channel allocation, link adaptation, and low reuse sizes add to the basic system features to facilitate adaptation into different carriers to interference ratios required by different services. In addition, the W-DTMA concept supports adaptive antennas, TDD operation, as well as joint detection of other-cell interference suppressions [RANTA96].

The main limitation associated with the system is considered to be the range with respect to low bit-rate services. This is due to the fact that in a TDMA-based operation, the slot duration is, at a minimum, only 1/64th of the frame timing, which results in either very high peak power or a low average output power level. This means that for large ranges with, for example, speech, the W-TDMA concept would not be competitive on its own, but would require a narrowband option as a companion.

Hybrid CDMA/FDMA [ETSI-UMTS].   FRAMES FMA 1 with spreading is based on the hybrid CDMA/TDMA concept, also called joint detection [KLEIN93, NAβHAN93, JUNG94]. The role of CDMA is to multiplex the different channels within a time slot, thus facilitating the transmission of low and high bit-rate services within the same RF bandwidth. Its TDMA burst structure introduces a midamble for channel estimation whereas the CDMA concept is applied on top of the TDMA structure for additional flexibility.

Original parameters of joint detection CDMA—frame length of 6 ms and 12 slots per frame [KLEIN93]—have been aligned with FMA1 without spreading to be backward compatible into GSM, i.e., 4.615 ms and 8 slots per frame, respectively. Specifically, short orthogonal spreading codes of length 16 are used. FMA1 with spreading uses either multicode or multislot to increase the user bit rate. With each slot maximum 16 codes can be pooled together for one user. However, typically only 10 of the 16 codes can be used due to interference limitation [OJANPE97].

Additional features of TDMA/CDMA include adaptive antennas support, operation in the TDD mode, and *dynamic channel allocation* (DCA).

Orthogonal frequency division multiple access (OFDMA) [ETSI-UMTS]. The OFDMA group was based on the OFDMA technology with inputs mainly from Telia, Sony, and Lucent. The system concept was shaped by the discussions about OFDMA in other forums such as the Japanese standardization forum, ARIB.

The basic system operation makes use of slow frequency-hopping with TDMA and OFDM multiplexing, whereas a 100-kHz wideband slot from the OFDM signal is used as the basic resource unit. Higher rates are achieved by allocating several bandslots, creating a wideband signal. System diversity is also provided by dividing the information among several bandslots over the carrier.

The OFDMA basic system is enhanced with transmit diversity, multiuser detection for interference cancellation, and adaptive antenna solutions.

OFDM is a method of digital modulation in which a signal is split into several narrowband channels at different frequencies. The difference between OFDM and other FDD systems lies in the way in which the signals are modulated and demodulated. Priority is given to minimizing the interference, or crosstalk, among the channels and symbols comprising the data stream. Less importance is placed on optimizing individual channels.

The introduction of OFDM into the cellular world has been driven by two main benefits:

- *Flexibility.* Each transceiver has access to all subcarriers within a cell layer.

- *Easy equalization.* OFDM symbols are longer than the maximum delay spread resulting in a flat fading channel which can be easily equalized.

The main drawback of OFDM is the high peak-to-average power, which is of main importance for the mobile station as well as in long-range applications [OJANPE98].

Opportunity driven multiple access (ODMA). Opportunity driven multiple access, proposed by Vodafone is basically a relaying protocol, not a pure multiple access as such. ODMA was later integrated in the W-CDMA and W-CDMA/FDMA concept groups and was not considered in the selection process as a concept on its own.

Briefly, in ODMA a terminal beyond the reach of the cell coverage uses another terminal as a relay to transmit its traffic to the base station. The ODMA operation has been considered to be feasible with the TDD operation where both reception and transmission are in the same frequency band. If it was desired to implement ODMA with FDD, it would require terminals either to receive in their normal transmission band or to transmit in their normal reception band. From an implementation point of view such a requirement makes ODMA undesirable for FDD.

All these proposed technologies were basically able to fulfill the UMTS requirements. It soon (January 1998) became evident in the selection process that W-CDMA and TDMA/CDMA were the strongest candidates [ETSI-UMTSb]. For the UTRA air interface [ETSI-UMTSa], W-CDMA was selected for operation with the paired frequency bands, i.e., for the FDD[4] operation, and TDMA/CDMA for operation with unpaired spectrum allocation, i.e., for the TDD[5] operation. This third-generation mobile system's standardization procedure proceeded within ETSI until the work was handed over to the third-generation partnership project (3GPP) in early 1999.

The creation of 3GPP was an initiative to create a single forum for UMTS standardization for achieving a common UMTS specification, taking into account the parallel standardization efforts in several regions around the world. 3GPP is the joint standardization project of the national standardization bodies from Japan (ARIB), Europe (ETSI), Korea (Committee T1), Japan (Telecommunications Technology Committee-TTC), and the United States (T1P1), which have all agreed to work together for the development of a third-generation telecommunications mobile system based on the evolved GSM core networks [HJELM00]. Figure 4.18 from Ref. [3GPP] symbolizes the 3GPP's partnership.

3GPP also includes market representation partners—GSM Association, UMTS Forum, Global Mobile Suppliers Association, IPv6 Forum, and Universal Wireless Communications Consortium (UWCC). In Ref. [3GPP] there are up-to-date links to all participating organizations.

Worldwide, the process of standardizing third-generation radio access has proceeded fastest in Japan. The Japanese standards body ARIB selected wideband CDMA (W-CDMA) to be the Japanese candidate scheme for the IMT-2000 air interface. Furthermore, the biggest Japanese operator NTT DoCoMo ordered prototypes for a W-CDMA trial system. Japan was the first to select and announce specific plans to introduce wideband radio networks based on W-CDMA technology. In fact, in Japan NTT DoCoMo was very successful with its *i-mode* technology (see Sec. 4.5.7) and later with its 3G services using W-CDMA.

---

[4]Frequency Division Duplex (FDD) uses two separate frequencies, one for the uplink and one for the downlink communications. This method causes less interference but it requires a pair of separated frequency bands to be allocated which may not be possible since bandwidth is limited in wireless communication.

[5]Time Division Duplex (TDD) uses the same frequency for uplink and downlink communications. The two ends take turns to transmit and hence the subscriber and base station must agree upon the time slot for uplink and downlink transmission. Since we want to focus more on W-CDMA, we will not discuss much about TDD.

Figure 4.18   Third-generation partnership project.

The U.S. Telecommunications Industry Association (TIA) was a little bit slower to react but faster to agree and decided on an IS-95 backward compatible W-CDMA. They called their W-CDMA's view cdmaOne, which is also referred to as cdma2000. Korea, on its part, is still considering two wideband CDMA technologies, one similar to W-CDMA and the other similar to cdma2000 (TTA—Telecommunications Technology Association—II and I respectively) [OJANPE98]. See Fig. 4.19.

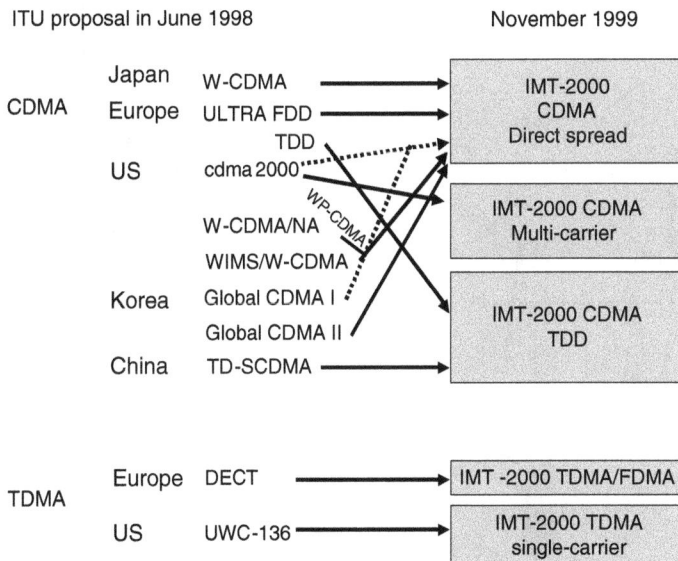

Figure 4.19   Results of harmonization (GSM World Focus 2001).

4.5.2   Radio access network: the UTRAN architecture

UTRAN has been defined as an access network. This means that the radio interface-independent functions are outside the scope of UTRAN specifications and handled by the core network.

The UTRAN consists of one or more radio network subsystems (RNSs), which in turn consist of base stations (called node Bs) and radio network controllers (RNCs) (see Figs. 4.20 and 4.21). A node B may serve one or multiple cells. Mobile stations are termed *user equipments* (UEs) and they are likely to be multimode to enable handover between FDD and TDD modes, and GSM as well.

The main function of node B is to perform the air interface L1 processing (channel coding and interleaving, rate adaptation, spreading, and the like). It also performs some basic radio resource management operation as the inner-loop power control. It corresponds to the GSM base station.

The UE consists of two parts—the *mobile equipment* (ME) used for radio communication over the Uu interface, and the *UMTS subscriber identity module* (USIM) which is a smart card that holds subscriber

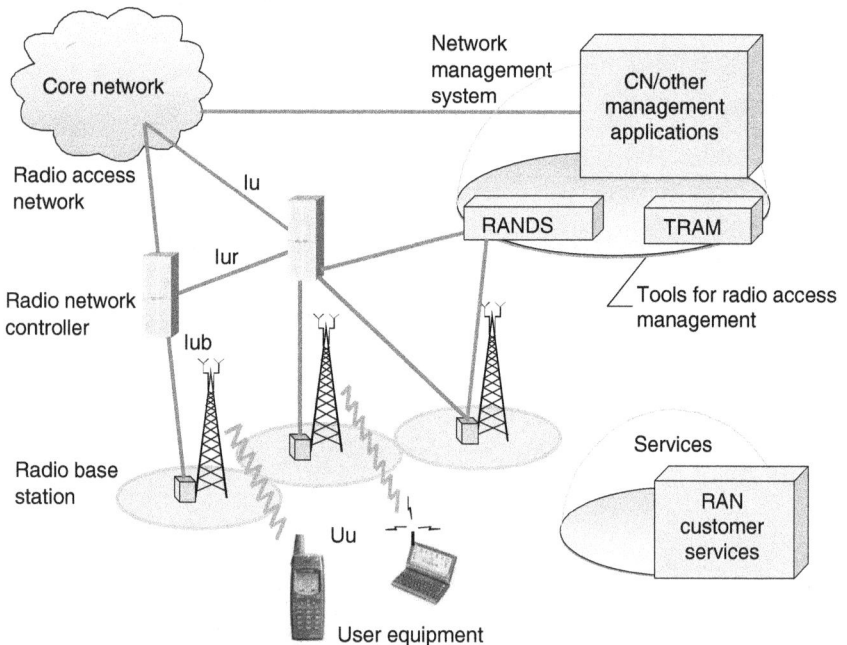

Figure 4.20   Radio access architecture. (*Source: www.ericsson.com.*)

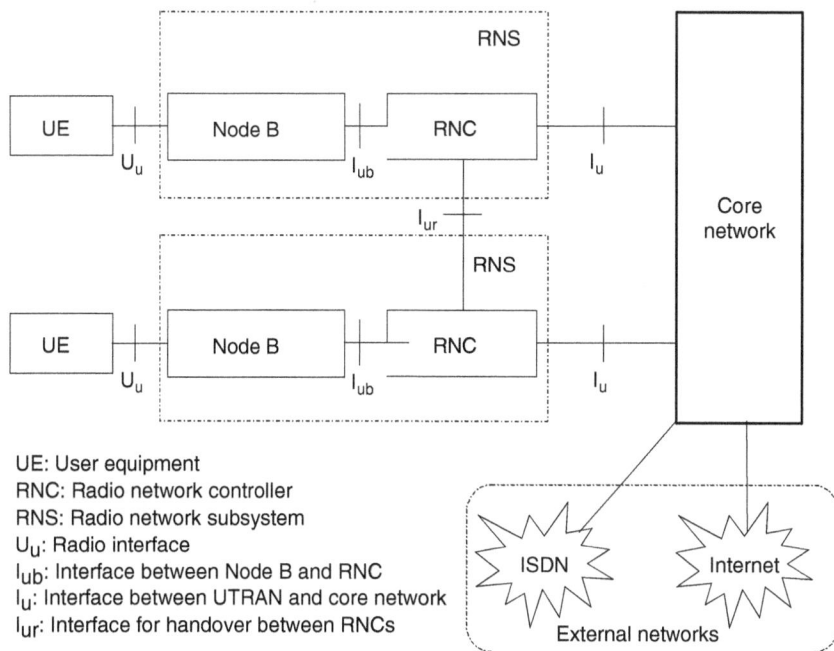

UE: User equipment
RNC: Radio network controller
RNS: Radio network subsystem
U_u: Radio interface
I_{ub}: Interface between Node B and RNC
I_u: Interface between UTRAN and core network
I_{ur}: Interface for handover between RNCs

Figure 4.21   UTRAN system architecture.

identity and performs authentication algorithms and stores authentication and encryption keys.

The protocols within UTRAN include (Fig. 4.22)

- *Node B application protocol (NBAP).*   It is responsible for the allocation and control of radio resources to node Bs.

- *Radio network subsystem application protocol (RNSAP).*   It is responsible for the coordination of radio resources between node Bs in the neighboring RNCs (i.e., in support of links during soft-handover)

- *Radio access network application protocol (RANAP).*   It is used to support signaling across the Iu interface. In particular, it supports the transfer of layer 3 messages between the UE and the core network. It is also used during the establishment of layer 3 connections between the UTRA and the core network.

The standardization of UTRA is scheduled to progress by 3GPP on the basis of annual releases. 3GPP release 99 specifies an ATM implementation on the Iub interface but it is likely that in the next release a migration to IP will be adopted.

Figure 4.22 shows a simplified version of the protocols running between a UE and the UTRAN. The transport channels carry control

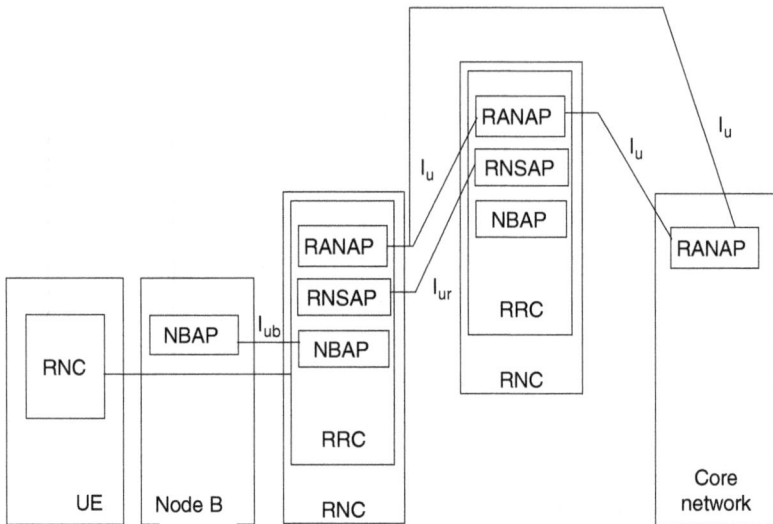

Figure 4.22 Resource control signaling protocols.

plane or user plane data between the UE and RNC, mapping onto physical channels on the air (Uu) interface allocated to the *radio resource control* (RRC) and ATM AAL2 connections over the Iub interface. The MAC layer and RLC reside in the RNC. The *frame protocol* (FP) is responsible for the relaying of transport channels between UE and RNC via the node B. This protocol stack is common to both FDD and TDD modes with some minor differences.

4.5.3 Core network

The core network is responsible for switching and routing calls and data connections to external networks (see Fig. 4.20). It comprises transport "pipes" for information flow—nodes that route traffic and nodes that provide call control intelligence.

There are three basic solutions for the core network to which W-CDMA radio access networks can be connected (see Fig. 4.23). The basis of the second-generation systems has been either the GSM core network or one based on IS-41. Both will naturally be important options in 3G systems. An alternative is the general packet radio service with an all-IP-based core network. The most typical connections between the core networks and the air interfaces are shown in Fig. 4.23. Other connections are expected to appear in the future. It is expected that the operators will remain with their second-generation core network for voice services and will then add packet data functionalities on top of that. Later, it will be possible to use IP-based core networks for all services.

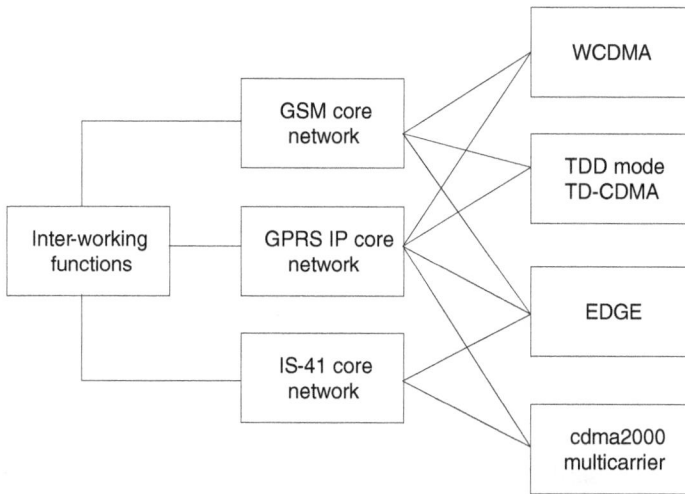

Figure 4.23   Core network relation to third-generation air-interface alternatives.

4.5.4   Radio interface protocols

The radio interface protocols are needed to set up, reconfigure, and release radio bearer services (including the UTRA FDD/TDD service).

In the UTRA FDD radio interface, the layer 2 (i.e., data link layer) is split into sublayers (see Fig. 4.24). The control plane layer 2 contains two sublayers—MAC and RLC. In the user plane, in addition to MAC and RLC sublayers, two additional service-dependent protocols exist—*packet data convergence protocol* (PDCP) and *broadcast/multicast control* protocol (BMC). Layer 3 (i.e., network layer) consists of one protocol, *radio resource control* (RRC), which belongs to the control plane. The PDCP, which exists only in the user-plane, contains compression methods that are needed to get better spectral efficiency for services requiring IP packets to be transmitted. The BMC, which exists only in the user-plane, is designed to adapt broadcast and multicast services on the radio interface.

The overall radio interface protocol architecture is shown in Fig. 4.24.

- The physical layer offers service to the MAC layer via transport channels that are characterized by how and with what characteristics are data transferred.

- The MAC layer offers services to RLC by means of logical channels that are characterized by the type of data transmitted. In the MAC layer, the logical channels are mapped onto transport channels. The MAC layer is also responsible for selecting an appropriate format for each transport channel according to the instantaneous source rate of the logical channels. Other functions of the MAC layer are: priority

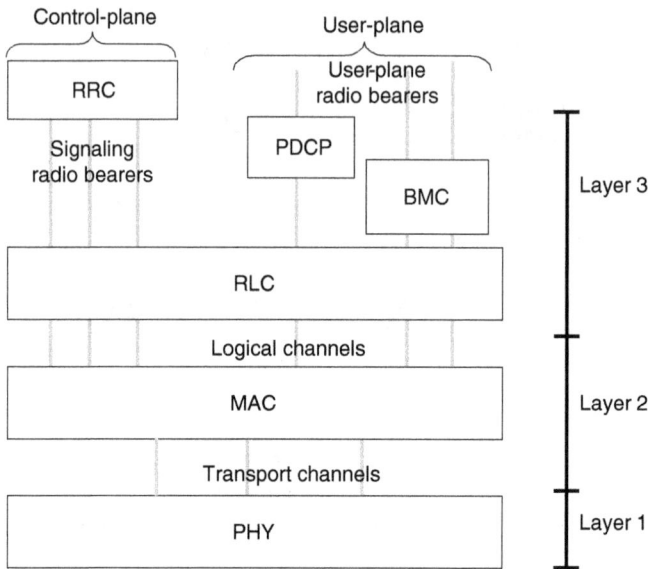

Figure 4.24 UTRA FDD radio interface protocol architecture.

handling between data flows of one UE, priority handling between UEs by means of dynamic scheduling, identification of UEs, multiplexing/demultiplexing of higher protocol data units, traffic volume monitoring, ciphering, and transport channel type switching.

- RLC offers services to higher layers that describe how RLC handles the data packets. The RLC protocol can operate in three modes—transparent, unacknowledged, and acknowledged mode. In the transparent mode, no overhead is added to the higher layers, erroneous protocol data units can be discarded if marked erroneous. In the unacknowledged mode, no retransmission protocol is in use and data delivery is not guaranteed. In the acknowledged mode, an automatic repeat request mechanism is used for error correction. The main functions of the RLC layer are: segmentation and reassembly of variable-length PDUs of higher layers, padding, transfer of user data, error correction, protocol error detection and correction, in-sequence delivery of higher layer PDUs, duplicate detection, flow control, sequence number check, ciphering, suspend/resume of data transfer.

- The RRC layer offers services to higher layers (to the nonaccess stratum) used by the UE side and by RANAP in the UTRAN side (see Fig. 4.21). RRC messages carry all parameters required to setup, modify, and release layer 2 and its protocol's entities. RRC messages carry all higher layer signaling and the mobility of user equipment in the connected mode is controlled by RRC signaling (measurements, handovers, cell updates, and the like). The RRC has a long list of functions to perform—broadcast of system information, paging, establishment, maintenance and release of RRC connections between UEs and UTRAN, control of radio bearers, control of security functions, integrity protection of signaling messages, UE measurement reporting and its control, RRC connection mobility functions, open-loop power control, cell broadcast related functions, and the like.

The data transfer services of the MAC layer are provided on the logical channel. Each logical channel is defined by the type of information transferred. There are control channels and traffic channels:

- *Broadcast control channel (BCCH).* A downlink channel for broadcasting system control information

- *Paging control channel (PCCH).* A downlink channel that transfers paging information

- *Dedicated control channel (DCCH).* A channel that transmits dedicated control information between UEs and the network

- *Common control channel (CCCH).* A channel for transmitting control information between UEs and the network

- *Dedicated traffic channel (DTCH).* A point-to-point channel dedicated to one UE for the transfer of user information

- *Common traffic channel (CTCH).* A point-to-multipoint unidirectional channel for the transfer of user information for a group of UEs

The control interfaces between RRC and all the low-layer protocols are used by the RRC layer to configure characteristics of the lower-layer protocol entities, including parameters for the physical, transport, and logical channels. See Figs. 4.24, 4.25, and 4.26.

Overview of the CDMA technology. The CDMA technology [PICKHO82, VITERBI, SOSL85a, Hol82, Dix84, PSM82, SH85] is used in commercial cellular 2G and 3G communications networks to make a better use of the radio spectrum than other technologies. Its unique features enable many more users to share the common RF spectrum than do alternative technologies. In CDMA each user is assigned a unique code sequence which is used to encode the information bearing signal. These codes are wideband, noise-like (pseudo-noise) signals. To allow multiple simultaneous transmissions from different users, code sequences should be pairwise orthogonal (the inner product of any two distinct sequences is 0).

The rate of the code signal, known as the *chip rate* (CR), changes at a much faster rate than the information bits. The bandwidth of the code signal is also chosen to be much larger than the bandwidth of the

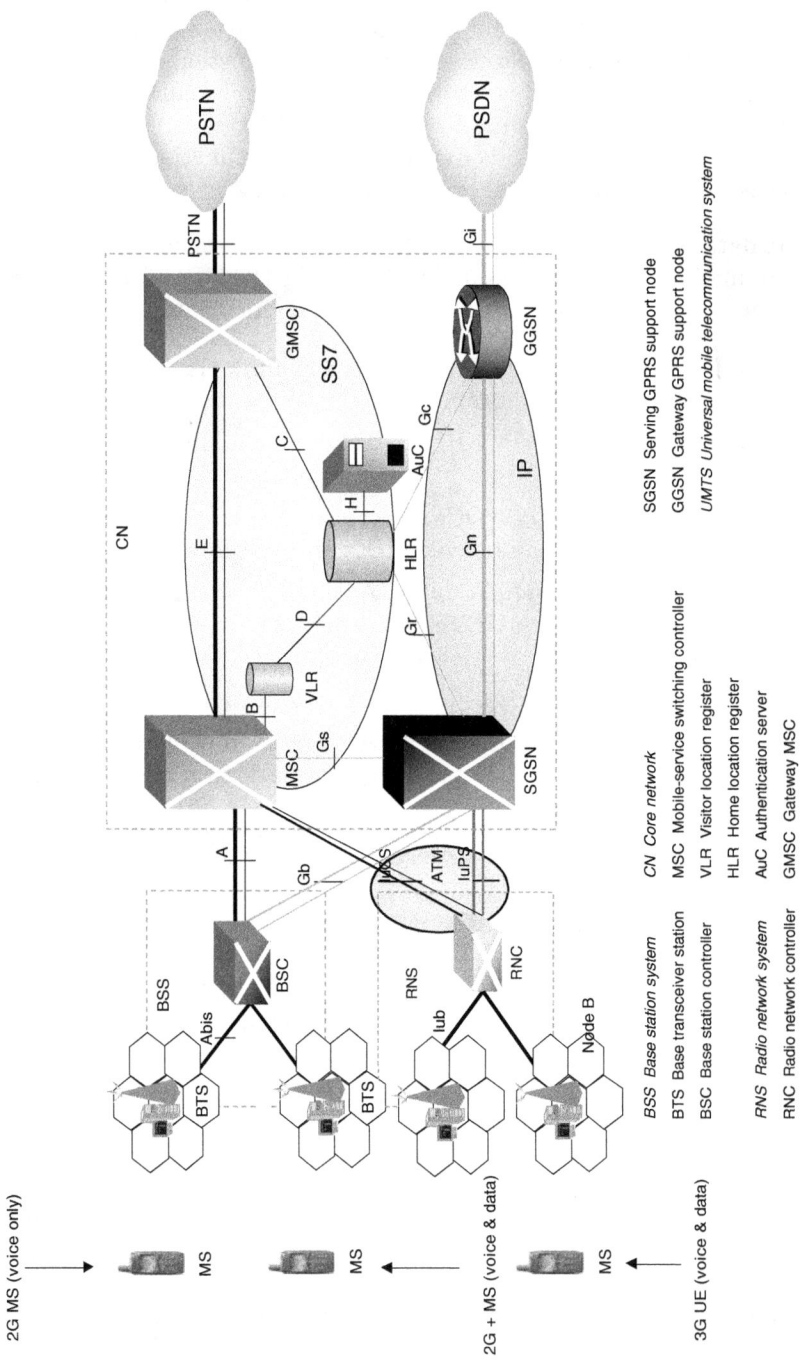

Figure 4.25 3G rel99 architecture (UMTS)—3G radios.

SGSN Serving GPRS support node
GGSN Gateway GPRS support node
UMTS *Universal mobile telecommunication system*

CN Core network
MSC Mobile-service switching controller
VLR *Visitor location register*
HLR *Home location register*
AuC Authentication server
GMSC Gateway MSC

BSS Base station system
BTS Base transceiver station
BSC Base station controller

RNS Radio network system
RNC Radio network controller

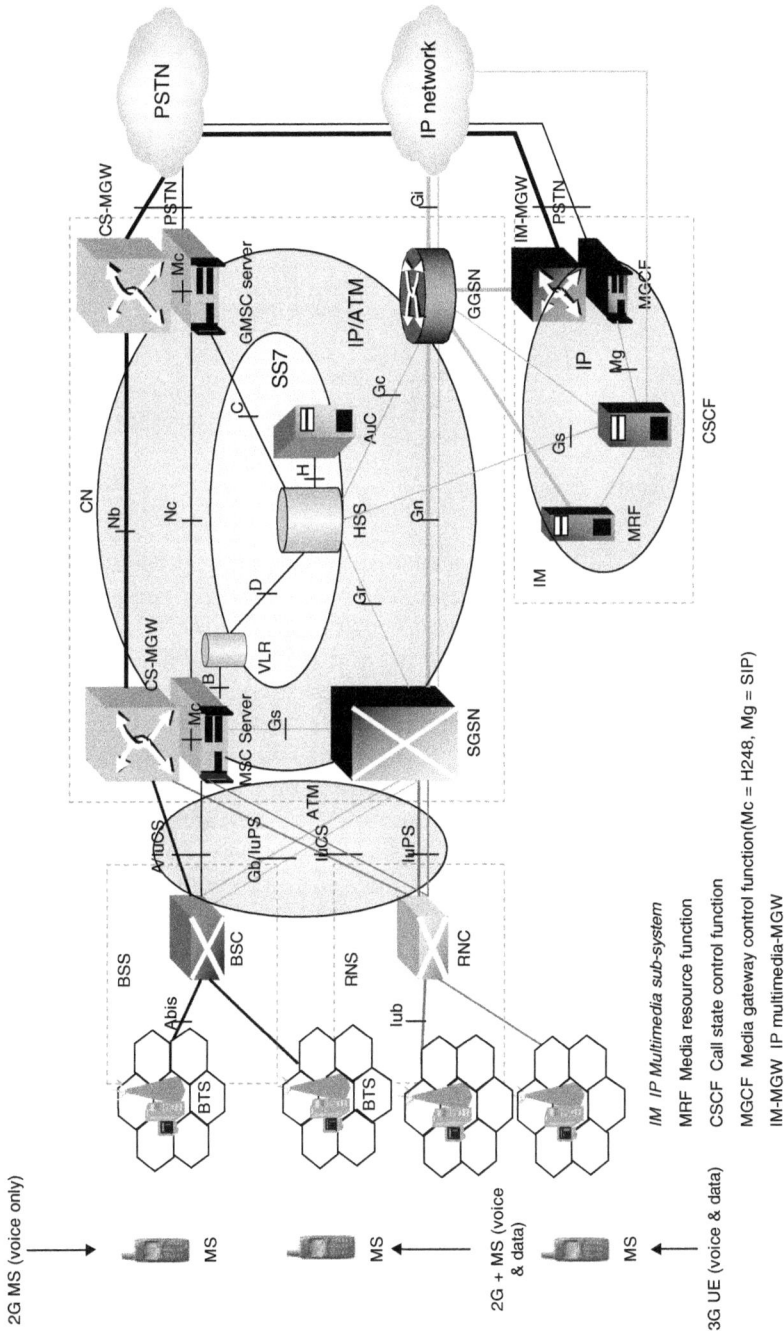

Figure 4.26 3G rel5 architecture (UMTS)—IP multimedia.

IM IP Multimedia sub-system
MRF Media resource function
CSCF Call state control function
MGCF Media gateway control function(Mc = H248, Mg = SIP)
IM-MGW IP multimedia-MGW

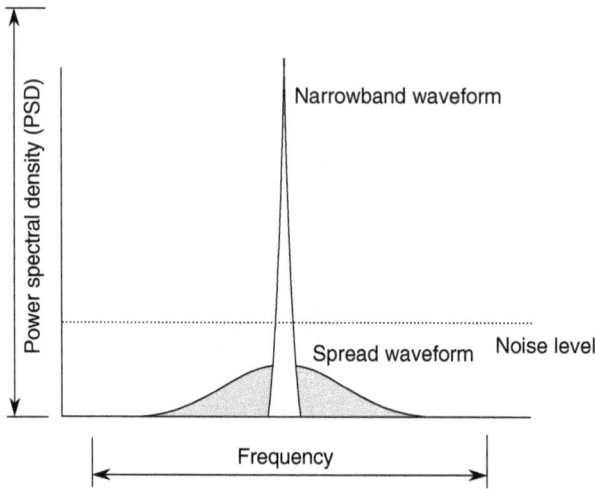

Figure 4.27 SS compared to narrow band.

information-bearing signal. As a result, the encoding process enlarges (spreads) the spectrum of the signal (see Fig. 4.27). This process is commonly known as *spread-spectrum modulation*. The resulting signal is also called a spread-spectrum signal, and this is why CDMA is often referred to as *spread-spectrum multiple access* (SSMA). Most SS systems transmit a radio frequency signal bandwidth as wide as 20 to 254 times the bandwidth being sent. (Some SS systems have even employed radio frequency bandwidths 1000 times their information bandwidth.)

The code lengths are limited to minimize the cell search time and facilitate the implementation of advanced multiuser receiver techniques. The choice of the chip rate is made on a few parameters such as spectrum deployment scenarios, dual mode terminal implementation and, of course, maximum data rate desired. In W-CDMA, for example, the CR was first set to 4.096 megachips per second (Mcps) mainly for backward compatibility with GSM, producing a minimum channel separation of 4.99712 MHz [OJANPE98]. The CR has now been relaxed to 3.84 Mcps in order to simplify the development of the RF part of multimode terminals (DS and multicarrier FDD, TDD).

A fundamental aspect of spread spectrum systems is the *processing gain*. Processing gain is defined as the ratio of transmission and information bandwidth:

$$G_p = \frac{BW_{\mathrm{RF}}}{R_{\mathrm{INFO}}}$$

where BW_{RF} = radio frequency bandwidth and R_{INFO} = information rate.

In practice, processing gain is used to describe the received signal fidelity gained. The numerical advantage is obtained from Claude Shannon's equation:

$$C = W \times \log\left(1 + \frac{S}{N}\right)$$

where C = channel capacity
$\quad W$ = bandwidth
$\quad S$ = signal power
$\quad N$ = noise power

From this equation the result of increasing the bandwidth becomes apparent. By increasing W in the equation, S/N may be decreased without a decreased performance. The process gain is what actually provides increased system performance without requiring a high S/N.

The processing gain is basically the "spreading factor," which gives CDMA systems the robustness against self-interference that is necessary for frequency reuse over geographically close distances.

For spread spectrum systems, it is therefore advantageous to have a processing gain as high as possible, at the cost, however, of an increased transmission bandwidth (by the amount of the processing gain).

Types of spread spectrum communications. There exist different techniques to spread a signal—direct-sequence (DS), frequency-hopping (FH), time-hopping (TH), multicarrier CDMA (MC-CDMA) [YEE94], or combinations of them (e.g., hybrid DS/FH). Each of these schemes is discussed in more detail in the following sections, with focus on the two popular ones—direct-sequence and frequency-hopping.

Direct-sequence. Direct-sequence is the most popular spread spectrum technique. The *direct-sequence spread spectrum* (DSSS) process is achieved by multiplying a radio frequency carrier with a pseudo-random noise code (PNcode). Each bit time is subdivided into m short intervals called chips (e.g., 64 or 128 chips per bit). A PNcode is a unique m-bit chip sequence valued −1 and 1 (polar) or 0 and 1 (nonpolar) and has noise-like properties. Pseudo-noise is produced at a much higher frequency than the data that are to be transmitted. Since the code has a higher frequency, it has a large bandwidth that spreads the signal in the frequency plane (i.e., spreads its spectrum).

The generation of PNcodes is relatively easy; a number of shift-registers is all that is required. When the length of shift-registers is n, the following can be said about the period N_{DS} of the previously mentioned code-families:

$$N_{DS} = 2^n - 1$$

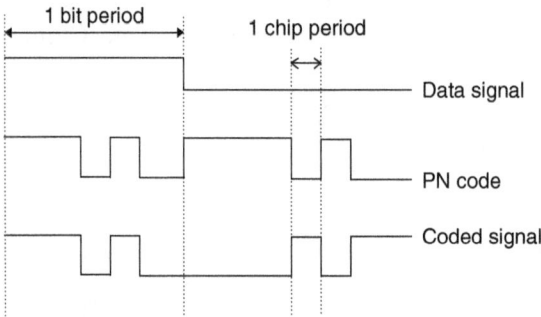

Figure 4.28 Direct-sequence spreading.

Signals generated with this technique appear as noise in the frequency domain. The wide bandwidth provided by the pseudo-noise code allows the signal power to drop below the noise threshold without losing any information. For this reason it is easy to introduce a large processing gain in direct-sequence systems.

In DS systems the length of the code (N_{DS}) is the same as the spreading-factor. As a consequence:

$$G_p \,(DS) = N_{DS}$$

This can also be seen from Fig. 4.28, which illustrates how the PNcode is combined with the data-signal (in this example $N_{DS} = 7$). The bandwidth of the data signal is multiplied by a factor of N_{DS}. The power contents however remain the same, which results in lower power spectral density levels, as depicted in Fig. 4.29.

At the receiver end, the received signal is multiplied again by the same (synchronized) PNcode. This is basically the dispreading of the signal. Dispreading completely removes the code from the signal and the original data-signal is recovered.

To put this into context, we demonstrate the basic operations of spreading and dispreading for a DS-CDMA system, using a simple example, as follows.

Figure 4.30 shows a simple, idealized CDMA encoding/decoding example. Suppose that the rate at which original data bits reach the CDMA encoder defines the unit of time; that is, each original data bit to be transmitted requires one bit-slot time. Let d_i be the value of the data bit for the ith bit slot. Each bit slot is further subdivided into M minislots; in Fig. 4.30, $M = 8$, although in practice M is much larger. The CDMA code consists of a sequence of M values c_m, $m = 1,...,M$, each taking a +1 or −1 value. In the example in Fig. 4.30, the 8-bit CDMA code being used by the sender is (1, 1, 1, −1, 1, −1, −1, −1).

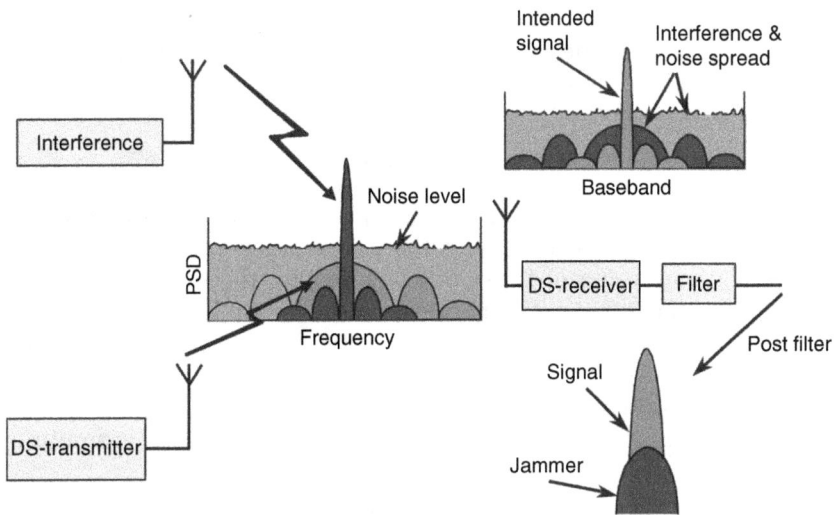

Figure 4.29   DS-concept before and after dispreading.

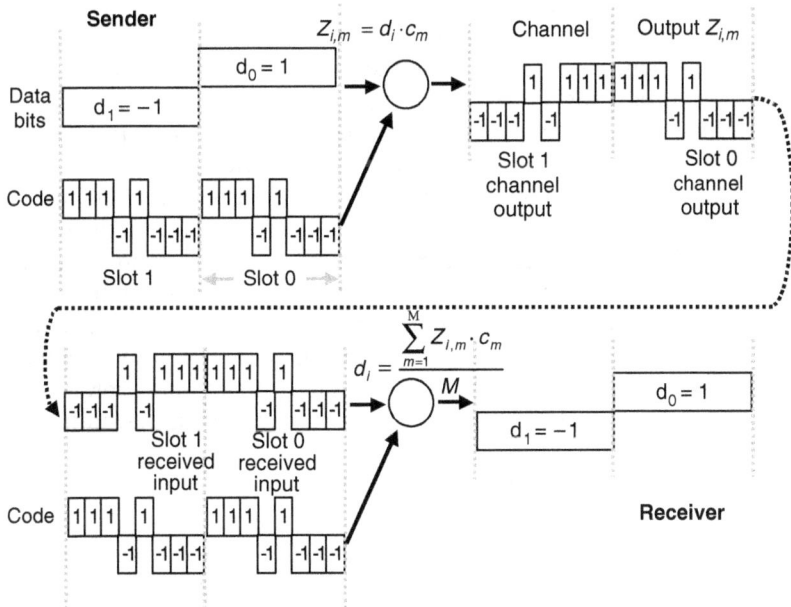

Figure 4.30   A simple CDMA example: sender encoding, receiver decoding.

To illustrate how CDMA works, let us focus on the ith data bit, d_i. For the mth minislot of the bit-transmission time of d_i, the output of the CDMA encoder, $Z_{i,m}$, is the value of d_i multiplied by the mth bit in the assigned CDMA code, c_m:

$$Z_{i,m} = d_i \times c_m \qquad (4.1)$$

During dispreading we multiply the spread user data/chip sequence, bit duration by bit duration, with the very same 8-code chips we used during the spreading of these bits. As shown, the original user bit sequence has been recovered perfectly, provided we have (as shown in Fig. 4.30) also perfect synchronization between the spread user signal and the (di)spreading code.

In a simple world, with no interfering senders, the receiver would receive the encoded bits, $Z_{i,m}$, and recover the original data bit, d_i, by computing:

$$d_i = \left(\frac{1}{M}\right) \sum_{m=1,M} (Z_{i,M} \times c_m) \qquad (4.2)$$

The reader might want to work through the details of the example in Fig. 4.30 to see that the original data bits are indeed correctly recovered at the receiver using Eq. (4.2).

The world is far from ideal, however, and as noted earlier, CDMA must work in the presence of interfering senders that are encoding and transmitting their data using a different assigned code. But how can a CDMA receiver recover a sender's original data bits when these data bits are being tangled with bits being transmitted by other senders? CDMA works under the assumption that the interfering transmitted bit signals are additive, e.g., if three senders send a 1 value, and a fourth sender sends a −1 value during the same minislot, then the received signal at all receivers during hat minislot is a 2 (since $1 + 1 + 1 − 1 = 2$). In the presence of multiple senders, sender s computes its encoded transmissions $Z_{i,m}^s$ in exactly the same manner as in Eq. (4.1). The value received at a receiver during the mth minislot of the ith bit slot, however, is now the sum of the transmitted bits from all N senders during that minislot:

$$Z_{i,m}^* = \sum_{s=1,N} Z_{i,m}^s$$

If the codes of the senders are chosen carefully, each receiver can recover the data sent by a given sender out of the aggregate signal simply by using the sender's code in exactly the same manner as in Eq. (4.3):

$$d_i = \left(\frac{1}{M}\right) \sum_{m=1,M} (Z_{i,m}^* \times c_m) \qquad (4.3)$$

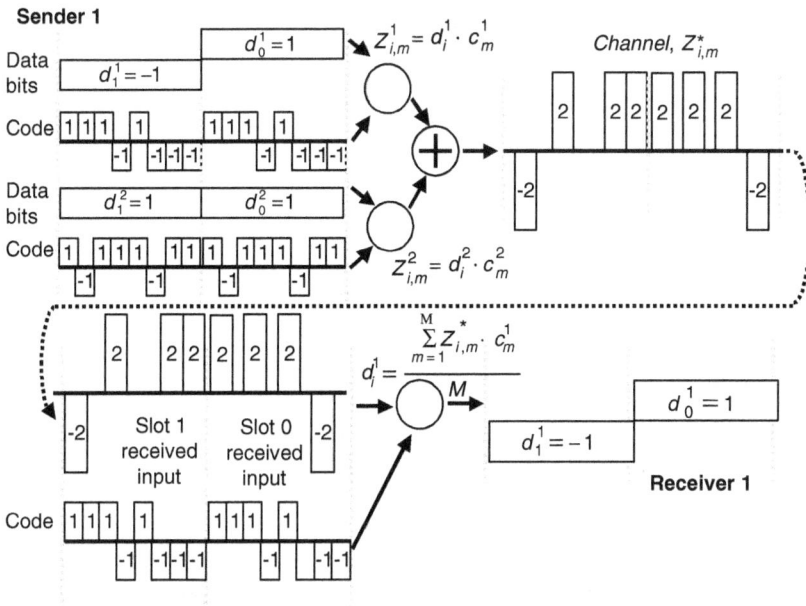

Figure 4.31   A two-sender CDMA example.

Figure 4.31 illustrates a two-sender CDMA example. The M-bit CDMA code being used by the upper sender is $(1, 1, 1, -1, 1, -1, -1, -1)$, while the CDMA code being used by the lower sender is $(1, -1, 1, 1, 1, -1, 1, 1)$. Figure 4.31 illustrates a receiver recovering the original data bits from the upper sender. Note that the receiver is able to extract the data from sender 1 despite the interfering transmission from sender 2.

Power-control in DS-CDMA systems.   The main problem with applying direct-sequence spreading is the so-called *near-far* effect (illustrated in Fig. 4.32). This effect appears when an interfering transmitter (MS_1 in Fig. 4.32) is much closer to the receiver than the intended transmitter (MS_2). Although codes in general are designed with low cross-correlation values (we cover the code properties in more depth in App. E "Pseudo-Random Noise Codes"), in our example the correlation between the received signal from the interfering transmitter can be higher than the correlation between the received signal from the intended transmitter. The result is that proper data detection may not be possible.

The solution to this problem is the proper control of transmitting power. In Fig. 4.32, it may happen that MS_2 at the cell edge suffers a path loss, say 60 dB above that of MS_1, which is near the base station BS. If there was no mechanism for MS_1 and MS_2 to be power-controlled to the same level at the base station, MS_2 could easily overshout MS_1

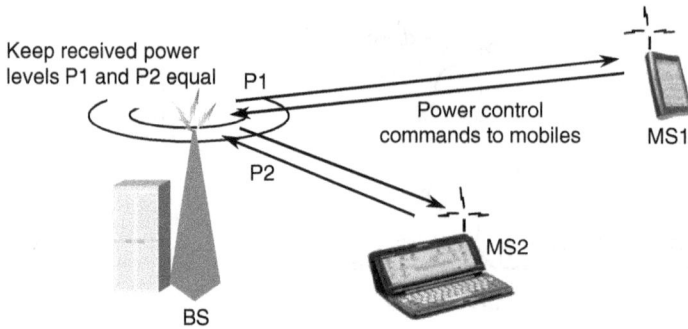

Figure 4.32 Near-far effect illustrated.

and thus block a large part of the cell, giving rise to the so-called near-far problem of CDMA. Accurate power control is therefore vital to the performance of a CDMA system, in particular on the uplink (base to mobile link) transmissions.

The main task of power control is to ensure that the base station receives all mobiles with equal strength. Besides, using minimum power level to obtain reliable communication with the most distant mobile in a single cell maximizes the overall network capacity.

Three techniques are commonly used for power control in the uplink—open-, closed-, and outer-loop. In open loop, mobiles measure the power received from the base station and adjust their transmit power accordingly—response time is a few microseconds. Open-loop power control mechanisms make, in essence, a rough estimate of path loss by means of a downlink beacon signal. Such a method would be, however, far too inaccurate. The prime reason for this is that fast-fading is essentially uncorrelated between the uplink and the downlink [SAUNDER99], which is due to the large frequency separation of the uplink and downlink band of the FDD mode. Open-loop power control is used in some CDMA systems to provide mobiles only a coarse power setting at the beginning of a connection.

In closed-loop power control, the base station performs frequent estimates of the received *signal-to-interference ratio* (SIR) from a specific mobile and compares it to a target SIR. If the measured SIR is higher than the target SIR, the base station will command the mobile station to lower its transmitting power; if it is too low the base station will command the mobile station to increase its power. Provided the mobile station has enough headroom to ramp the power up, only very little residual fading is left and the channel becomes an essentially nonfading channel as seen from the base station receiver. The measure-command-react cycle is executed at a rate of 1500 times per second (1.5 kHz) for each mobile station. It thus operates faster than any significant change of path loss that could possibly occur and even faster than the speed of fast-fading (Rayleigh) for low to moderate mobile speeds. Thus, closed-loop power control will prevent any power imbalance among all the uplink signals received at the base station.

The same closed-loop power control technique is also used in the downlink. Here the motivation is different—on the downlink there is no near-far problem due to the one-to-many scenario. All the signals within the cell originate from the base station to all mobiles. Though the fading removal is highly desirable from the receiver point of view, it comes at the expense of an increased average transmit power at the transmitting end. This means that a mobile station in a deep fade, (which will then use a large transmission power) will cause increased interference to other cells. It is thus desirable to provide a marginal amount of additional power to mobile stations at the cell edge as they suffer from increased other-cell interference.

In contrast to open- and close-power control mechanisms, outer-loop power control adjusts the target SIR setpoint in the base station according to the needs of the individual radio link and aims at a constant quality, usually defined as a certain target *bit error rate* (BER) or *frame error rate* (FER). Why should there be a need for changing the target SIR setpoint? The required SIR for, say, FER = 1 percent depends on the mobile speed and the multipath profile. If we were to set the target SIR setpoint for the worst case, i.e., high mobile speeds, we would waste much of the capacity for those connections at low speeds. Thus, the best strategy is to let the target SIR setpoint float around the minimum value that just fulfils the required target quality. The outer-loop power control is illustrated in Fig. 4.33.

Outer-loop control is typically implemented by having the base station tag each uplink user data frame with a frame reliability indicator,

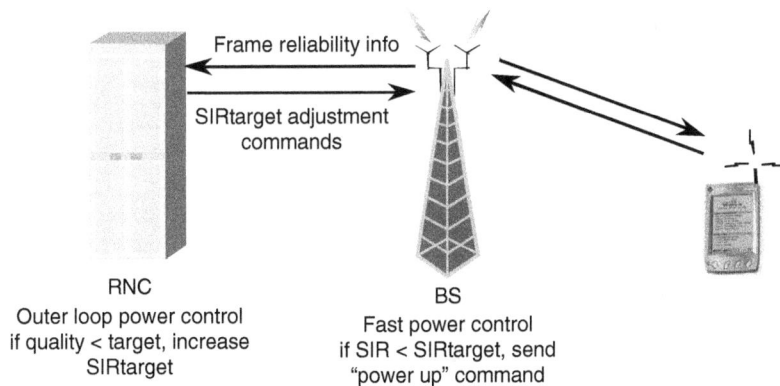

Figure 4.33 Outer-loop power control.

such as a CRC, and check the result obtained during the decoding of that particular user data frame. Should the frame quality indicator indicate to the system (*radio network controller*—see Sec. 4.5.2 for details) that the transmission quality is deteriorating, the system will in response command the base station to increase the target SIR setpoint by a certain amount.

Frequency-hopping. The bandwidth in frequency-hopping systems is divided into many possible broadcast frequencies. In FH systems, the wideband signal is generated by transmitting at different frequencies, hopping from one frequency to another according to a unique code sequence (an FH sequence of length N_{FH}). At any particular moment in time, the code determines the frequency at which the mobile shall transmit or receive. Hence, the only way for a user to transmit/receive is to use the proper code sequence. The process of frequency-hopping is illustrated in Fig. 4.34.

One can easily observe that both DS and FH systems generate wideband signals controlled by the code sequence generator. For DS systems, the code is the direct carrier modulation (direct-sequence) whereas in FH systems the code sequence dictates the carrier frequency (frequency-hopping).

A downside of FH, as opposed to DS, is that obtaining a high processing gain is hard. The processing rate is directly related to the hopping rate—the faster the "hopping-rate," the higher the processing gain.

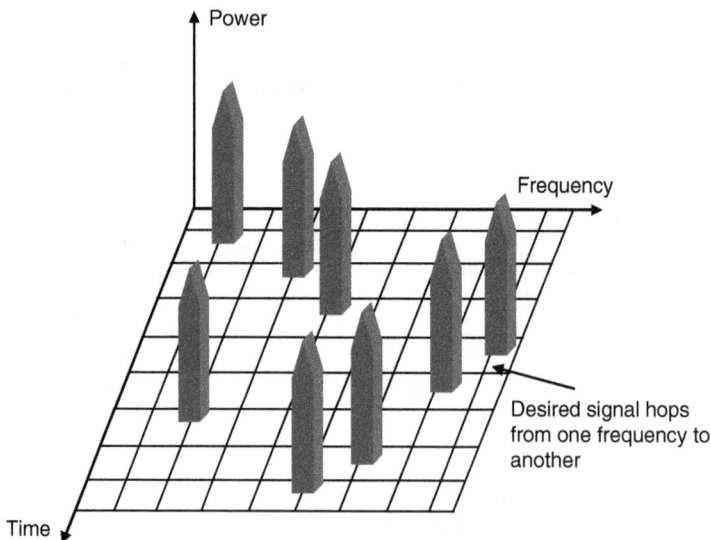

Figure 4.34   Illustration of the frequency-hopping concept.

Therefore, there is a need for a frequency synthesizer to perform fast-hopping over the carrier-frequencies.

On the other hand, FH is less affected by the near-far effect compared to a DSSS system. Frequency-hopping sequences have only a limited number of "hits" with each other. This means that if a near-interferer is present, instead of the whole signal only a limited number of "frequency-hops" will be blocked. From the "hops" that are not blocked it should be possible to recover the original data message.

In addition, due to the large number of frequencies used (quick hops), deciphering of the code is nearly impossible.

Hybrid system DS/FH.   The DS/FH SS technique is a combination of direct-sequence and frequency-hopping. One data bit is divided over frequency-hop channels (carrier frequencies). In each frequency-hop channel, one complete pseudo-noise code is multiplied with the data signal (Fig. 4.35).

As the frequency-hop sequence and the pseudo-noise codes are coupled, an address is a combination of pseudo-noise codes and the frequency-hop sequence. To bound the hit-chance (the chance that two users share the same frequency channel in the same time), the frequency-hop sequences are chosen in such a way that two transmitters with different FH-sequences share at most two frequencies at the same time (time-shift is random).

The principles of spread spectrum communication.   The spread spectrum technology allows for a number of unique features, such as

- Protection against interference. Coding enables a bandwidth trade for processing gain against interfering signals.

- Noise-effect reduction. Error-detection and correction codes can reduce the effects of noise and interference.

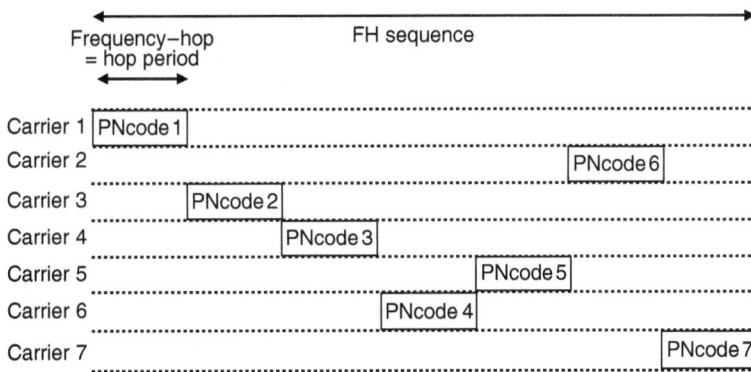

Figure 4.35   Hybrid DS/FH coding.

- Frequency, space, and time diversity (see "Smart Antennas and rake Receivers" in Sec. 4.6.1).

- Provision for privacy. Coding enables protection of signals from eavesdropping.

These properties of SS signals are further analyzed as follows:

Capacity gains. With existing networks nearing saturation, increased minutes of use, capacity-intensive data applications, and the steady growth of worldwide wireless subscribers, mean operators will have to find effective ways to accommodate increased wireless traffic in their networks at the lowest possible cost. Deploying new cell sites, for example, is not the most economical or efficient means of increasing capacity. Wireless carriers have begun to explore new ways to maximize the spectral efficiency of their networks and improve their return on investment. Smart antennas and CDMA have emerged as two of the leading spectrally-efficient technologies for achieving highly efficient networks that maximize capacity and improve quality and coverage. CDMA can infact outperform other non-CDMA-based technologies by a factor of 2, or better, on the number of users supported at any given time [VITERBI].

Universal frequency reuse. The goal of any wireless communication system is to deliver the desired energy (signal) to a designated receiver and also to minimize the undesired energy (interference) that it receives. This goal can be achieved by providing disjoint slots (frequency or time) to each user of a cell. But once these are fully assigned, the only way to provide more user capacity is by creating more cells. Then users in adjacent cells must also be provided disjoint slots; otherwise, their mutual interference will become intolerable for narrowband (nonspread) transmission. This leads to limited frequency reuse, where typically a slot is reused only once every nth cell (see Ref. [GSM92]). It also requires frequency plan revision and user channel reallocation every time a new cell is introduced.

With spread spectrum, universal frequency reuse applies not only to all users in the same cell but also to those in all other cells. The processing gain together with the wideband nature suggests a frequency reuse of 1 between different cells of a CDMA wireless system. Insertion of new cells as traffic intensity grows does not require a revision of the frequency plans of the existing cells. Equally significant, as more cells are added, is the fact that the network's ability to insert and extract energy at a given location is enhanced. Hence, the transmitted power levels of both the mobile user and the base station can be reduced significantly by exploiting the power control capabilities of both cells. In short, since the allocated resource of each user's channel is energy rather than time or frequency, interference control and channel allocations merge into a single approach.

CDMA's single cell frequency reuse capability and noncontiguous bandwidth requirement, together with its extended coverage range, simplifies RF planning and implementation. This allows an operator to invest in fewer cell sites, with faster deployment, ultimately giving the service provider increased and quicker access to revenues.

Multiple users can use the same band.   CDMA is a "spread spectrum" technology; as its name implies the use of different binary sequences allows several SS systems to operate independently of each other within the same band. If the systems parameters are chosen judiciously and also the right conditions exist, conventional users in the same band space will experience very little interference from SS users. This allows more signals to be packed into a band; yet with some extra interference added to all users, which is induced from each additional signal, conventional or SS. As we described earlier, since the signal is spread over a large frequency-band, the power spectral density becomes very small; hence, other communications systems do not suffer from this kind of an occurrence.

Hard to intercept.   The very foundation of the spreading technique is the code used to spread the signal. Without knowing the code it is impossible to decipher the transmission. Also, because the codes are so long, simply viewing the code would still not make it possible to resolve the code, hence interception is very hard to achieve.

Good antijam performance.   One of the most important features of SS is its ability to reject interference. At the first glance, it may be considered that SS would be most affected by interference. However, for any signal which is spread in the bandwidth, the bandwidth signal is equal to its original bandwidth plus the bandwidth of the local interference. Hence, the wider the interference bandwidth, the wider the output signal. Similarly, the wider the input signal, the less is its effect on the system. The latter property is attributed to the fact that the power density of the signal (after processing) is lower, and less power falls in the band pass filter.

When considering frequency-hopping in particular, it should be noted that carrier wave (CW) interference is the least effective jamming technique (opposite to direct-sequence). This is because the CW interference interferes with only one of the frequencies of the channel. It thus needs to only be e greater than that of the desired signal to cause an error. Here e is the decision threshold of the receiver's bit decision device. Hence, for a single CW, the expected errors would average one every n transmissions, where n is the number of available channels.

Hard to detect.   Spread spectrum signals are much wider than conventional narrow band transmissions (as stated earlier, of the order of 20 to 254 times the bandwidth of narrow band transmissions). Since the communication band is spread, SS signals can be transmitted at a much lower spectral power density than narrow band transmitters without being deterred by the background noise. This allows a spread spectrum and a narrow band signal to occupy the same bandwidth but with little to no interference (see antijammering characteristics in the following text).

When a signal is broadcast at such low power, it blends into the noise. An observer therefore would overlook it, and thus the signal is not subject to eavesdropping. Since the applied codes are, in principle, unknown to a hostile user, this also means that it is hardly possible to detect a message of another user.

Soft handoff.   Universal frequency reuse makes it possible for a mobile user to simultaneously "receive from" and "send to" two different base stations the same call. In the case of reception by the user, the signals from the two base stations can be combined to improve performance, as with multipath combining (see the next paragraph). In fact, one could regard the second base station signal as a delayed version of the first, generated actively and on purpose, rather than as a delayed reflection of the first caused by the environment.

For the reverse link, the two different base stations will normally decode the signals independently. Should they decode a given frame or message differently, it will be the switching center that receives inputs from all base stations and connects the cellular network to the landline-switched network, to arbitrate. In general terms, it is enough to state that the more reliable signal from the two base stations will prevail. Qualitatively, as a user moves from one cell to an adjacent one, this feature provides more reliable handoff between the two base stations. Quantitatively, through proper power control, soft handoff can result in more than double its capacity as well as its coverage area when the network is lightly loaded.

Receiver diversity: rake receiver and smart antennas.   The basic operation of the correlation receiver for CDMA is shown in Fig. 4.30. As shown, the correlation receiver integrates (i.e., sums) the resulting products (data × code) for each user bit. The result of multiplying the interfering signal with the own code and integrating the resulting products leads to interfering signal values lingering around 0. As can be seen, the amplitude of the own signal increases on an average by a factor of 8 relative to that of the user of the other interfering system, i.e., the correlation detection has raised the desired user signal by the spreading factor—in the example, 8—from the interference present in the CDMA system.

Both base stations as well as mobiles for W-CDMA use essentially this type of a correlation receiver. However, due to multipath propagation (and possibly multiple receive antennas), it is necessary to use multiple correlation receivers in order to recover the energy from all paths

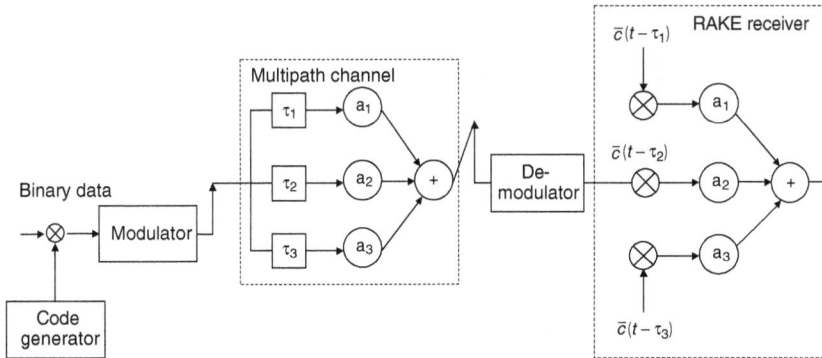

Figure 4.36   Principle of the rake receiver.

and/or antennas. Such a collection of correlation receivers, termed "fingers," is what comprises the *CDMA rake receiver* [CDMAWEB].

After dispreading by correlators, the signals are combined using, for example, maximum ratio combining. Since the received multipath signals are fading independently, diversity order and thus performance are improved. Figure 4.36 illustrates the principle of the rake receiver. After spreading and modulation, the signal is transmitted and passes through a multipath channel that can be modeled by a tapped delay line (i.e., the reflected signals are delayed and attenuated in the channel). In Fig. 4.36 we have three multipath components with different delays (τ_1, τ_2, and τ_3) and attenuation factors (a_1, a_2, and a_3), each corresponding to a different propagation path. The rake receiver has a receiver finger for each multipath component. In each finger, the received signal is correlated by a spreading code, which is time-aligned with the delay of the multipath signal. After dispreading, the signals are weighted and combined. In Fig. 4.36 maximum ratio combining is used, that is, each signal is weighted by the path gain (attenuation factor). Due to the mobile movement, the scattering environment changes, and thus, the delays and attenuation factors change as well. Therefore, it is necessary to measure the tapped delay line profile and to reallocate rake fingers whenever the delays change by a significant amount. Small-scale changes, less than one chip, are taken care of by a code tracking loop that tracks the time delay of each multipath signal.

Smart antenna technology is one step further in the evolutionary path toward capacity improvements and interference mitigation. Base station antennas have till date been omnidirectional or sectored. Omnidirectional transmissions waste a lot of power since most of it is radiated in other directions than toward the user. In addition, the power radiated in other directions will be experienced as interference by other users. The idea

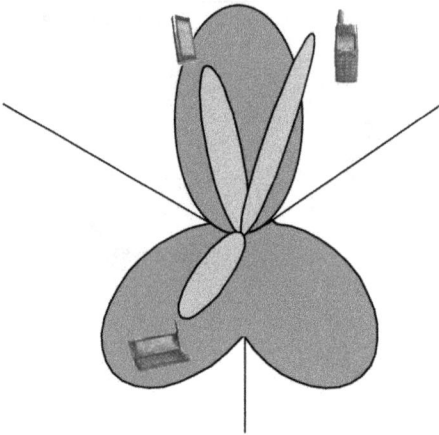

Figure 4.37 Illustration of the difference between a traditional base station radiation pattern and a smart antenna base station.

of smart antennas is to use base station antenna patterns that are not fixed, but adapt to the current radio conditions. This can be visualized as the antenna directing a beam toward the communication partner only. The difference between the fixed and the smart antenna concept is illustrated in Fig. 4.37.

Traditionally, whereas users communicating via the same base station have been separated by frequency, as in FDMA; by time, as in TDMA; or by code, as in CDMA, smart antennas, on the other hand, add a new way of separating users, namely by space, through *space division multiple access* (SDMA). In SDMA users of the same cell can use the same *physical communication channel*. (The *physical communication channel* is defined as a combination of the carrier frequency, time slot, and spreading code.)

This is achieved by using multiple antennas to provide more accurate directional targeting as well as fine-tune antenna coverage patterns that match the traffic conditions and complex radio frequency environments. This capability of smart antennas to adjust their patterns to the changing network traffic or RF conditions provides network operators maximum flexibility in controlling and customizing the sector antenna pattern beamwidth and azimuthal orientation over that of standard sector antennas. Thereby, adaptive directional reception and transmission are achieved on the uplink and adaptive directional transmission on the downlink. At the same time, less interference is received from other directions on the uplink, or transmitted toward the other directions on the downlink. Hence, more users can be accommodated in the system and, at the same time, a corresponding increase in the spectral efficiency is achieved.

TABLE 4.9   Key Features of the W-CDMA Proposal

Channel bandwidth	5, 10, 20 MHz
Downlink RF channel structure	Direct spread
Chip rate	3.84 Mcps
Frame length	10 ms/20 ms (optional)
Spreading modulation	Complex quadrature spreading
Coherent detection	User-dedicated time multiplexed pilot
Base station synchronization	Asynchronous operation
Multirate	Variable spreading and multicode Time-multiplexed simultaneous services
Spreading factors	4-256 (4.096 Mcps)
Power control	Open and fast closed loop (1.6 kHz)
Spreading	Variable length orthogonal sequences for channel separation, Gold sequences for cell and user separation
Multiuser detection, smart antennas	Supported by the standard, optional in the implementation
Service multiplexing	Multiple Services with different quality of service requirements multiplexed on one connection

Summary of main parameters in W-CDMA.   Table 4.9 summarizes the main parameters related to the W-CDMA air interface.

The main system design parameters of W-CDMA are highlighted as follows:

W-CDMA is a wideband direct-sequence code division multiple access (DS-CDMA) technology; user information bits are spread over a wide bandwidth by multiplying the user data with quasi-random bits (chips) derived from CDMA spreading codes. The chip rate of 3.84 Mcps is used. In order to support very high bit rates (up to 2 Mb/s), the use of a variable spreading factor and multicode connections is supported.

Besides, in order to achieve the very high data rates, W-CDMA requires a wide frequency band around 5 MHz of high capacity with 50 to 80 voice channels (per 5 MHz carrier) compared to eight channels per 200 kHz carrier for GSM.[6] The actual carrier spacing can be selected on a 200-kHz grid between approximately 4.4 and 5 MHz, depending on the interference between the carriers. The choice is motivated by the necessity to balance the high data-rate demand of third-generation systems

---

[6]DS-CDMA systems with a bandwidth of about 1 MHz, such as IS-95, are commonly referred to as narrowband CDMA systems. Subject to the operating license, the network operator can deploy multiple such 5 MHz carriers to increase capacity, possibly in the form of hierarchical cell layers.

with the scarcity of the bandwidth. Another reason to choose the large 5-MHz bandwidth is that it can resolve more multipaths than narrower bandwidths, increasing diversity and thus improving performance. W-CDMA can then provide up to eight times more traffic per carrier compared to a narrowband CDMA carrier.

With 5 MHz, 144, and 384 kbit/s data rates can be achieved easily while 2 Mb/s peak rate can be provided under specific conditions.

W-CDMA channeling. W-CDMA defines five logical channels [OJAN-PER]—two dedicated channels and one common channel in the uplink (UL—mobile to base station), and three common physical channels in the downlink (DL—base to mobile). Here are the so-defined logical channels:

- The broadcast control channel (BCCH) carries system- and cell-specific information (DL).

- The paging channel (PCH) for messages to the mobiles in the paging area (DL).

- The forward access channel (FACH) for messages from the base station to the mobile in one cell (UL).

- The dedicated control channel (DCCH) covers the two dedicated control channels—the stand-alone dedicated channel (SDCCH) and the associated control channel (ACCH) (UL/DL).

- The dedicated traffic channel (DTCH) for point-to-point data transmission in the uplink and downlink (UL/DL).

The primary and secondary common control physical channels (CCPCH) carry the downlink common control logical channels (BCCH and PCH) while the synchronization channel (SCH) (see Fig. 4.38, from Ref. [OJANPER]) provides timing information and is used for handover measurements by the mobile station.

In the uplink, user data are transmitted on the *dedicated physical data channel* (DPDCH), which carries user data and is I/Q modulated with the control information. The latter is sent out on the dedicated physical control channel (DPCCH) and is needed to transmit pilot symbols for a coherent reception, power control signaling, and rate information for rate detection. Time-multiplexed pilot symbols are used in the uplink of W-CDMA to perform coherent detection. In comparison with second-generation CDMA approaches that use noncoherent detection, the addition of the pilot is used to increase the performance of the uplink by a factor depending on the proportion of the pilot signal power to data signal power. In fact, this improvement can generally be estimated around 3 dB, which is a significant factor in wireless communications. Figure 4.39 (from Ref. [OJANPER]) depicts the structure of W-CDMA uplink's dedicated channels. In addition, W-CDMA uses a

Downlink – Time multiplexed control and data

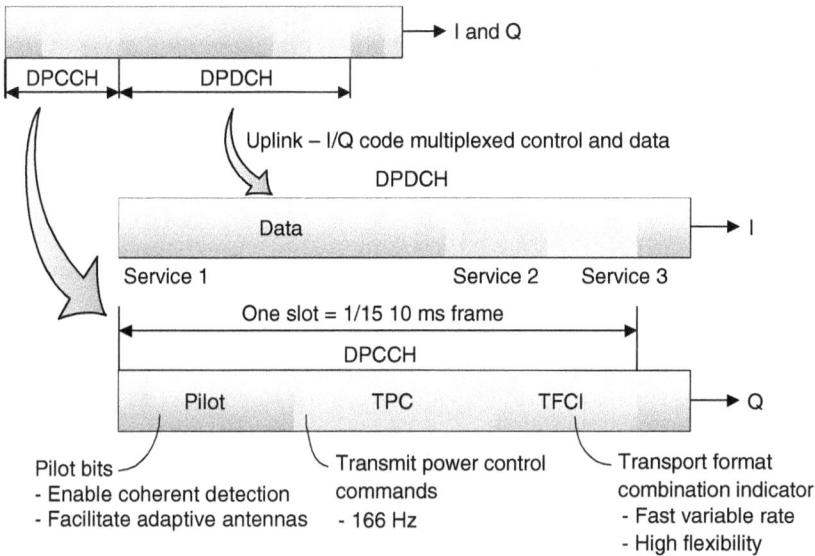

Figure 4.38 Structure of the W-CDMA physical layer.

combined I/Q and code multiplexing solution (dual-channel QPSK) for multiplexing the DPDCH with the DPCCH to avoid *electromagnetic compatibility* (EMC) problems with *discontinuous transmission* (DTX).

In the downlink, the dedicated channels (DPDCH and DPCCH) are time-multiplexed and so are the pilot symbols that are primarily used for coherence detection. Since these are user-dedicated, i.e., they have to take the same path as the data signal, they can be used for channel estimation, even with adaptive antennas beamforming. In addition, fast

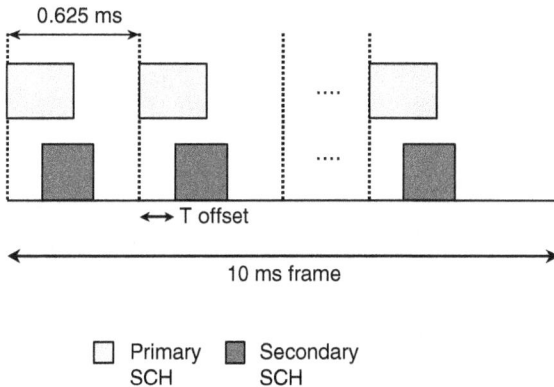

Figure 4.39 Synchronization channels in the downlink.

power control is used in the downlink to improve its transmission performance. This is achieved in two ways—first, it decreases multipath fading effects, and second, it increases the multiuser interference variable within the cell. This is so because the orthogonality between users is imperfect due to the multipath channel [OJANPE98, OJANPER]. Note that pure QPSK is used in the downlink as opposed to dual-channel QPSK in the uplink.

Furthermore, W-CDMA uses different spreading codes for cell separation in the downlink and user separation in the uplink. In fact, the FDD component applies a two-layered code structure consisting of spreading codes and scrambling codes. In the uplink, the spreading codes are first applied to each transmission to distinguish between different mobile stations and then the DPDCH and DPCCH are QPSK-modulated. In the downlink, spreading is performed by channelization codes for each DPCH depending on the transmission service required. Afterward, a cell-specific scrambling code is applied to distinguish between different cells. The data modulation used is also QPSK. The *orthogonal variable spreading factor* (OVSF) codes are employed for the spreading codes in order to preserve the orthogonality between different rates and spreading factors in both the uplink and the downlink.

The choice of those coding schemes leads to parallel transmission of two channels since the In phase/quadrature (I/Q) multiplexing is used. For this reason, attention must be paid to modulated signal constellation and related peak-to-average power ratio. Using the complex spreading circuit, shown in Fig. 4.40, the peak-to-average power is greatly reduced and, as a consequence, the power efficiency is improved [OJANPE98].

Multirate.   Multirating is the multiplexing of multiple services of the same connection on one DPDCH or the multiplexing of several connections of different QoS requirements on the same DPDCH. This concept is illustrated in Fig. 4.41. Multirate is necessary to provide flexible data rates and this is of great importance since the bandwidth is needed to be allocated only on demand (bandwidth on demand requirement of UMTS) primarily because of its scarcity and also because it needs to be provided if demanded. Each user is allocated frames of 10 ms duration, during which the user data rate is kept constant.

The general W-CDMA strategy is to use a single code transmission for small data rates and a multicode transmission for higher ones.

If we are dealing with two services involved in the same transmission (like speech and video in video-conferencing, for example), time multiplexing can be used either before or after the inner or outer coding, as illustrated in Fig. 4.42 [OJANPER]. After service multiplexing and channel coding, the multiservice data stream is mapped to one DPDCH. Since the total bit rate can be almost arbitrary it is necessary to permit the service provider to allocate several DPDCHs

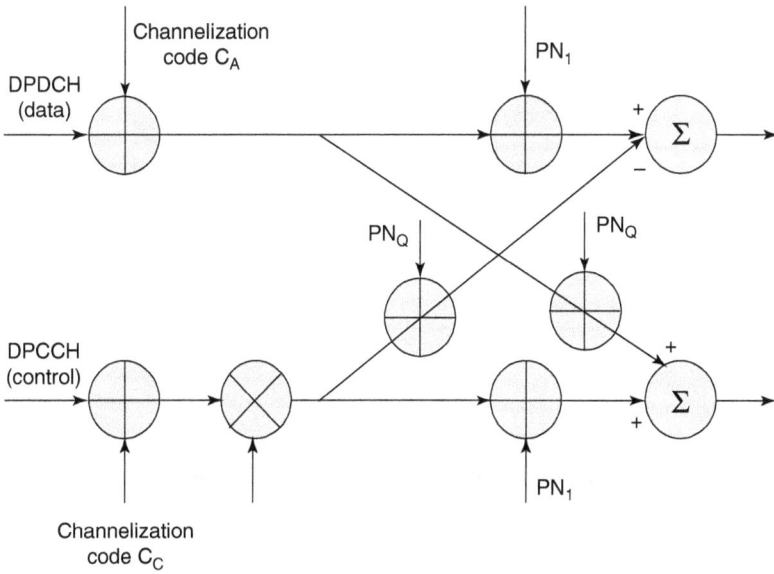

Figure 4.40 Complex spreading.

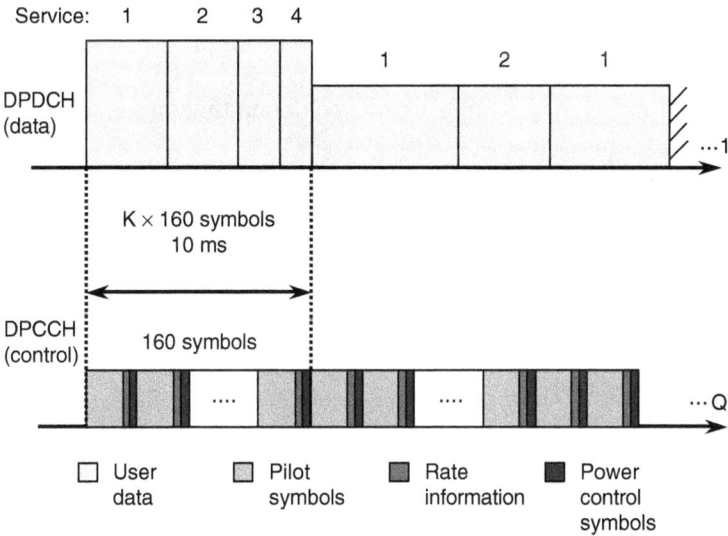

Figure 4.41 W-CDMA uplink multirate transmission.

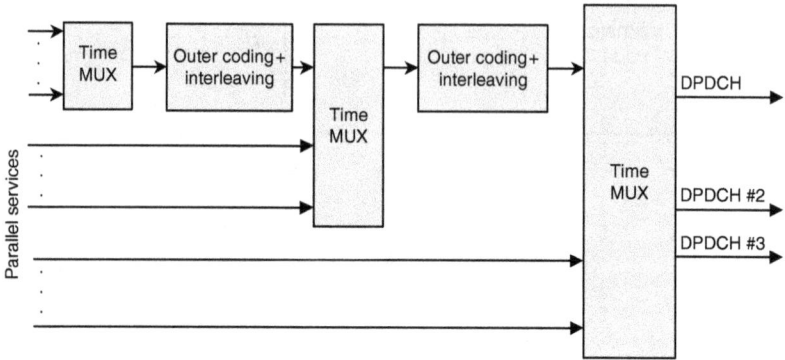

Figure 4.42 Service multiplexing in W-CDMA.

if the total rate exceeds the upper limit for single-code transmission. The main advantage of this first technique is that time multiplexing avoids multicode transmission thus reducing peak-to-average power of transmission.

In a second scheme, in the case where we are dealing with multiple transmissions at the same time, parallel services could be mapped to different DPDCHs using multicode coding/interleaving of distinct channels. With this alternative scheme, both the power requirement and the QoS of each channel can be separately and independently controlled. The drawback of code multiplexing is obviously the need for multicode transmission, which will increase the mobile station complexity, set higher requirements for the power amplifier linearity in transmission, and finally ask for more correlators in reception. Different coding schemes can be employed depending on the bit error rate and delay requirements of different services. As an example, for services with a BER of 10^{-3}, convolutional coding of 1/3 can be used for a relatively low bit rate and convolutional coding of 1/2 for higher bit. For higher-quality service, up to the 10^{-6} BER level classes, outer Reed-Solomon coding can be used and turbo codes have also been proposed for transmissions with a bit rate higher than 32 kbit/s [OJANPE98].

To summarize, a certain multirate scheme is first chosen to suit the needs of the different transmissions, thus resulting in the use of one or several DPDCHs. Then each transmission is assigned a particular spreading code selected to distinguish the actual mobile station from the others. After that, each of the several transmissions is time/code multiplexed on its assigned DPDCH before the latter is QPSK-modulated with the corresponding dedicated control channel.

W-CDMA supports the operation of asynchronous base stations, so that, unlike the synchronous systems (e.g., IS-95), there is no need for a global time reference, such as a GPS. (Deployment of indoor and micro base stations is easier when no GPS signal is needed.) In line with our discussions in Sec. 4.4.3, one of the key issues in IMT-2000's vision is global roaming. Therefore, W-CDMA must include certain specifications that allow it to achieve this. The first thing to consider is the structure of W-CDMA asynchronous base stations. These base stations are considered when designing soft and softer handover algorithms. The second aspect is the interfrequency handover support; this regards the utilization of *hierarchical cell structures* (HCSs), an important concept of third-generation systems as shown in Fig. 4.15. Interfrequency hard handovers can be used, for example, to hand a mobile over from one W-CDMA frequency carrier to another. One alternative for this is high-capacity base stations with several carriers. Third—intersystem hard handovers that take place between the W-CDMA FDD system and another system, such as W-CDMA TDD or GSM.

Each of these aspects is covered in more detail in the following paragraphs.

During soft handover, illustrated in Fig. 4.43, a mobile station is in the overlapping cell coverage area of two sectors belonging to different base stations. The communications between the mobile station and the base station take place concurrently via *two* air-interface channels from each base station separately. This requires the use of two separate codes in the downlink direction so that the mobile station can distinguish the signals. Both the channels (signals) are received at the mobile station by maximal ratio combining rake processing.

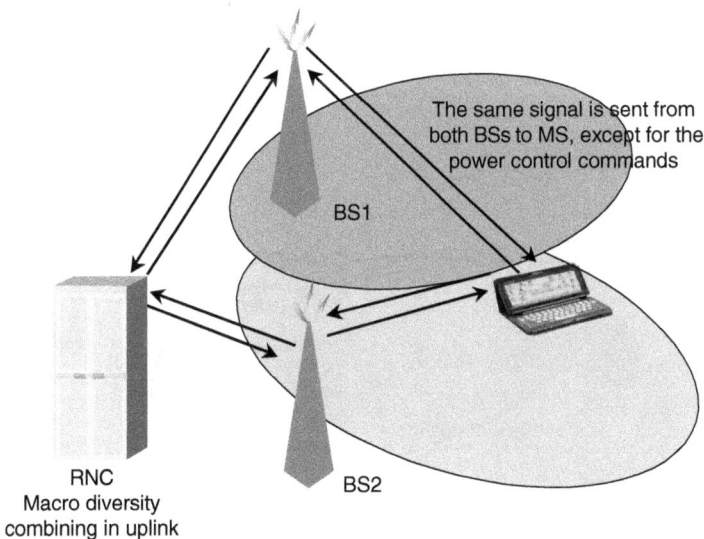

The same signal is sent from both BSs to MS, except for the power control commands

BS1

RNC
Macro diversity
combining in uplink

BS2

Figure 4.43   Soft handover.

In order to enter soft handover, the observed timing differences of the downlink synchronization channels (from the two base stations) are measured by the mobile station, which then reports it to its present base station. The latter is therefore able to adjust the timing of the new downlink soft handover connection with a resolution of one symbol. This enables the mobile rake receiver to collect the macrodiversity energy from the two base stations and then achieve the timing adjustments of dedicated downlink channels without losing the orthogonality of downlink codes but still reaching a resolution of one symbol [OJANPER].

In the uplink direction a similar process takes place at the base station—the code channel of the mobile station is received in each sector, then routed to the same baseband rake receiver and the maximal ratio combined there in the usual way.

As in the soft handover, during softer handover, a mobile station is in the overlapping cell coverage area of two adjacent sectors of a base station. The communications between the mobile station and the base station take place concurrently via two air-interface channels, one for each sector separately. The two signals are received at the mobile station by means of rake processing, very similar to multipath reception, except that the fingers need to generate the respective code for each sector for the appropriate dispreading operation. Figure 4.44 shows the softer handover scenario.

Soft/softer handovers are an essential interference-mitigating tool in W-CDMA, as with fast power control. Without soft/softer handover there would be near-far scenarios of a mobile station penetrating from one cell deeply into an adjacent cell without being power-controlled by the latter. Very fast and frequent hard handovers could largely

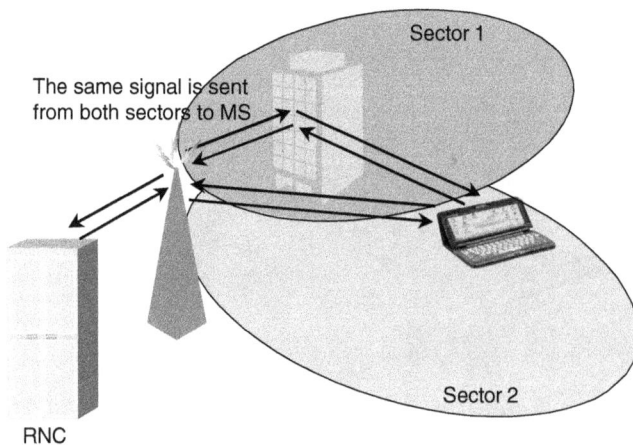

Figure 4.44 Softer handover.

avoid this problem; however, they can be executed only with certain delays during which the near-far problem could develop.

Although soft and softer handovers seem similar in the downlink direction (BTS to mobile), in the uplink direction they differ significantly—the code channel of the mobile station is received from both the base stations, but the received data are then routed to the RNC for combining. This is typically done so that the same frame reliability indicator, as provided for outer-loop power control, is used to select the better frame between the two possible candidates within the RNC. This selection takes place after each interleaving period, i.e., every 10 to 80 ms. Note that during soft handover, two power-control loops per connection are active, one for each base station.

Interfrequency handover, on the other hand, is needed for utilization of hierarchical cell structures (macro-, micro-, and picocells) since the cells belonging to different layers are likely using different frequencies. Several carriers and interfrequency handovers may also be used for taking care of high-capacity needs in hot spots as well as to provide handovers with second-generation systems, like GSM in Europe or IS-95 in the United States. A key requirement for seamless interfrequency handover is the ability of the mobile station (the handset) to carry out cell search on a carrier frequency different from the current one without affecting ordinary data flow [OJANPE98].

Various methods of making measurements on other frequencies while still having the connection running on the current frequency are considered for interfrequency measurements in W-CDMA. The first approach, dual receiver, is considered suitable for mobile stations with receiver diversity. The concept is to temporarily reallocate one of the receiver branches for measurements on a different carrier while the other keeps receiving from the current frequency. The main advantage of the dual receiver technique is that there is no break in the current frequency connection.

The second idea, the slotted mode, is presented in Fig. 4.45 (from Ref. [OJANPER]). This approach is proposed for single-receiver mobile stations (no antenna diversity) to allow interfrequency measurements for interfrequency handover. The information normally transmitted during a 10-ms frame is compressed in time (either by code puncturing or by changing the forward error correction's rate) thus leaving unused time for the mobile station to perform measurements on other frequencies [OJANPE98, OJANPER].

W-CDMA employs coherent detection on the uplink and the downlink based on the use of pilot symbols or common pilot. While already used on the downlink in IS-95, the use of coherent detection on the uplink

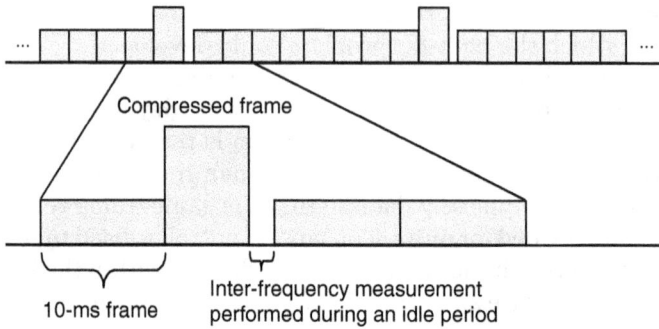

Compressed frame

10-ms frame

Inter-frequency measurement
performed during an idle period

Figure 4.45 Slotted mode structure.

is new for public CDMA systems and results in an overall increase in the coverage and capacity on the uplink.

The W-CDMA air interface has been crafted in such a way that advanced CDMA receiver concepts, such as multiuser detection, can be deployed as a system option to increase capacity and/or coverage.

Second-generation CDMA receivers, based on the rake receiver principle, are interference limited because they consider other users' signals as interference. In practice this means that when a new user, or interferer, enters the network, the service quality of other users can go below the acceptable level. Optimally, all signals would be detected jointly or interference from other signals would be removed by subtracting them from the desired signal. That would enhance the network interference resistance thus boosting the number of users who can be served at the same time. *Multiuser detection* (MUD), also called joint detection and *interference cancellation* (IC), provides a means of reducing the effect of multiple access interference, and hence increases the system capacity [OJANPER]. In addition to capacity improvement, MUD alleviates the near-far problem typical to DS-CDMA systems. If a user's mobile station is too close to the base station, it may block the whole cell traffic by using too high a transmission power which would result in a high-level interference for others. However, if this user is previously detected and his or her signal subtracted from the input signal, the other users will not see any interference.

Such a strategy is quite simple to implement on a system using short spreading codes. This is because cross-correlation between different signals does not change every symbol compared to long ones. W-CDMA, however, uses long spreading codes which in fact complicate the deployment of MUD in UMTS's air interface. The most feasible approach that has been proposed is the regenerative parallel interference cancellation algorithm. This technique exercises the

| RACH burst | USER packet | Time between packets | RACH burst | USER packet |

Common channel without fast power control

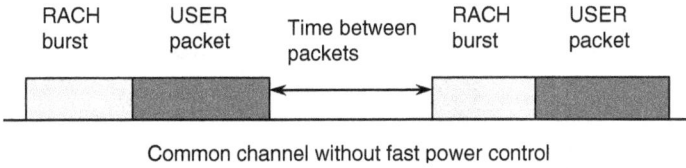

Figure 4.46   Packet transmission on the common channel.

interference cancellation at the chip level, avoiding, in that way, an explicit calculation of the cross-correlation between spreading codes from different mobile stations [OJANPER]. Complexity issues impose that there must be different ways of using MUD in the uplink and the downlink. Base stations have to demodulate the signal of all users, while mobile stations only have to take care of one single signal. For that reason, a simpler interference suppression scheme can be used in the mobile stations than the one used in the base stations [OJANPE98].

Packet data.   W-CDMA uses two different strategies for data packet transmission. The first one, called common channel packet transmission, is used for short infrequent packets. The idea is to add short data packets directly to a random access burst that is 10 ms long and is transmitted with a fixed power, which is based on the slotted Aloha scheme, in order to avoid having to maintain a dedicated channel where there is no real point in doing so. In the meantime, the delay associated with the transfer to a dedicated channel is avoided. Taken from Ref. [OJANPER], Fig. 4.46 illustrates packet transmission on a common channel.

The second strategy is used to send more frequent packets through a dedicated channel. For large packets, a single-packet scheme is used. The channel is reserved only for the packet transmission duration and release immediately afterward. If some packets need to be transmitted more often, a multipacket scheme can be used. For that option, the dedicated channel is maintained by transmitting power control and synchronization information between subsequent packets [OJANPE98].

2G CDMA technology.   2G CDMA networks, branded cdmaOne, were commercially launched in 1995, and provided roughly 10 times more capacity than analog networks—far more than TDMA or GSM.

cdmaOne: The family of IS-95 CDMA technologies.   cdmaOne describes a complete wireless system based on the TIA/EIA IS-95 (Telecommunications Industry Association/Electronic Industries Association Interim Standard-95) CDMA standard, including IS-95A and IS-95B revisions. cdmaOne provides a family of related services including cellular, PCS [COX92], and fixed wireless (WLL).

IS-95, developed in the 800 and 1900 MHz radio bands, offers two different rates—8 and 13 kbit/s. The standard supports 55 users per channel (1.25 MHz) per base station (55 out of 64 different spreading codes for users). More capacity is possible with sector antennas. Guard bands are needed to protect AMPS services in the same frequency range. Adjustable rate vocoder is also used based on the *QualComm code excited linear prediction* (QCELP) algorithm.

IS-95A: The first 2G CDMAcellular standard.   TIA/EIA IS-95 was first published in July 1993. The IS-95A revision was published in May 1995 and is the basis for many of the commercial 2G CDMA systems around the world. IS-95A describes the structure of the wideband 1.25 MHz CDMA channels, power control, call processing, handoffs, and registration techniques for system operation. In addition to voice services, many IS-95A operators provide circuit-switched data connections at 14.4 kbit/s. IS-95A was first deployed in September 1996 by Hutchison.

IS-95B: 2.5G.   The IS-95B revision, also termed as TIA/EIA-95, combines IS-95A, ANSI-J-STD-008, and TSB-74 into a single document. The ANSI-J-STD-008 specification, published in 1995, defines a compatibility standard for 1.8 to 2.0 GHz CDMA PCS systems. TSB-74 describes interactions between IS-95A and CDMA PCS systems that conform to ANSI-J-STD-008. Many operators that have commercialized IS-95B systems offer 64 kbit/s packet-switched data, in addition to voice services. Due to the data speeds that it is capable of supporting, IS-95B is categorized as a 2.5G technology. cdmaOne IS-95B was first deployed in September 1999 in Korea and has since been adopted by operators in Japan and Peru.

Differences between W-CDMA and 2G air interfaces.   In this section we illustrate the main differences between the third- and second-generation

TABLE 4.10   Main Differences between W-CDMA and GSM Air Interfaces

	W-CDMA	GSM
Carrier spacing	5 MHz	200 kHz
Frequency reuse factor	1	1-18
Power control frequency	1500 Hz	2 Hz or lower
Quality control	Radio resource management algorithms	Networks planning (frequency planning)
Frequency diversity	5 MHz bandwidth allows multipath diversity with rake receiver	Frequency-hopping
Packet data	Load-based packet scheduling	Time-slot-based scheduling with GPRS
Downlink transmit diversity	Supported for improving downlink capacity	Not supported by the standard

air interfaces. GSM and IS-95 (the standard for cdmaOne systems) are the second-generation air interfaces considered here.

Table 4.10 lists the main differences between W-CDMA and GSM. In this comparison only the air interface is considered. The differences in the air interface reflect the new requirements of the third-generation systems. For example, the larger bandwidth of 5 MHz is needed to support higher bit rates.

Transmit diversity is included in W-CDMA to support the asymmetric capacity requirements between the downlink and the uplink. Transmit diversity is not supported by the second-generation standards.

Table 4.11 lists the main differences between W-CDMA and IS-95. Both W-CDMA and IS-95 utilize direct-sequence CDMA. The higher chip rate of 3.84 Mcps in W-CDMA gives more multipath diversity than the chip rate of 1.2288 Mcps in cdmaOne, especially in small urban cells. The importance of diversity for system performance is discussed in "Overview of the CDMA technology" in Sec. 4.5.4. Most importantly, increased multipath diversity improves the coverage. The higher chip rate also results in a higher trunking gain, especially for high bit rates, than do narrowband second-generation systems.

W-CDMA has fast closed-loop power control in both the uplink and the downlink, while IS-95 uses fast power control only in the uplink. The downlink fast power control improves link performance and enhances the downlink capacity. It requires new functionalities in the mobile, such as SIR estimation and outer-loop power control, which are not needed in IS-95 mobiles.

The IS-95 system was targeted mainly at macrocellular applications. The macrocell base stations are located on masts or rooftops where the GPS signal can be easily received. IS-95 base stations need to be synchronized and this synchronization is typically obtained via GPS. The need for a GPS signal makes the deployment of the indoor and microcells more problematic since GPS reception is difficult without a line-of-sight connection to GPS satellites. Therefore, W-CDMA is designed to operate with asynchronous base stations where no synchronization from GPS is needed.

Interfrequency handovers are considered important in W-CDMA to maximize the use of several carriers per base station. In IS-95 interfrequency measurements are not specified, making interfrequency handovers more difficult.

Experience from second-generation air interfaces has been important in the development of the third-generation interface, but there are many differences, as listed above. In order to make the optimum use of the capabilities of W-CDMA, a deep understanding of the W-CDMA air interface is needed, from the physical layer to network planning and performance optimization. For the sake of completeness, in the following

TABLE 4.11   W-CDMA vs cdmaOne (cdma2000)

	W-CDMA	cdmaOne
Channel bandwidth	5 MHz	1.25 MHz
Downlink RF channel structure	Direct spread	Direct spread or multicarrier
Chip rate	3.84 Mcps	1.2288 Mcps
Base station Synchronization	Not needed	Yes, typically obtained via GPS
Roll-off factor for chip shaping	0.122	Similar to IS-95
Frame length	10 ms, 20 ms (optional)	20 ms for data and control/ 5 ms for control information on the fundamental and dedicated control channel
Spreading modulation	Balanced QPSK (downlink) Dual channel QPSK (uplink) Complex spreading circuit	Balanced QPSK (downlink) Dual-channel QPSK (uplink) Complex spreading circuit
Data modulation	QPSK	QPSK
Coherent detection	User dedicated time multiplexed pilot (downlink and uplink); no common pilot in downlink	Pilot time multiplexed with PC and EIB (uplink) Common continuous pilot channel and auxiliary pilot (downlink)
Channel multiplexing in uplink	Control and pilot channel time multiplexed I&Q multiplexing for data and control channel	Control pilot, fundamental, and supplemental code multiplexed I&Q multiplexing for data and control channels
Multirate	Variable spreading and multicode	Variable spreading and multicode
Spreading factors	4-256 (3.84 Mcps)	4-256 (3.6864 Mcps)
Power control	Open and fast closed loop (1.6 kHz)	Uplink: 800 Hz, control Downlink: slow power
Spreading (downlink)	Variable length orthogonal sequences for channel separation Gold sequences 2^{15} for cell and user separation (truncated cycle 10 ms)	Variable length Walsh sequences for channel separation, M-sequence 2^{15} (same sequence with time shift utilized in different cells, different sequence in I&Q channel)
Spreading (uplink)	Variable length orthogonal sequences for channel separation Gold sequences 2^{41} for user separation (different time shifts in I and Q channel truncated cycle 10 ms)	Variable length orthogonal sequences for channel separation, M-sequence 2^{15} (same sequence for all users, different sequences in I&Q channels; M-sequence 2^{41} for user separation different time shifts for different users)

TABLE 4.11 W-CDMA vs cdmaOne (cdma2000) (*Continued*)

	W-CDMA	cdmaOne
Handover	Soft handover Interfrequency handover	Soft handover Interfrequency handover
Packet data	Load-based packet scheduling	Packet data transmitted as short-circuit-switched calls
Downlink transmit diversity	Supported for improving downlink capacity	Not supported by the standard

text we highlight various issues involved in the complex, yet of great significance, task of UMTS network planning.

4.5.5 UMTS network planning

The high price of auction-based 3G licenses is putting enormous pressure on manufacturers and operators to configure the next-generation mobile networks in the most cost-effective way possible. This means taking great care of the UMTS radio planning and ensuring terminal and base station compatibility between different vendors are prerequisites for keeping 3G costs down and improving revenue prospects.

UMTS network planning is a very complex area involving optimal site location, sectoring, and handover issues. It requires the use of both advanced planning and analysis tools, coupled with tremendous skill and effort from planning engineers in order to achieve a high-capacity preoptimized network.

One key factor that influences the available UMTS network capacity is the choice of a soft handover margin. This factor determines the number of subscribers that would enter soft handover. Experience shows that, while a large margin can provide a certain reduction in the uplink and downlink intercell interference—and therefore an improvement in capacity—additional downlink connections from other base stations must be made available - thus reducing overall downlink capacity. On the other hand, a lower soft handover margin reduces the load on additional downlink resources but increases the uplink and downlink intercell interference. An optimum soft handover margin therefore needs to be set depending on where the traffic is located with respect to the base stations.

In addition to the soft handover margin problem in CDMA-based 3G systems, an important factor that leverages 3G systems to hit higher capacities (many users and high data rate) is the proper management of RF resources, with power control being the primary issue. By effectively deploying a preoptimized UMTS network, an operator may be able to avoid or delay the need to install additional base stations due to capacity limitation issues and manage the capital expenditure more efficiently. As described earlier, this is important because in a W-CDMA system, all users appear as noise to other users in the network. To achieve maximum capacity minimum noise is required, and so the effect of other users must be limited. This is explained by Shannon's law—if a higher data rate is required at the same signal to noise ratio, with the same available bandwidth, then more power is required. Therefore the high data rate user must transmit at a higher power output. This creates more "noise" for other users. These other users must then increase output power to overcome the higher noise floor, and then the high data rate user must increase power to overcome the higher power of other users. This is a positive feedback system that can very quickly run out of control with all users trying to run at full power to overcome the noise of other users.

Hence the power control algorithm, which runs in node B, is critical to ensuring that the correct balance of transmit powers between all users is enabled, so that they can all get through above the noise while still keeping the noise floor at a minimum level.

An important issue here is a user near the outer edge of a cell. Such a user is running at full power to overcome the noise floor and atmospheric propagation loss. This is acceptable and a call can be placed and held. If a high data-rate user comes into the cell, he or she will raise the noise floor of the entire cell, possibly to such an extent that the distant user's call cannot be maintained. If the high data-rate user has a guaranteed service level agreement for certain data rates, he or she will require to be connected, and so the other user will be dropped.

But how will it be accepted if one user has to be dropped to accept another "more important" user? What happens if both have guaranteed SLAs? The first (distant user) is on a call, and so would not expect to be thrown off the network. The second (close, high data rate) is close to the node B base station and knows that he or she gets good coverage and so will expect the call to connect.

The effect of the noise floor rising (and hence cell radius shrinking) is called *cell breathing* (Fig. 4.47). In effect, the cell coverage is a function of the number of users and the type of service being used within the cell. So if a cell is "mapped" by the operator tracing the coverage of a "typical" call, this will not represent the true coverage with many other users on the call. Also, customers may find that on one day they have a satisfactory coverage at a certain location but a short time later from the same location they cannot connect. This problem can be managed by the fact that the level of "interference," or noise floor, is a function of the data rate being offered to the user. So a user can be accepted at a certain data rate, but that rate can be lowered dynamically to support other users coming on.

Radio resource management (RRM) is the process defined within 3GPP, which controls how users are admitted onto the network, what

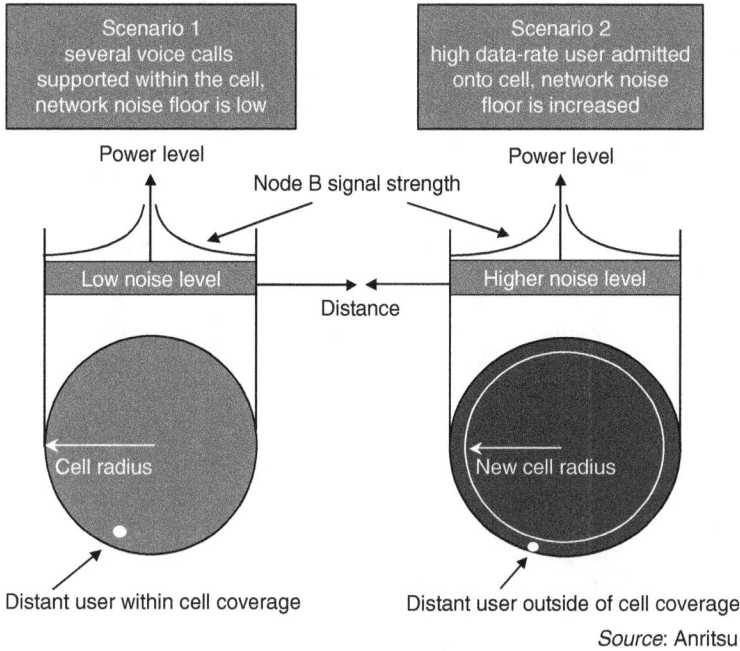

Figure 4.47   The "cell breathing" effect [TEL-INTER02v36a].

types of services are being used at any time by the network, and the impact on other users. UMTS network simulators are being used to help evaluate some of the basic RRM algorithms that need to be implemented for the initial network rollouts. These include considering scenarios such as real-time and non-real-time 3G services, variable bit rates up to 384 kb/s, dedicated or common or shared channels, and switching and multiplexing between voice and data transport channels.

Since third-generation networks rely predominantly on packet-switched traffic, the management of which is more complex and requires RRM to be able to control the setting up and control of many different services simultaneously. As network traffic changes, the allocation of resources needs to change; hence simulating measurement evaluations of real-world environments including loading and interference need to be pursued.

4.5.6   3G services and applications

The billions of dollars spent on 3G licenses and infrastructure have led to huge debts and the biggest challenge that the mobile operator industry has ever had to face as it tries to make its investments materialize into positive returns. The key to the future success of 3G lies with the applications and services that operators will offer to their subscribers, and within the underlying network, optimized to fully support the end-user services. Customers do not buy technology; customers buy services which add value to their professional or personal lives and are affordable too. Attractive services are the livelihood of the UMTS economy and an indispensable factor in UMTS's competitive and sustainable growth. To this avail, the main issue that the industry has to grapple with is introducing compelling applications and services [3GSM01, 3GSM02].

The best-known new feature of UMTS is that higher user bit rates—on circuit-switched connections 384 kb/s and on packet-switched connections up to 2 Mb/s—can be reached. Higher bit rates naturally facilitate new services, such as mobile e-services, advanced location-based services, and context-aware services. These types of services tend to involve several key characteristics, which are equally important in a digital mobile multimedia environment:

- Ubiquitous, real-time, multimedia communications. The only hope for dramatically increased fidelity, akin to communicating in person, is high-speed access and transport for any medium, anytime, anywhere, and in any volume.

- More "personal intelligence" distributed throughout the network. This includes applications that can access personal profiles of users (e.g., subscription information and personal preferences), learn from their behavior patterns, and perform specific functions on their behalf (e.g., "intelligent agents" that notify them of specific events).

- More "network intelligence" distributed throughout the network. This includes applications that know about, allow access to, and control network services, content, and resources. It can also perform specific functions on behalf of a service or network provider (e.g., "management agents" that monitor network resources, collect usage data, provide troubleshooting, or broker new services/content from other providers).

- More simplicity for users. This shields users from the complexity of information gathering, processing, customization, and transportation. It allows them to more easily access and use network services/content, including user interfaces that allows for natural interactions between the users and the network. It involves providing context-sensitive options/help/information, transparently managing interactions among multiple services, providing different menus for novices and experienced users, and providing a unified environment for all forms of communication.

- Personal service customization and management. This involves the ability of users to manage their personal profiles, self-provision network services, monitor usage and billing information, customize their user interfaces and the presentation and behavior of their applications, and create and provision new applications.

- Intelligent information management. This helps users manage information overload by giving them the ability to search for, sort, and filter content, manage messages or data of any medium, and manage personal information (such as calendar and contact list).

Mobile e-services.   Information technologies are an integral part of our lives, businesses, and society. The wide acceptance of Internet standards and technologies is helping us to build global computer networks capable of connecting everything and reaching everyone [PIS99]. Until now, the Internet has been mainly about delivering information - often too much of it and with few effective means of finding, sorting, and selecting the details most relevant to the user. E-services unlock the true potential of the Internet by turning it from a simple repository of knowledge into a framework for creating functionality. *E-services* are Web-based services that dynamically interact with each other to deliver true value-added service propositions to the end user. They make the Internet work for the user, rather than forcing the user to work the Internet, and effectively transform the Internet into a powerful back-end for delivering services and solutions that add value in a range of real-life scenarios. In fact, in the future, many Web pages will be consulted by other computers that have been instructed to search for relevant suppliers, compare prices, and even make purchases on behalf of their human masters. Mobile e-services thus aim at bringing a new generation of dynamic, interactive service solutions to consumers, enterprises, retailers, and service providers anytime, anywhere; thus enabling users to appreciate the full potential of the Internet for businesses and consumers alike.

A flexible, always-on infrastructure ensures that e-services can be rapidly deployed, easily managed, and scaled to meet the needs of the exploding mobile market. This service infrastructure envisages a world transformed by "service-centric computing," changing both the way businesses operate and serve their customers, and the way individuals live their lives on the go. However, for today's highly mobile population, delivering a new generation of Web-based service solutions on its own is not enough. People on the move, for instance, cannot always find a place to plug in their modems in order to access the Internet. *Mobile e-services* are the convergence of two of the most significant industrial developments—mobile telephony and e-services. They combine the latest developments in mobility with innovative e-service to bring new service solutions to all types of digital appliances, wherever they are being used. Working dynamically together, e-services will deliver a host of service offerings that can be personalized and tailored to the specific needs of each individual user. And it is not only the end user that stands to benefit. Enterprises of all kinds are able to take full advantage of the Internet for reaching new markets, increasing revenues, improving productivity, and delivering better customer service. Almost daily, new wired or wireless e-business models emerge—e-shop, e-procurement, e-auction, e-mall, third party marketplace, virtual community, value chain service provider, value chain integrator, collaborative platform, and information brokerage and trust [TIMMER98]. For someone walking down a street, for example, the most useful mobile experience might be e-services deployed to the user's mobile phone [3GSM02]. Once the user gets into a car, however, that same mobile phone could join forces with the car's powerful on-board computer to provide far richer services, specific to that environment. When the user enters a client's office, that same mobile phone could join forces with the local net-enabled printer to again deliver value-added services that meet specific needs in that new environment.

In addition, context-specific e-services can be proactively delivered and can transparently adapt themselves to the capabilities of the appliances in that environment—wired or wireless, smart or not so smart. To this avail, mobility is not really about putting the Internet in your pocket, since many users have neither the time nor the desire to browse the Web on their PDA. What mobile users benefit most from is proactive, context-specific e-services, offering them what they need, when they need it. Such services allow them to automatically and transparently interact with the environment in which they find themselves. Their mobile device knows what information they need and automatically delivers it to them. For example, their mobile device could announce their presence (e.g., at a hotel), gather information (room number), provide information (credit card number), customize their room (program the hotel room TV or stereo), and then check them out of the hotel (automatically picking up a copy of the bill). Leveraging then such key technologies as WAP, GPRS, Bluetooth, and e-speak, mobile e-services succeed in combining the convenience of mobility with the power of e-services [3GSM01]. See Fig. 4.48.

Location-based services.   A subsequent successor of mobile e-services is *location-based services* (LBS). Common examples of location-based services include discounted calls in a certain area, broadcasting of a service over a limited number of sites (broadcasting video on demand), and

Mobile banking	Mobile trading	Mobile ticketing	Mobile shopping	Mobile betting
• Bank accounts • Balances • Statements • Transfers • Lines of credit • Bill payments • Special orders • Post-dated pay	• Balance • Portfolios • Holdings • Quotes • Ratings • Trading • Open orders • Warrants	• Ticket availability • Orders • Reservations • Confirmations • Coupons • Bank account • Prepaid • Ticket inspection	• Product search • Product selection • Shopping cart • Alerts • Orders • Confirmations • Check–out • Bank accounts • Loyalty schemes • Prepaid	• Bet types • Selections • Ratings/odds • Bets • Confirmations • Bank accounts • Prepaid

Shopping

Trading

Ticketing

Banking

Betting

Solution platform

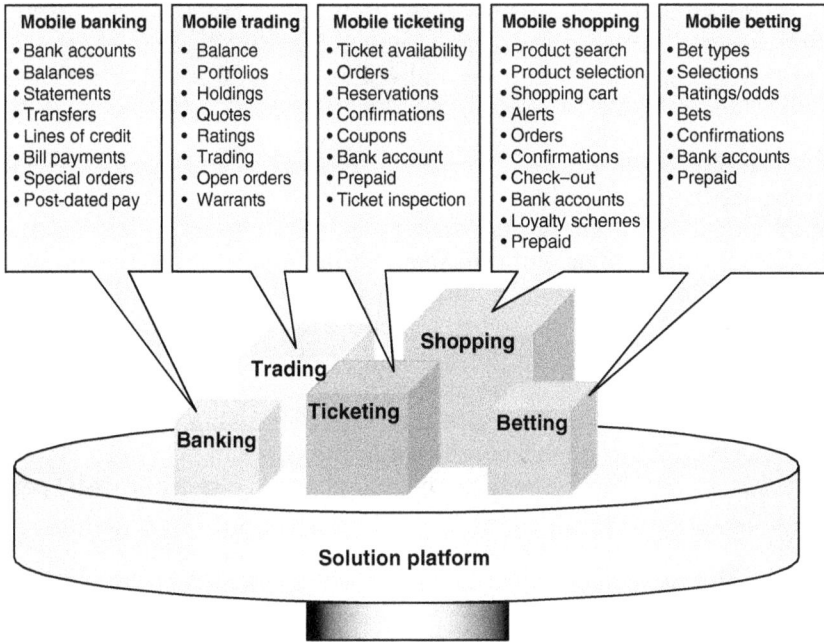

Figure 4.48   Solutions platform for mobile e-commerce applications.

retrieval and display of location-based information, such as the location of the nearest gas stations, hotels, restaurants, and so on.

Location-based services and applications are one of the basic dimensions in the 3G wireless mobile market. A number of reasons buoy the success of LBS over 3G:

- 3G license holders are under obligation to build UMTS services to provide a minimum level of coverage of approximately 50 percent by the end of 2005 and 80 percent by the end of 2007 (coverage obligations vary slightly from country to country). The new network must have LBS capabilities built-in to comply with the European Emergency Services' location-reporting directive for mobile devices (E 112). This directive is due to come into force by spring, 2008.

- Regulators want to see license terms enforced and 3G used. LBS are one area in the telecoms space where regulators are almost forcing the pace rather than enforcing complex and limiting rules and regulations. However, operators want extra time to rollout 3G services, so progress may be slower than envisaged—instead of a near-national coverage by 2007 it is probably more realistic to expect a slow, incremental growth over the next 10 years.

- LBS are already a proven technology. Consumers are using and paying for LBS applications over existing networks. 3G will make LBS more prevalent and more accurate and enable an altogether better user experience.

A location-based service is provided either by a teleoperator or by a third-party service provider that uses available information on the terminal location. The service is either push (e.g., automatic distribution of local information) or pull type (e.g., localization of emergency calls). Depending on the service, the data may be retrieved interactively or as background. For instance, before travelling to an unknown city abroad one may request night-time download of certain points of interest from the city. The downloaded information typically contains a map and other data to be displayed on the top of the map. By clicking the icon on the map, one gets information from the point. Information to be downloaded as background or interactively can be limited by certain criteria and personal interests.

Besides, location-based services can bring together groups of people with mobile handsets. A "friend-finder" service based on buddy lists within the phone would enable one to locate friends with similar interests who were in his or her immediate geographical location. Once this was established, we could go into something like an instant messaging/chat-type application and all this would be done via SMS.

Mobile operators have a good starting point for LBS as most GSM/GPRS-based providers already offer general services via their portals. These are often location-enabled tourist services, which use cell ID and voice or WAP-based content delivery. If you travel abroad for example, you now get an SMS from the foreign national operator with a contact number to dial for a location-based audio-guide to tourist attractions. This is a genuinely useful application that will only get better as multimedia PDAs become more prevalent.

For operators, the positive news is that people are already happy to pay for this type of content. Latest predictions (Fig. 4.49) report 80 percent of mobile subscribers by 2005, a total market worth of US$32.9 billion!

While they will have a place, unsolicited "pull" applications such as "where shall I eat?" and "what's on the menu?" may not draw in many customers. The focus should be on opt-in "push" applications, such as "did you know this is on tonight/tomorrow night?" or 'half price tickets to the Lion King tonight only plus links to further information, booking services, maps, and opt-in marketing.' Push services are far more valuable to operators, providing scope for a wide range of revenue-sharing partnerships. They are also more valuable to consumers who can select when and what information they receive direct to their mobile device,

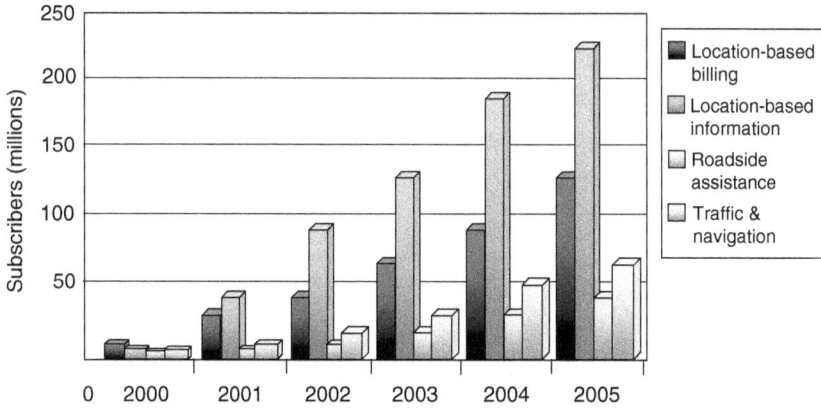

Source: The strategic group, March 2000

Figure 4.49   Market pull for LBS value-added services in Europe.

with minimum effort. The challenge is to implement the right business models to make LBS equitable to all parties and to rollout strict regulations to protect users from unwanted push marketing.

On the business side, LBS present more real and varied applications, such as field service information, logistics tracking and—for emergency services—call prioritization and navigation. The business applications are ideally suited to mobile workforce automation, where they have particular appeal to fast response businesses needing location tracking, route optimization and real-time service-call prioritization. Industry figures reflect this [TEL-INTER03v36]. Business data revenues, particularly from telemetry (wireless monitoring) and logistics applications stand to be far greater than consumer data revenues.

Content-aware services.   Considering the dominant role of Internet access in today's life, we could image the same need for information access or Web browsing in 3G mobile networks. The driving force for 3G services is the content [3GSM01]. Content can be considered as information, products, and services stored remotely on servers and applications platforms and delivered to the end users on request. Such servers could be owned and operated by either the operator, or a content provider, or both. Content generally includes news and information, games and entertainment, customer-generated information such as a corporate intranet, and applications such as "find my nearest" or m-commerce.

In their general context, content-aware services can be broadly divided into six main categories:

- Communication and community: email, calendar, and chat
- Information: news, weather, directories
- Lifestyle: listings of events, restaurants, movies, and games
- Travel: hotel listings, direction assistance, and timetables
- Transaction: banking, stock trading, purchasing, and auctions
- Other: This includes information about personalization, location-based services, device dependence and advertising, and also about the openness of the mobile portal, billing, and target group.

Content is not new to GSM. In the mid 1990s there were voice-based weather and traffic reports and, more recently, there have been SMS and WAP-based information services. In our days, the growth of mobile portals represents an important development area, heralding many of the personalized applications and services seen as holding the key to the successful realization of 3G in the future [3GSM01]. In a 3G world, mobile users are increasingly becoming reliant on their phones to do more than just communicate with their family, friends, and business associates [MOBCOM03]. European networks are engaging mobile users to vote on their favorite television shows via a touch of their mobile phone. Providers and carriers worldwide are realizing that sending customized messages, such as an SMS when a user's football team scores a goal, can build loyal and long-lasting relationships with their customers. Besides, consumers have increased their appetites for downloading ring tones and screen savers, sending MP3 files, emailing photos to friends and family, and voting via mobile phones. And by increasing their interactions with the tools available on mobile devices today's, operators are quick to seize the revenue opportunities associated with these new services.

Initial market findings from mobile operators and industry forums, such as PayCircle—the consortium that sets industry standards for mobile payment—indicate that the initial wave of content services is seeing fast adoption rates. In fact, a recent report by Strategy Analytics notes that with the quick adoption of these content services, the global wireless industry will grow from US$46 billion in 2002 to about US$114 billion by 2008. In addition, the report predicts that carriers will retain almost 60 percent of the 2008 revenue total as they continue to dominate the value chain.

As these next-generation services continue to penetrate the European market, service providers are asking themselves how they can deliver value to their customers while increasing revenue. With consumers using their mobile accounts as payment instruments, carriers are finding new ways to drive revenue by offering content-aware services.

4.5.7 NTT DoCoMo's i-mode technology

I-mode is a packet-based wireless data service for mobile phones first offered by Japan's NTT DoCoMo in 1999 [TEL-INTER02v36a]. I-mode has shown just how popular such services over wireless can be, where games, ringtones and logos, horoscopes, and jokes have attracted relatively high volumes of traffic. In terms of subscriber numbers, the mobile data service has been phenomenally successful—30 million subscribers to date—and enables users to do telephone banking, make airline reservations, conduct stock transactions, send and receive e-mails, and have access to the Internet via color screens.

The reason why i-mode has worked in Japan is largely attributed to its ability to attract content providers through revenue-sharing deals (where the mobile operator takes only a small percentage of the money). And consumers are drawn to the services, which are cheap and charged to the phone bill of customers by the mobile operator on behalf of the content provider. I-mode's favorable outcome is additionally due to the fact that it is marketed with an emphasis on services and the overall user experience with no reference to the actual technology.

Considering the business model's success in Japan, it is not surprising that mobile operators are now taking i-mode into Europe (supported over their GPRS networks). But, if mobile operators are to successfully run i-mode in Europe for the long term, it will be despite very real and significant market variations. Firstly, NTT DoCoMo in Japan has an extremely powerful position in the value chain, which means that it has been able to dictate network equipment and handset specifications to manufacturers and impose its will on market players. In Europe, the links between operators and handset manufacturers are not as close and operators do not control the rights to wireless Internet standards in the same way as NTT DoCoMo owns the i-mode standard—cHTML (i-mode's simplified HTML version).

Secondly, when i-mode was introduced in Japan, Internet penetration was low (22 percent) and the new service was the only method of accessing the Internet for users without PCs. In contrast, wireless Internet offerings in Europe have to compete with a large installed base of PCs which already offer a rich Internet experience. This means that the wireless Internet is only an add-on for most people, rather than a primary source of access in Europe. Because of this, European operators have to convince consumers that the wireless Internet "extension" is something that customers really need when they are mobile.

Thirdly, i-mode handsets carrying Asian brand names are not as popular as other makes in Europe, especially among i-mode's primary target groups (young and fashion-conscious mobile customers). Handset replacement cycles are also much longer in Europe than in Japan.

Finally, running an i-mode service over GPRS—while combining cHTML and WAP in one browser—may prove to be a major technical challenge to the Asian device manufacturers (only Toshiba and NEC are trying to do this at present).

KPN Mobile became the first European mobile operator to introduce i-mode services in Germany (with its partner NTT DoCoMo) through its E-Plus subsidiary in March 2002. It has already rolled out i-mode in the two other countries where it has European wireless interests (the Netherlands and Belgium) and has already planned to reach more than one million i-mode users across its three European markets.

The operator sold 27,000 handsets during the first two months after its i-mode launch. E-Plus has signed up with 85 content partners and is following the NTT DoCoMo business model of giving a large proportion of the revenue to the content supplier. E-Plus keeps a monthly basic subscription fee of €3 and takes only 14 percent of the revenues itself for bill management and credit risk, along with additional revenues from the traffic. The services from the content providers are priced at up to €2 per month. The same revenue approach is being used in the Netherlands and Belgium, and other European mobile operators launching i-mode will roughly follow the same revenue distribution path.

There are, at present, two other operators that intend to launch i-mode in Europe. The first is Bouygues Telecom (France), which concluded an i-mode license and technology transfer agreement with NTT DoCoMo in April 2002. The second, Spain's Telefónica that reached an agreement with NTT DoCoMo to rollout i-mode by the first quater of 2003 and intended to introduce the service under the operator's mobile Internet brand name—e-moción. At the end of June 2004, there were 6.5 million i-mode subscribers in Europe [WIELAND04].

There's definitely space in the European market for I-mode to continue and expand as an effective means for operators to deliver and for subscribers to access mobile content.

But a number of issues still have to be resolved before i-mode can succeed in Europe. These include handsets being GPRS, WAP, and i-mode-enabled with colored screens, a wide selection of handsets for customers to choose from and the involvement of western manufacturers. I-mode handsets and contents also have to be significantly cheaper than existing services, and the introduction of a prepay version—along with an increase in NTT DoCoMo alliances and effective i-mode marketing—needs to be satisfactorily addressed.

4.5.8   3G versus W-LAN

Third-generation wireless mobile systems and W-LANs are presently the two main emerging technologies in the wireless broadband access landscape. 3G is a radio communications technology that supports mobile wireless voice as well as high-speed mobile wireless access to Internet-based services. W-LAN, on the other hand, is essentially a technology that provides high-speed wireless access in limited coverage areas called *hot spots*, such as airports, corporate offices, coffee shops, and hotels.

The key differences between 3G and W-LANs are the following: 3G uses a licensed spectrum [UMTS-RPRT], in which cellular operators have invested in, is costly, and forces cellular operators to conform to certain regulations. By mid-2002 more than 110 frequency licences were granted for 3G network deployments, guaranteeing exclusive frequency spectrum in the order of approximately 30 MHz (duplex) per operator. In addition, CEPT has designated license-exempt spectrum for the use of UMTS/TDD technology in the bands from 2010 to 2020 MHz.

W-LANs, on the other hand, operate in the free, unlicensed, and unregulated spectrum (2.4 GHz or 5 GHz) [IEEE-STDRD] resulting in low barriers to entry but, at the same time, hindering operators from providing a guaranteed quality of service due to interference issues beyond their control.

Furthermore, 3G is positioned as a high-speed service with a data rate of 2 Mb/s (urban) and 100 to 300 kb/s (elsewhere), whereas 2.5G can deliver theoretical data rates of about 171 kb/s with GPRS and about 144 kb/s with cdma1xRTT. W-LAN technologies can get end-user rates of around 2 to 5 Mb/s in hot spots (theoretical maximum is 11 Mb/s for IEEE 802.11b—see Sec. 1.7.1 for details on the IEEE 802.11 standard), with a promise (to date) of a maximum rate of 54 Mb/s (using IEEE 802.11a).

Table 4.12 summarizes the key differences between the two technologies, W-LAN and 3G.

Given the previously mentioned semantics of 3G and W-LAN technologies, it becomes evident why a number of mobile professionals saw the proliferation of comparatively cheap wireless LANs as a mortal threat to the whole business premise for 3G. Perhaps this is why so many incumbent mobile operators were for a time seemingly slow to get moving. The industry view is changing, with wireless LAN being increasingly seen as a technology ally for the near-3G systems that mobile operators invested so heavily in.

To begin with, it has been a very entrepreneurial market with small start-ups (like Boingo and Wayport) rapidly dotting local regions with hot spots whilst the mobile operators watched with both interest and anxiety, fearing that most of their potential 3G customers would find the scattered wireless LAN service good enough. However, business models for hot sspot specialists remain uncertain. For mobile operators, the vision has been to bring these together with 3G over time, with much work being invested in making them far more seamless. However, it is not yet apparent whether the current crop of business models for the new

TABLE 4.12   Comparison of the Two Technologies W-LAN and 3G

	3G	W-LAN
Data Rates	144 kb/s (outdoor) up to 2 Mpbs (indoor). 2.5G can deliver theoretical data rates of about 171 kb/s with GPRS and about 144 kb/s with cdma1xRTT. However, the practical data rates will be about half the theoretical ones	High data rates: 802.11b standard up to 11 Mb/s and 802.11a up to 54 Mb/s. The practical data rate is about half the theoretical data rate.
Handoff	Implemented as integral part of the specification.	Currently not implemented, however possible in the future with solutions like MobileIP or STP bridging.
Cost of Network	Expensive to install: 3G base station costs about $250,000.	Cheap to install: access point costs anywhere from $1000 to $5,000
Coverage Area (per BS/AP)	~5 mi radius	100–300 ft radius
Spectrum	Licensed	Unlicensed (2.4 GHz and 5 GHz)
Supported Media	Voice and data	Primarily data. VoIP in future
Connectivity	Seamless (anytime, anywhere)	Interrupted; only possible in limited locations (i.e., hot spots)
Security	Provides a more secure environment than W-LAN	Currently security provided with WEP (wired equivalent privacy) security mechanism using RC4 encryption which is very insecure. 802.11TGi will improve security using various mechanisms like 802.1x, RADIUS, KDC, and TKIP/AES.

breed of hot spot service providers will succeed. Simply put, will people pay between US$25 and US$60 a month for patchy wireless Internet access in coffee shops, airports, and hotels? We do not really know yet, but what is becoming clearer is that rather than eroding the demand for 3G and near-3G services by skimming the cream off the market, mobile operators are coming to see public access wireless LAN as a way to promote the concept of connecting to the Internet through the air, and actually drive demand for wireless data services; indeed, wireless LAN may prove to be the optimal way to meet heavy service demands that would overwhelm 3G networks in crowded places such as airport terminals, train stations, and convention centers.

Besides, it is always worth remembering that services for mobile users will not be the same as those targeted at the desktop. In this

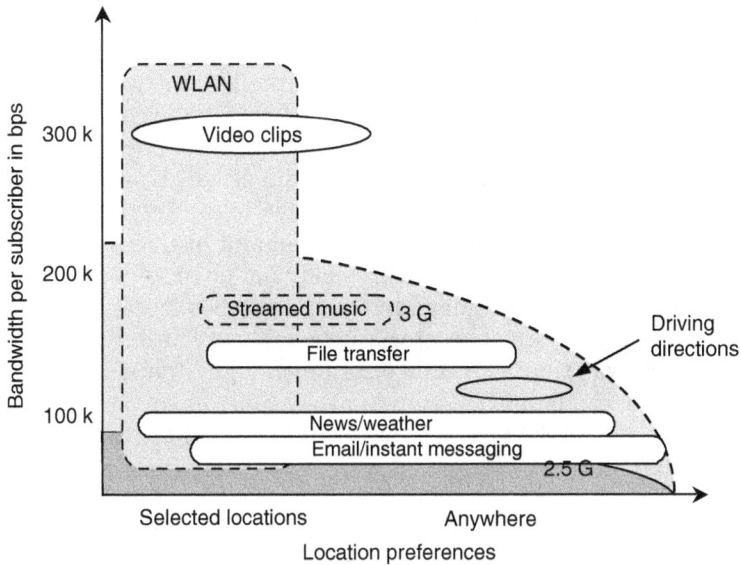

Figure 4.50 Application requirements compared to provisioning alternatives [HENKEL02].

sense, two different access preferences, i.e., anytime/anywhere versus selected locations, identify customer preferences for certain applications for 3G versus W-LAN access. The most attractive applications for mobile devices include email, driving directions, instant messaging services, and streamed music. While these applications are mostly valued anytime/anywhere, users are still highly interested in these applications in limited locations. The different data rate requirements for each of these applications versus the respective location sensitivity are illustrated in Fig. 4.50. The dotted lines delimit areas where 2.5G, 3G, and W-LAN service provisioning is possible, depending on the technical specification of each of the standards. The fact that 3G and W-LAN support different types of applications, based on technological capabilities and customer preference, allows us to believe that these technologies serve different niche market segments (Fig. 4.50). In this regard, W-LANs should only be seen as an optional extra to 3G by end users and MNOs alike—it is not an alternative technology to 3G.

4.6 Systems Beyond Third Generation

Over the past years, the emphasis has been on utilizing means to extend and enhance coverage; simply put, how to achieve the desired level of coverage by employing the minimum of costly infrastructure equipment (base stations, base site controllers, switching centers, and the like). At the same time, customers are becoming more discerning; they are beginning to demand better quality in terms of intelligibility, clarity, and absence of the artifacts traditionally associated with radio communications systems. Today, the real goal of MNOs is to offer seamless, secure, and cost-effective range of data and multimedia services—across heterogeneous networks such as fixed, cordless, cellular, and satellite— anywhere, anytime, anyhow, and on an on-demand basis, even when roaming to networks where such features may not be provided.

When 3G was initially envisioned in the early 1990s it was not very clear what its breakthrough characteristics would be. The only self-evident requirement was that 3G would offer much higher data rates and higher traffic capacity than 2G. Soon after the initial deployment of 2G, the goals and targets of 3G were better understood and the research— and later standardization work—could really start. In the original vision of the third generation around 1990, the capabilities of different wireless access systems such as cellular, cordless, and data services should be supported in all radio environments by a single radio interface. During the definition and standardization phase of UMTS it turned out that there is no single radio technology that can be optimized for all applications. Therefore, the UTRA concept combined the FDD and TDD components to support the different symmetrical and asymmetrical service needs in a spectrum-efficient way [CHAUD99, PRASAD00].

Although 3G systems represent clear improvements over 2G systems in terms of spectral efficiency, peak data rates, QoS control, and the like, work is already ongoing on developing the systems even further. Now that the first 3G systems are being built, it is natural to start asking what could be the next big step; what might come after 3G is surely not quite the same as with 2G and its successor 3G 10 years ago.

The major driving forces for systems *beyond third generation* (B3G) will be the economic success based on user demands for new and advanced services, with high security and reliability, and economic advantages and business opportunities for the involved players, competition, costs, size and weight of terminals, and battery lifetime. The ratio between cost and performance will improve beyond third generation.

The available, emerging, and evolving access technologies have basically been designed in the classical vertical communication model that a system has to provide a limited set of services to users in an optimized manner. Systems beyond third generation will mainly be characterized by a horizontal communication model, where different access technologies like cellular, cordless, W-LAN type systems, systems for short-range connectivity, and wired systems will be combined into a common platform to complement each other in an optimum way for different service requirements and radio environments [WWRF]. These access systems will be connected to a common, flexible, and seamless core network. This contrasts with 3G, which merely focuses on developing new standards and hardware. B3G systems will support comprehensive and personalized services, providing stable system performance and quality service. Mobility management will be a part of a new media access system as the interface between the core network and the particular access technology to connect a user via a single number for different access systems to the network. This will correspond to a generalized access network. Global roaming for all access technologies is required. The interworking between these different access systems in terms of horizontal and vertical handover and seamless services with service negotiation including mobility, security, and QoS will be a key requirement, which will be handled in the newly developed media access system and the core network.

Figure 4.51 shows this vision of a seamless network including a variety of interworking access systems that are connected to a common IP-based core network.

Further characteristics of such systems will be:

- A variety of supported data rates according to second-generation cellular systems to third generation systems, broadband access (W-LAN

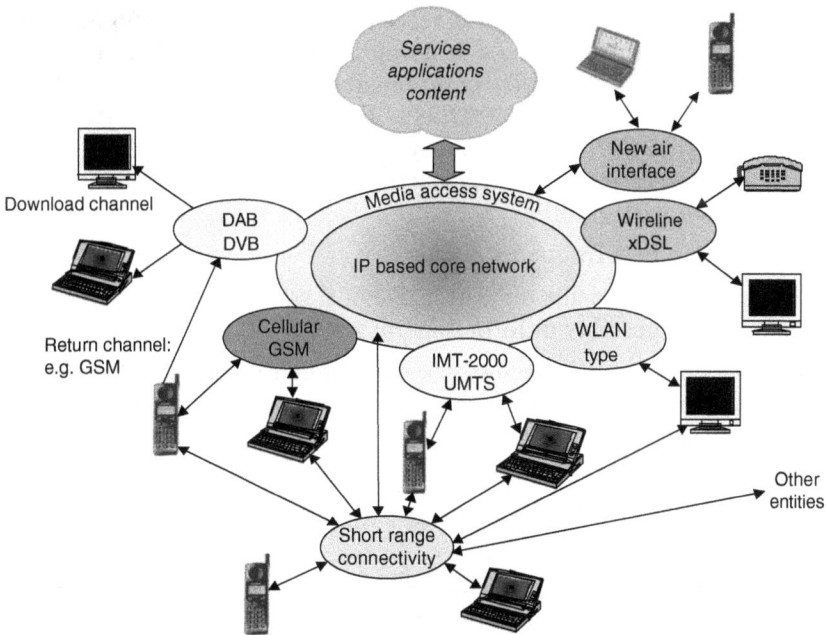

Figure 4.51 Seamless future network including a variety of interworking access systems.

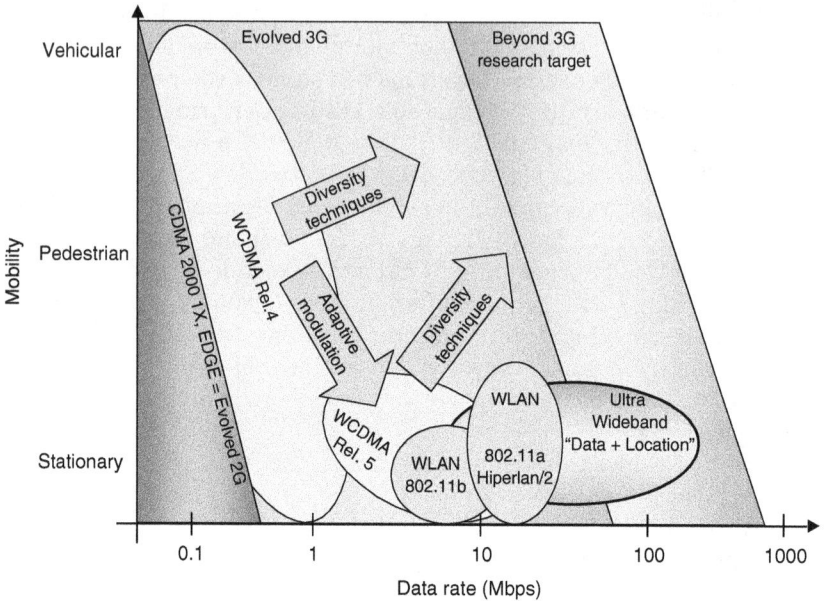

Figure 4.52 Mobile systems beyond IMT-2000.

type systems), short-range connectivity systems and wired systems as xDSL with maximum data rates up to 10, 20, and 155 Mb/s for current and future services and applications (ITU-R WP8F is currently discussing data-rate requirements for the new elements of systems beyond UMTS [ITU2002] (see Fig. 4.52). The data rates of up to 100 Mb/s for the mobile access and up to 1 Gb/s for the nomadic/local area wireless access are defined as the aggregate cell payload (i.e., the peak useful data rate))

- Current and new frequency bands
- Sharing of frequencies for better use of resources between different systems and possibly different operators
- New network types and network management like ad hoc and self-optimizing networks, automatic and dynamic network reconfiguration, and dynamic frequency allocation to support a variety of access systems on a common platform
- Support of symmetrical and asymmetrical services by FDD and TDD systems
- Core and radio access networks designed for efficient packet transmission by supporting improved QoS requirements for real-time services
- Optimization of transmission links with respect to asymmetrical traffic

- Core and radio access network based—from today's point of view—on IP due to lower infrastructure costs, faster provisioning of new features, and easy integration of new network elements

- Separation of the physical layer and different access technologies from the applications by means like JAVA virtual machines or *common object request broker architecture* (CORBA) to decouple the applied software from the hardware

Table 4.13 summarizes the main parameters of some typical access technologies.

4.6.1 B3G challenges

Researchers are currently developing frameworks for future B3G networks. Different research programs, such as Mobile VCE, MIRAI, and DoCoMo, have their own visions on B3G features and implementations. Some key features (mainly from the point of view of users) of B3G networks are stated as follows:

- High usability—anytime, anywhere, and with any technology

- Support for multimedia services at low transmission cost

- Personalization

- Integrated services

First, B3G networks are all IP-based heterogeneous networks that allow users to use any system at any time and anywhere. Users carrying an integrated terminal can use a wide range of applications provided by multiple wireless networks. Second, B3G systems provide not only telecommunications services, but also data and multimedia services. To support multimedia services, high-data-rate services with good system reliability will be provided. At the same time, a low per-bit transmission cost will be maintained. Third, personalized service will be provided by this new-generation network. It is expected that when B3G services are launched, users in widely different locations, occupations, and economic classes will use the services. In order to meet the demands of these diverse users, service providers should design personal and customized services for them. Finally, B3G systems also provide facilities for integrated services. Users can use multiple services from any service provider at the same time. Just imagine a B3G mobile user, Stella, who is looking for information on movies shown in nearby cinemas. Her mobile may simultaneously connect to different wireless systems. These wireless systems may include a global positioning system (for tracking her current location), a wireless LAN (for receiving previews of the movies in nearby cinemas), and a code division multiple access (for making a

TABLE 4.13 Main Parameters of Different Access Systems

System	Data rates	Technology	Range	Mobility	Frequency range	Original application area
GSM (including GPRS, HSCSD and EDGE)	9.6 kb/s up to 384 kb/s	TDMA, FDD	up to 35 Km in GSM lower for data	high	900, 1800, 1900 MHz	public and private environment
IMT-2000, UMTS UTRA)	max. 2 Mb/s	IMT-2000 family, W-CDMA (FDD) + TD-CDMA (TDD)	30 m –20 Km	high	2 GHz (ITU spectrum)	public and private environment
DECT/ DECTlink	max. 2 Mb/s	TDMA / TDD	up to 50 m	low	1880–1900 MHz	office and residential environment
Bluetooth	max. 721 kb/s	direct sequence or frequency-hopping	0.1–10 m	very low	2.4 GHz ISM band	cable replacement, SoHo environment
HIPERLAN 2	25 Mb/s	OFDM, TDD	50– 300 m	low	5 GHz	corporate environment, public hot spots
IEEE 802.11a	about 20 Mb/s	OFDM, TDD	50–300 m	low	5 GHz	corporate environment, public hot spots
HIPERACCESS	about 25 Mb/s 1.5 Mb/s	not yet specified OFDM	2 ... 10 km ≤100 km	no high	5–40 GHz	business access, feeder
DAB					e.g. 176–230 MHz 1452–1467.5 MHz	audio broadcasting
DVB-T	5–31 Mb/s per 8 MHz channel (mobile: 5–8, fixed 16–31)	OFDM	≤100 km	medium to high	TV bands below 860 MHz	video broadcasting
Cable modem	down <40 Mb/s up <10 Mb/s	FDD QAM / QPSK	5 to ~20 km	no	down ~60 to 860 MHz up 10 to ~40 MHz	residential environment
ADSL	down ≤6.144 (8) Mb/s up ≤0.640Mb/s	DMT (carrierless AM/PM CAP)	2–6 Km	no	base band	SoHo (Small Office – Home Office), SME, residential environment

telephone call to one of the cinemas). In this example, Stella is actually using multiple wireless services that differ in QoS levels, security policies, device settings, charging methods, and applications. It will be a significant revolution if such highly integrated services are made possible in B3G mobile applications.

The interworking of different access systems on a common platform and the necessary multimode or adaptive and multiband terminals for different access systems and a wide range of services are some of the desirable key characteristics of beyond 3G systems. Challenges are in several areas such as the radio interface, the radio access and core network, implementation issues, and services-related issues.

To migrate current systems to B3G with these features, we have to face a number of challenges. As the present evolution seems to be going toward 10 Mb/s/100 Mb/s peak data rates, any air interface for systems B3G should be clearly better in order to justify its technical and commercial feasibility. Figure 4.53 shows the mobility versus data rate both for different mobile system generations and also the research target for a possible revolutionary air interface for B3G systems.

Many technical challenges have to be solved by extensive research to make the vision of systems beyond third generation a reality. The interworking of different access systems on a common platform and the necessary multimode or adaptive and multiband terminals for different access systems and a wide range of services are some of the desirable key characteristics of beyond 3G systems. Challenges are in several areas as the radio interface, the radio access and core network, implementation issues, and services-related issues.

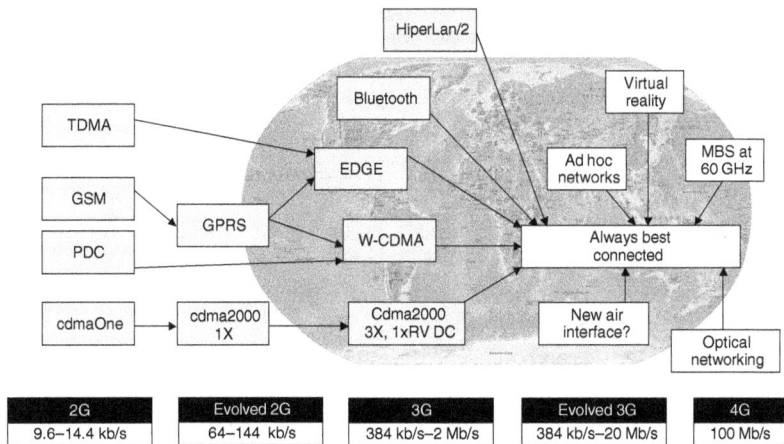

Figure 4.53 The path to future-generation mobile systems.

Spectrum efficiency is of paramount importance in future systems. Systems B3G have to use the frequency resources as efficiently as possible. Therefore, several physical-layer-related techniques have to be investigated:

- Optimization of evolving and emerging access systems by improved modulation and channel coding schemes for further enhancement of spectrum efficiency and system performance

- Advanced detection schemes as multiuser detection and interference cancellation to gain from the a priori knowledge about intra- and intercell interference signals

- Signal processing algorithms as the trade-off between performance gain and computing complexity

- Compression techniques for source coding to reduce the needed user data rate

Improved algorithms to support these physical layer issues are

- Link adaptation according to the channel conditions, traffic load, and services for better usage of the frequency resources and improved system performance

- Spectrum sharing between different systems and the investigation of coexistence conditions between different radio access systems

- Advanced antenna concepts to improve the link quality and channel capacity

These concepts are used to increase the channel capacity of the radio link. Diversity concepts reduce basically the impact of fading due to multipath transmission. Multiple antenna concepts are a further extension of diversity concepts, gaining from uncorrelated multipath transmission channels between the different antenna elements on the base station and the terminal side [IEEE-PERS98]. The basic idea is to reuse the same frequency band simultaneously for parallel transmission channels by space-time coding to increase the channel capacity.

Adaptive antenna (Fig. 4.54) concepts improve the link quality by reducing the co-channel interference from different directions and in the more advanced spatial division multiple access (SDMA) concept by reusing the same frequency channels simultaneously for different users in distinct directions. System aspects like common control channels and the signaling concept are an essential part of advanced antenna concepts in order to achieve the possible range extension for the common control channels as for the traffic channels. These are key concepts to use the scarce frequency spectrum as efficiently as possible without major impacts on evolving access systems.

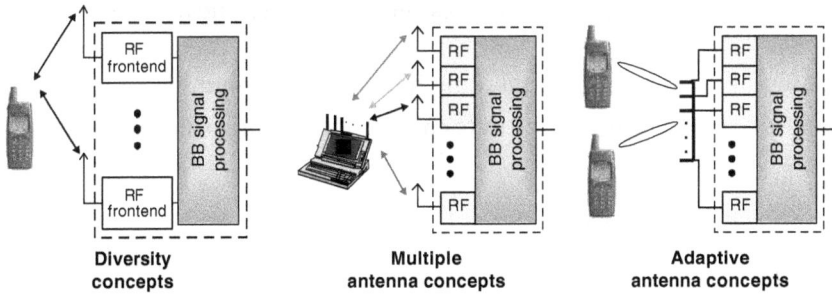

Figure 4.54 Advanced antenna concepts.

A prerequisite for improving link quality and channel capacity is the design and evaluation of a realistic wideband channel characterization for new frequency bands up to about 60 GHz. Based on wideband propagation, measurements channel models are needed for the international standardization process including models for the direction of arrival.

Higher layer protocols in the access network (medium access system) are an additional area to improve the system performance even further. These include:

- Self-optimizing networks and automatic network reconfiguration
- Resource allocation algorithms with respect to varying traffic load, services, bearer capabilities, radio environment, and channel conditions
- Dynamic frequency allocation
- Interworking of different access systems on higher layers via horizontal and vertical handover and service negotiation
- Network management

The seamless future network in Fig. 4.51 comprises several access systems with seamless interworking for different applications and radio environments. The interworking of different access systems on an IP-based common platform via the medium access system with horizontal and vertical handover results in several challenges to the core and the medium access system. Major topics are related to the improvement and extension of IP for mobile applications and the optimization for radio transmission as

- Support of real-time and non-real-time services with respect to service requirements like QoS and especially delay requirements for real time services

- Service negotiation for seamless services versus the available access systems and bearer capabilities
- Mobility management including handover and roaming
- Security mechanisms like authentication, authorization and accounting (AAA)

Besides, users will gain from this flexible concept only when multimode and multiband terminals with low power consumption and reasonable size are feasible. A variety of terminal types such as PDAs, notebooks, and handsets are supporting these applications. Other technical challenges are improved display techniques and battery technology.

Several concepts for the terminal implementation are under discussion (Fig. 4.55), which will also be applied to base station equipment. In its simplest form, several fixed modes are implemented in parallel. However, this straightforward concept is inflexible for future improvements. More advanced concepts apply a signal processing platform, where the parameter sets of different access systems are downloaded to the signal processing unit. In most advanced concepts a flexible and freely programmable signal processing unit can be adapted to the actual access system. The last two concepts are software-defined radios, which are the major technical challenge for the terminal implementation [IEEE-PERS99]. Such concepts will become feasible with the progress in semiconductor technology and the increasing available signal processing power.

From the user's point of view the man–machine interface has to be easy to use and self-explanatory to enable people—even handicapped and elderly people—to use easily advanced services. The content has to be adapted automatically to the actual bearer capability of the used access system.

For the sake of completeness, we further elaborate on some of the previously mentioned research areas which constitute the forefront research activities on the air interface of the systems beyond 3G. Our goal here

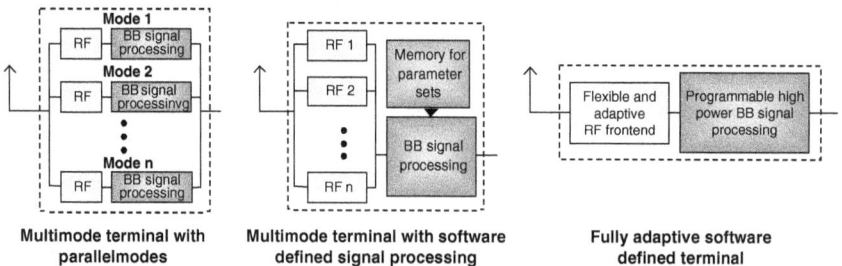

Figure 4.55   Concept for multimode and software-defined terminals.

is to develop an intuitive feel for the technological challenges and list some potential research areas that need to be covered as part of beyond 3G air-interface research.

Spatial domain solutions.   Smart antennas, transmit/receive(Tx/Rx) diversity, beamforming, and *multiple-input-multiple-output* (MIMO) systems.

Spatial domain solutions are very important if considerable improvements to wireless system performance are targeted. Techniques like transmit/receive diversity, beamforming and MIMO should be studied carefully both with short- (e.g., evolution of 3G) and long-term goals (possible beyond 3G solutions) in mind.

Smart antennas and beamforming techniques.   Future wireless applications will be required to provide significant improvements in the system capacity, and per-user data rates and operate with significantly-reduced costs per transmitted bit—as compared to 3G systems—in multiservice, multitechnology, networks and highly variable propagation scenarios. Smart antenna and MIMO processing techniques offer the capability of providing such improvements. The smart antenna technology encompasses a wide variety of techniques that can be used at both the base station and in the user terminal, in order to achieve significantly higher data rates, better link quality, and increased spectral efficiency. Moreover, the adoption of MIMO processing techniques in future wireless systems is expected to have a significant impact on the efficient use of the spectrum, the minimization of the cost of establishing new wireless networks, on the optimization of the service quality provided by those networks, and facilitate transparent operation across multitechnology wireless networks.

To achieve these goals, the spatial processing algorithms have to be optimized for the chosen air interfaces to achieve the best performance/cost trade-off. Different transmission strategies have been developed in order to achieve increased spectral efficiencies, such as space-time codes, which achieve space diversity (space-time trellis coding and space-time block coding) and coding gain (for space-time trellis coding), BLAST technology, which takes advantage of several independent spatial channels through which different data streams can be transmitted, as well as beamforming, which minimizes interfering power, and space-time techniques which optimally exploit the inherent diversity of frequency selective multipath channels. Beamforming techniques are generally used at the base station for uplink reception and downlink transmission with multiple antennas. Beamforming allows spatial access to the radio channel by means of different approaches, such as based on directional parameters or by exploiting the second-order spatial statistics of the radio channel.

These transmission strategies require efficient techniques to separate the signals of multiple users sharing the spectrum resources at the receiver and to cancel interference, under various interference scenarios. Some of them use pilot signals known by both the transmitter and receiver (nonblind techniques), while others employ the a priori knowledge of the received signals (blind/semiblind techniques). Depending on the multiple access technique, different strategies have been proposed, from theoretically optimal strategies to practical ones like parallel or successive interference cancellation (PIC or SIC), decorrelation, and joint detection. Reduced complexity schemes may use turbo techniques for performance enhancement.

Optimization of smart antenna and MIMO transceivers can be achieved for various air interfaces by developing robust (the best if possible) and reduced complexity baseband signal processing algorithms, which exploit the propagation environment (picocell, macrocell, and the like), the channel stochastic properties (narrowband, frequency selective, long-term, short-term), and user information (contextual, location-based, and the like), in order to enhance performance

- At the link level, by improving channel estimation (blind, semiblind, and the like), interference cancellation capability and multiuser detection

- At the system level, by optimizing resource allocation and increasing capacity

To summarize, the integration of smart antennas into mobiles results in many very attractive features, and their implementation is thus worthwhile even despite some difficult technical challenges:

- Due to the increasing number of subscribers, the limited resources of the air interfaces in the current cellular networks are measured to be too short for the future. Smart antennas will increase these resources by adding an additional degree of freedom to diversity. Integrated into handsets, they will allow the usage of this additional diversity for the uplink as well.

- Electromagnetic radiation can be significantly reduced by forming a "beam" in the opposite direction of the human body. Two different kinds of effects are suspected to cause harm—the conversion of the radiation into heat and a direct effect on biological tissues. Negative effects on the brain, on the ear, and on the intraocular pressure, caused by localized heating, are in discussion [NIEMEG01, DAM01]. The other direct effects of electromagnetic waves on biological tissues has not been understood completely yet. A damage of animal DNA strands by even low-level exposure of radio frequency radiation has been reported [WALKE01]. Obviously, the direction of suitable smart antennas can reduce the risk caused by both the effects of electromagnetic radiation.

- A smart antenna focuses its transmission power toward the base station. By this, either the range will be increased for fixed transmission power, or the power can be reduced for a fixed range. This will result in reduced power consumption.

The integration of smart antennas into mobiles is, however, a difficult task that has not been realized yet. A major challenge faced by technology experts is the limited space in handsets. The lack of space in a mobile terminal forces deviation from the golden rule that the spacing between the array elements should not be smaller than half of the used wavelength of the transmitted or received radiation. With a wide spacing, a better directivity can be obtained.

Reconfigurable and robust signal processing techniques.   B3G systems require signal processing techniques that would be capable of operating under highly variable scenarios, with respect to

- Propagation environment (indoor/outdoor, rich-scattering/specular, and the like)

- Traffic environment (hot spot/spotty, uniform/dense/sparse, and the like)

- Interference environment (intracell/intercell, same system/other system, and the like)

- System mobility (static/mobile users, speed of interference, and the like)

- Antenna configuration (number of antennas at the terminal/base, antennas correlation/bandwidth, antenna topology, and the like)

There are two main approaches to allow wireless communication transceivers to operate in a multiparametric, continuously changing environment:

- Reconfigurable, adaptive techniques for adjusting the structure and parameters of the transceivers to allow them to demonstrate the best performance in a variety of situations

- Robust techniques, which can demonstrate reasonable (required) performance in a variety of the unspecified situations

The first approach assumes that the particular scenario can be identified, the optimal solution can be known, and the required transceiver configuration can be provided. MIMO receivers, which are capable of reconfiguring themselves by switching automatically between a beam-forming and a spatial multiplexing technique [NIEMEG01, DAM01], can be considered as an example of the first approach.

The second approach can be illustrated by short-burst systems [WALKE01], which allow avoiding nonstationarity tracking.

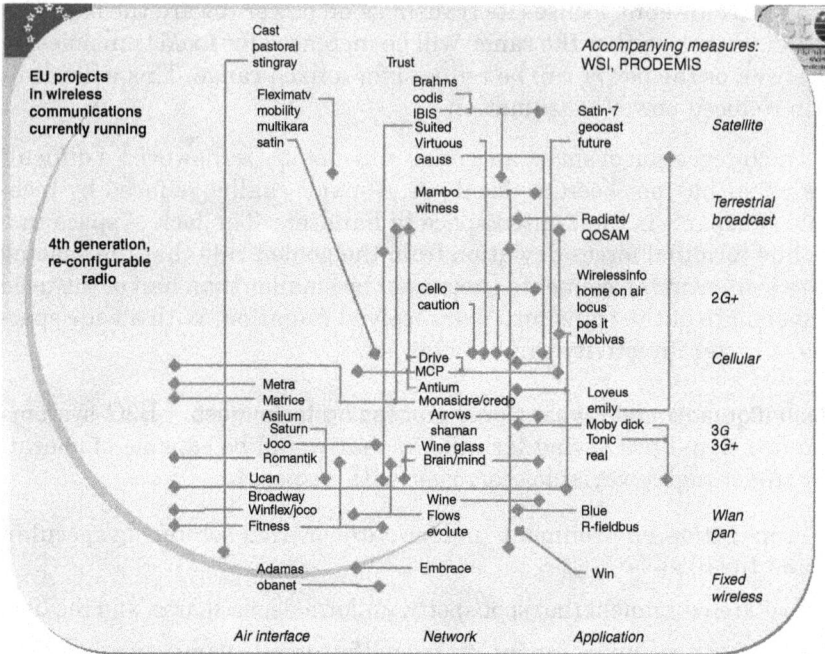

Figure 4.56 EU research projects in wireless communications (*www.cordis.lu/ist/ka4/mobile/index.htm*).

Other examples are MIMO and interference cancellation techniques [REMON01, CHEN97] based on the semiblind estimation algorithms without explicit estimation of the propagation channel for all signals received by an antenna array.

To meet these challenges extensive international research activities are necessary to solve technical issues and to prepare the consensus building for international standardization of new ideas, including the interworking of systems by vertical handover, global roaming, the optimization of evolving and emerging access systems, as well as the radio access and core network.

Figure 4.56 presents a synoptical map of the research European activities on 3G and (partly) on 3G+ systems, funded by the European Union IST FP5 and FP6 (framework programme) programmes (*www.cordis.lu/ist/fp5*).

4.6.2 Small-cell structures and the potential of multihop relaying in future-generation systems

The previously mentioned research issues dictate that the development of B3G mobile services and technology presents major and unprecedented technical challenges. Part of this technological "labyrinth" is the use of ultrahigh frequencies (see Fig. 4.53). On one hand, in high frequencies, higher bit rates are possible. On the other hand though, channel characteristics become more severe as the operation frequency gets higher, giving rise to system losses. To compensate for these losses, power is necessarily increased. These factors all probably demand a reduction in cell sizes for systems B3G.

It is evident that for a given service area with a fixed available frequency bandwidth, a wireless infrastructure using small cells can provide more capacity than a system using larger cells [COX95]. Besides, assuming that the capacity of a cell remains constant, the number of channels that can be provided within the service area increases as the cell size is reduced. The increase in available traffic channels is in fact proportional to the inverse of the square of the decrease in the size of the cells of the system. If the diameter of a cell is decreased by a factor N, the number of cells that cover the same service area increases by a factor N^2, and the number of available channels within the given service area increases by a factor N^2.

Small-cell systems trade system capacity with infrastructure costs and handover rates. As cell sizes decrease, users cross the cell boundaries more frequently, raising the flag of higher handover rates. In addition, to fully provide communications in a certain area, a higher number of base stations are required when small cells are deployed, significantly increasing infrastructure costs.

The multihop relaying technique scheme is a promising approach for enhancing the area coverage without significant additional infrastructure costs. In this approach, one or more mobile terminals relay transmission signals between an end-user terminal—with no direct link to the base station—and its serving base station. A mobile with a poor link to the base station need not increase its transmission power to compensate for the link's losses but instead hand over its call to another mobile terminal, which lies in a more advantageous position and can help as an intermediate (relaying) node of communication between the mobile terminal and the base station. The unconstrained on-demand connectivity coupled with fault tolerance has made wireless multihop networks to evolve largely during the past decade [EPHR87, JUB87, LAU95] and they are expected to play an important role in future-generation networks where mobile access to a wired network is either ineffective or impossible.

Relay-based deployment concepts for wireless and mobile broadband cellular radio [WALKE03a]. This section presents concepts of relay implementations and aims at pointing out the performance benefits that multihop relaying can provide in broadband networks when applied in certain

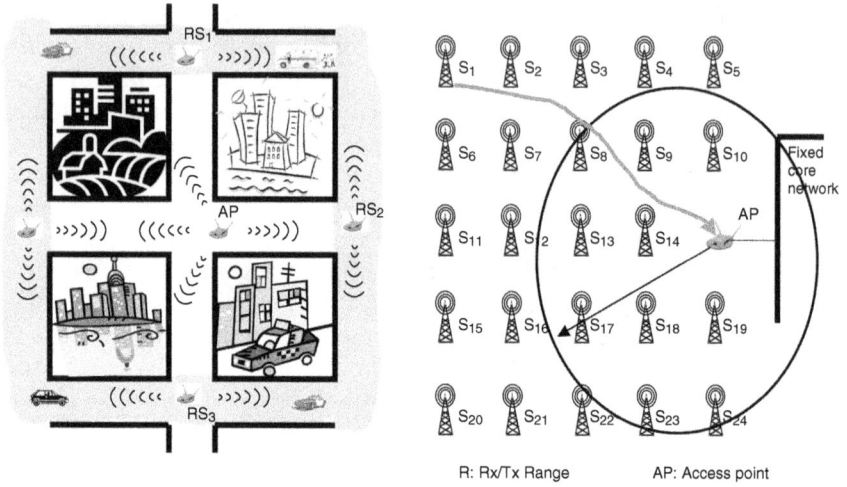

Figure 4.57   (a) City scenario with one AP (serving the white area) and four RS covering the shadowed areas "around the corners"(shown in gray). (b) Schematic of the scenario. (Source: Ref. [WALKE03])

scenarios [WALKE03a]. Figure 4.57 (a) shows a city scenario with one AP (providing radio coverage to the areas marked white) and four fixed-mounted *relay stations* (RS) to provide radio coverage to areas "around the corner" shadowed from the AP (shown in gray).

While the intersection can be covered well by the AP, the close-by streets can only be served if a line-of-sight (LOS) connectivity is available between mobile terminal and its serving station, owing to the difficult radio propagation conditions known for, e.g., the 5 to 6 GHz frequency band. The RSs allow extending radio coverage to these streets.

A schematic of this scenario—a multihop network with an access point—is illustrated in Fig. 4.57 (b) where the transmit/receive radius R is shown to be the parameter determining the connectivity of the nodes shown. A Fixed Relay (S_1) would have to route the traffic of the wireless terminals (not shown) it is serving via the intermediate Fixed Relay (S_8) using a low PHY-mode, or via S_2 and S_8 using a higher PHY-mode and thus a higher link capacity, and so forth from S_8 either via S_9 or directly to the AP. The interpretation of Fig. 4.57 (b) multihop network is that all the nodes shown are either Fixed Relays or APs and the mobile user terminals roaming in the area (not shown) are served by the nodes shown. The basic element of Fig. 4.57 (a) can be repeated to cover a wide area.

Figure 4.58 shows three examples of concepts for fixed or mobile relaying, see Ref. [WALKE03]:

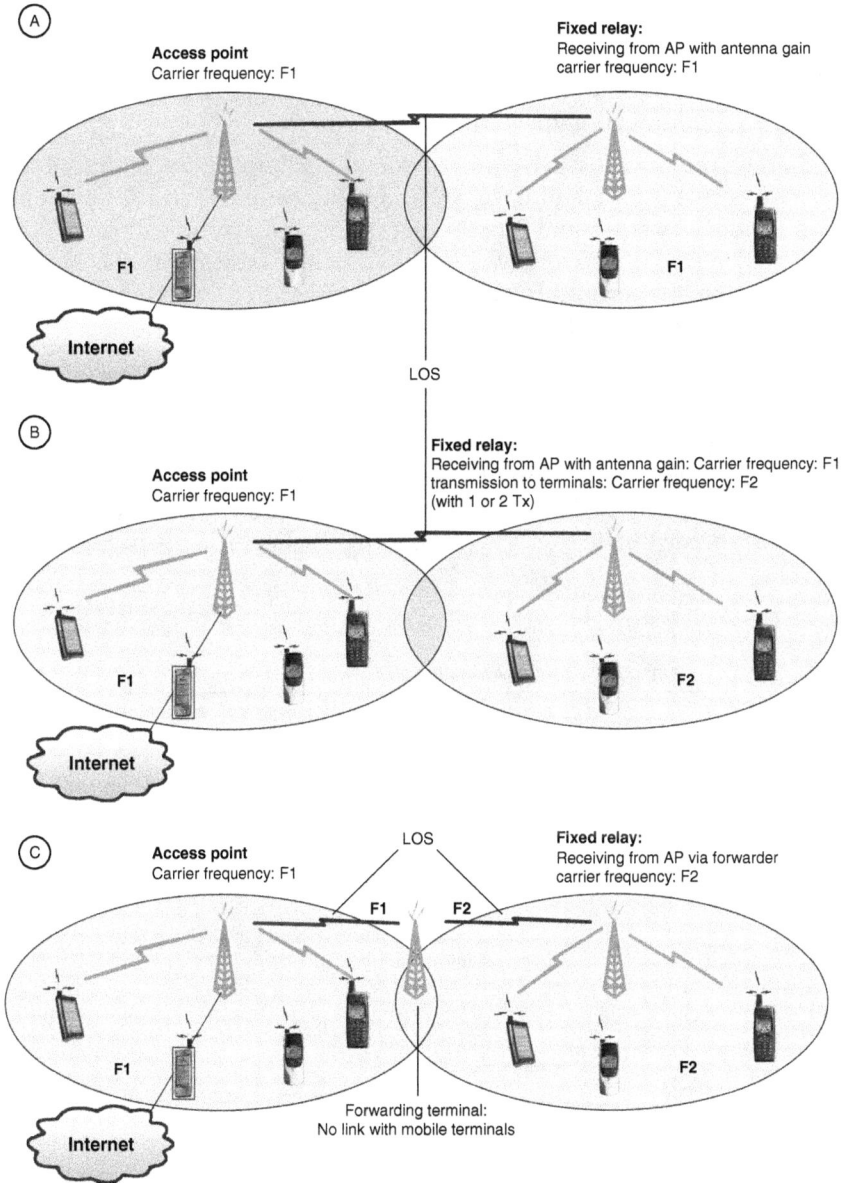

Figure 4.58 Example relay concepts (Source: Ref. [WALKE03]).

1. Relaying in the time domain with AP and Fixed Relay operating at the same carrier frequency F1

2. Relaying in the frequency domain with AP and Fixed Relay operating at different frequency carriers;

3. 2-stage relaying in the frequency domain, where a fixed mounted RS that is in the range of both AP and relay connects the AP and the second RS by store-and-forward operation and dynamically switches between frequency carriers F1 and F2. Unlike the second RS, which serves mobile terminals, the only role of the first RS is to bridge the distance between the AP and the second RS where this is not possible due to lack of LOS (see also Ref. [HABETHA02]).

In (a) and (b) the radio link is based on LOS radio with transmit and receive gain antennas at AP and relay.

An analytical estimation of the bit rate over distance from an AP that is supported by Fixed Relays to extend the radio range for the approach according to Fig. 4.58 (a) is shown schematically in Fig. 4.59.

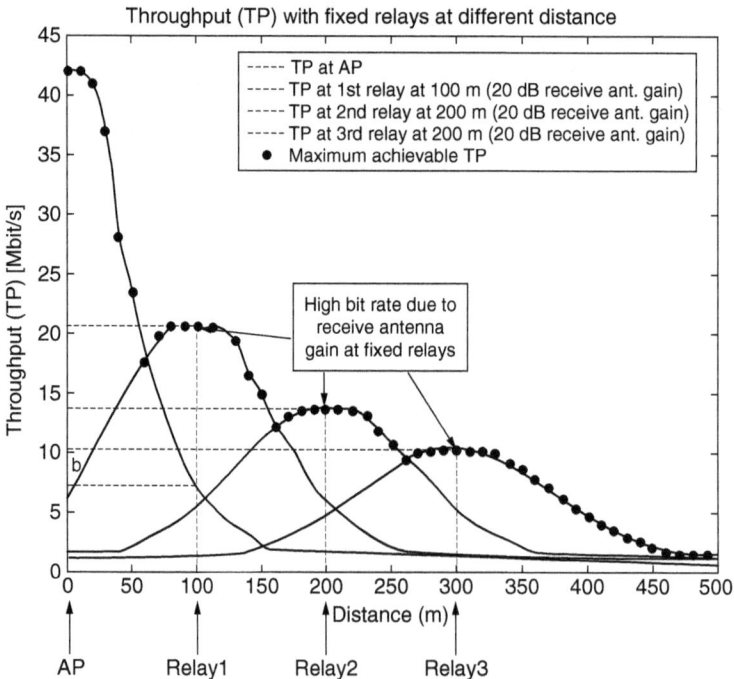

Figure 4.59 Analytical estimation of the extension of the radio range of an AP by Relays with receive antenna gain (Source: [WALKE03])

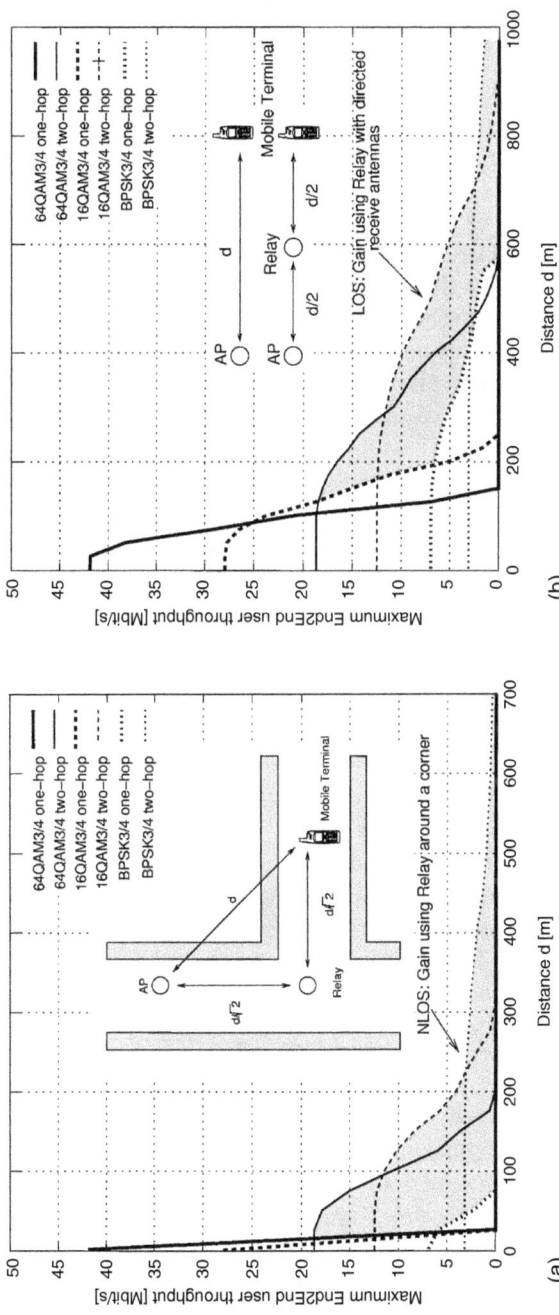

Figure 4.60 (*a*) Fixed Wireless Router at an intersection to extend the radio range of an AP "around the corner" into a shadowed area to serve a remote mobile terminal. (*b*) Maximum end-to-end throughput vs. distance for forwarding under LOS conditions with directed receive antennas having a gain of 11 dB (Source: Ref. [ESSEL02]).

Without receive antenna gain, Relay 1 would have available only a bit rate equivalent to the value b [Mbit/s], while with receive antenna gain it achieves a value a [Mbit/s]. As also shown in Fig. 4.59, the throughput decreases with an increasing number of hops.

Similar considerations apply to Relays 2 and 3.

The relaying function could be performed according to the ISO/OSI reference model either in layer 1 (physical) "repeater," 2 (link) "bridge," or 3 (network) "router." Figures 4.59 and 4.60 show analytical and simulation results according to the solution of a layer 2-relay as described in Refs. [ESSEL00] [WALKE01A] for the HiperLAN/2 standard.

Figure 4.60 (a) shows the concept introduced in Fig. 4.58 (a) the simulated end-to-end throughput between an AP and a mobile terminal that is located at a distance d for different modulation and coding schemes known from the air interface of HiperLAN/2 or IEEE 802.11a [ESSEL02]. The terminal is shadowed by a building at the street corner and is therefore connected "around the corner" with the help of a Fixed Relay. The shaded area under the curves shows the gain in terms of throughput possible from the use of the Fixed Relay without which the MT could not be connected to the AP. It can be seen that the range extension resulting from using the Fixed Relay is substantial. It is worth noting that the smart antenna technology at the AP or mobile terminal cannot provide radio coverage around an obstacle. A relay is currently the only way to achieve this.

It has been stated that the capacity of an AP when using a modem as standardized for HiperLAN/2 and IEEE 802.11a might be excessive compared to the rate requirements of mobile terminals roaming in its pico cell area that is formed by omnidirectional or sectorized antennas and a maximum transmit power of 1 W EIRP [MOHR02].

The Fixed Relay concept can also be used to extend the range of an AP to nonshadowed areas beyond the regulatory EIRP limits as shown from the simulation results for HiperLAN/2 (Fig. 4.60 (b)) according to the scenario in Fig. 4.58 (a). It can be seen that the radio range can be dramatically increased, especially when using receive antenna-gain at both ends. The use of smart antennas and beamforming to reduce the path loss between AP and relays and to connect multiple relays at the same time has been known for a long time [WALKE85].

Relay-based implementations in existing cellular radio systems [WALKE03a].
No smart relaying concept has been adopted in existing cellular systems so far. Solely bidirectional amplifiers have been used in 2G systems and are introduced for 3G systems. Yet, these analog repeaters increase the noise level and they suffer from the danger of instability due to their fixed gain, which has limited their application to specific scenarios.

Most existing and standardized systems were designed for bidirectional communication between a central base station and mobile stations directly linked to them. The additional communication traffic between a mobile terminal and a relay intermediately inserted into a link between the mobile terminal and the base station requires additional radio resources to be allocated—one of the reasons that hitherto have hindered the deployment of smart relay concepts.

TDMA-based systems are especially well-suited to introduce relaying, as this scheme allows for an easy allocation of resources to the mobile-to-relay and relay-to-BS links. The first system based on TDM and relaying to connect fixed and mobile was proposed in 1985 [WALKE85]. Another method proposed for F/TDMA systems is to reuse a frequency channel from the neighboring cells [SRENG03]. The ETSI/DECT standard in 1998 was the first specifying Fixed Relays (called wireless base stations) for cordless systems using TDM channels for voice and data communications. The ETSI/TETRA standard specifies a dual-watch function allowing the aggregate traffic of a number of mobile terminals connected in direct mode to be relayed by TDM channel-switching to a dispatcher panel connected to a BS. Relaying in cellular CDMA-based systems has been investigated by Zadeh et al. [ZADEH01]. Uplink and downlink are separated using frequency division duplex, as it is done in IS-95 and UTRA FDD. All these concepts can easily be extended to packet-based systems [ESSEL00] [WALKE01A].

A completely different approach is considered in Ref. [AGGE01], incorporating an additional ad hoc mode into the GSM protocol stack to enable relaying with the goal to achieve ubiquitous coverage and more efficient and robust utilization of network resources. With a negligible increase in the mobile station's complexity, relaying enhancements on GSM could improve robustness against radio-link failures as well as offset the difficulties of high data rate transmission over significant distances, and associated intercell interference, by lowering transmission powers. Besides, the decentralized nature of network control in wireless multihop networks provides additional robustness against coverage limitations and/or wireless channel impairments. Figure 4.61 illustrates different scenarios of a GSM cell using multihop relaying to transparently extend communication at dead spot locations (such as subway train platforms, indoor environments, and basements) with the goal to improve indoor as well outdoor coverage.

Squeezing more out of a legacy network and existing subscriber base is an excellent strategy for GSM operators to evolve in a cost effective way to the B3G business. Conceptually, the radio access part of an enhanced ad hoc GSM (A-GSM) system [AGGE01] comprises two segments—the

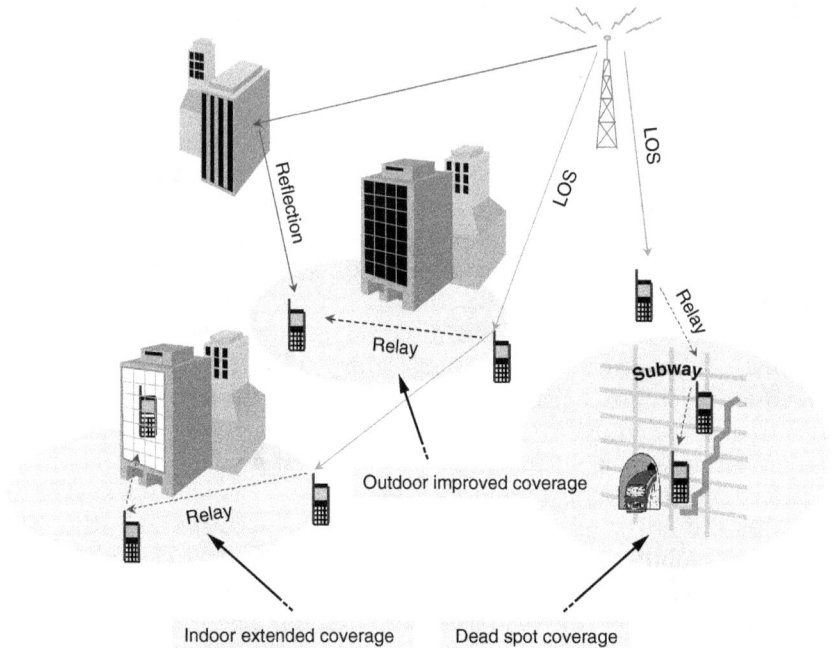

Figure 4.61 GSM scenarios enhanced with multihop relaying features.

GSM radio access and the wireless multihop (ad hoc) access part. The relay nodes in the range of BTS will toggle between A-GSM and GSM modes. The basic idea is similar to that followed by Ericsson's proposal called the ODMA TDMA/WB-CDMA proposal [WBCD].

Similar to [AGGE01], the framework proposed in Ref. [WU01] employs relaying stations to divert traffic from possibly congested areas of a cellular system to cells with a lower traffic load. These relaying stations utilize a different air interface for communication among themselves and with mobile stations that could, for example, be provided by a wireless LAN standard.

ETSI-BRAN/HiperLAN/1 and IEEE 802.11x contain the elements to operate ad hoc networks. ETSI/HiperLAN/2 [HLAN-SPEC2] in the *home extension* contains an ad hoc mode of operation that allows the nodes to agree on a *central controller* (CC) to take the role of an AP in a cluster of nodes, but no multihop functions are specified so far in any W-LAN standard. Multihop operation based on wireless relays that operate alternating on different frequency channels to connect neighboring clusters has been proven workable in Ref. [HABETHA02]. In the HiperLAN/2 basic mode (using an AP) it has been shown that multihop operation via forwarding mobile terminals is easy to perform within the framework of the standard [ESSEL00] [WALKE01A].

In the following sections, we describe in detail the architectural and functional building blocks of the approach considered in Ref. [AGGE01]. We will primarily focus on the GSM protocol enhancements as well as the networking requirements for enabling ad hoc relaying in GSM.

4.6.3   GSM evolves to B3G: toward ad hoc GSM communications

Radio channel structure in A-GSM.   The A-GSM channel structure follows the timeliness of the GSM physical channel structure. As in the GSM system, TDMA is applied to the A-GSM radio path. Eight time slots are used for each frequency band. Figure 4.62 illustrates the mapping of A-GSM frame structure onto the GSM frame structure. We covered the GSM channel structure in Chap. 3.

The channel allocation process in an A-GSM system inherits the rules and restrictions imposed by the channel allocation process in the GSM system. Specifically, A-GSM multihop nodes within the same cell must be assigned distinct "orthogonal frequencies/time slots" (i.e., channels).

In addition, an A-GSM end-to-end multihop connection must be assigned different channels on each link along the multihop path. Even if perfect power control is assumed so that channels within the same cell can be reused from nodes of this cell, the highly dynamic nature of the network may create concerns on whether the power control, which is a distributed process performed by all multihop nodes along the multihop path, can react and adapt fast enough to the frequent topological changes.

Furthermore, whereas mobiles in GSM do not have to contend for a channel during the channel assignment phase, in an A-GSM system, however, a distributed medium access layer is required to ensure a collision-free operation. The difficulties and intricacies of MANET MAC layer are discussed in detail in Sec. 2.4.

The following example highlights various channel layer issues for a call setup and forwarding along an A-GSM path.

Let us assume that MSa (see Fig. 4.63) has a call established with its serving base station (BTS) and during its call, its connection to BTS is heavily shadowed and is thus severely degraded. To salvage its ongoing call, MSa triggers a GSM-to-A-GSM connection handover from its direct connection through MSb. Upon triggering of a handover, the nodes along the multihop path perform the following procedures:

1. If not synchronized, nodes must first synchronize themselves for the signaling exchange and setup of the new signaling and data paths

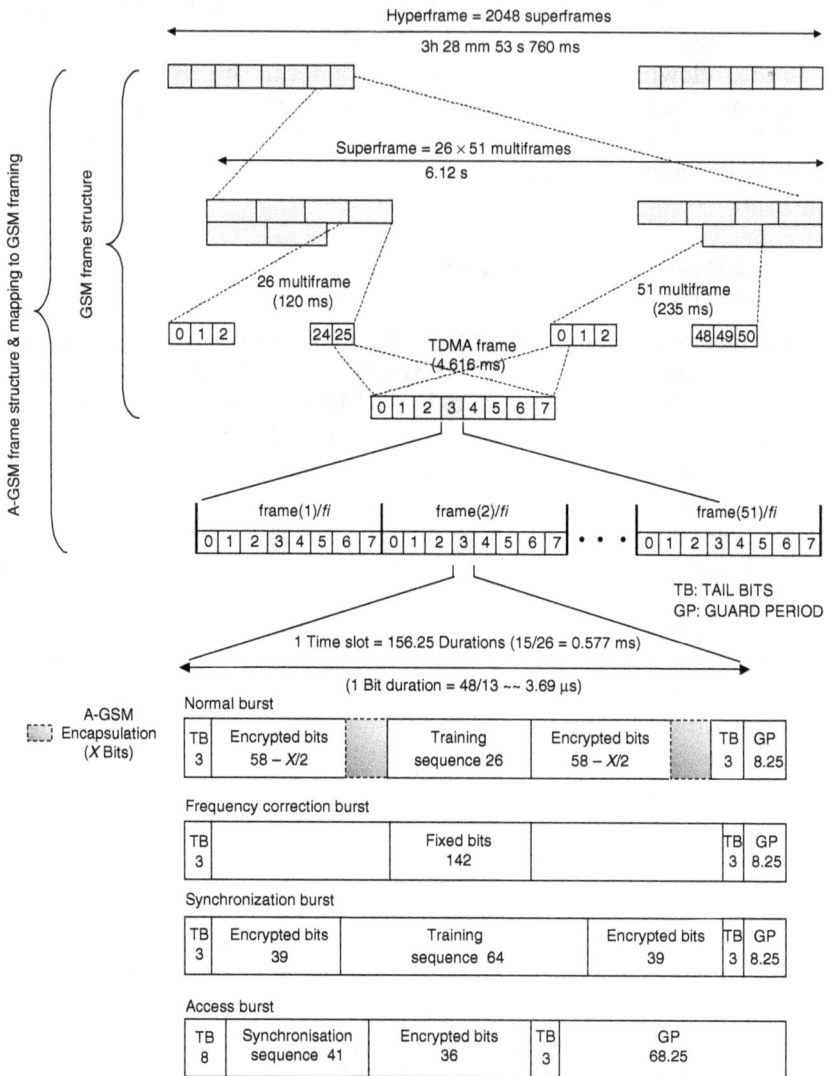

Figure 4.62 Hierarchy of frames in the A-GSM physical channel structure.

2. Switch to multihop frequency for transmission/reception

3. Transmitters should decrease their transmitting power to minimize interference

4. Relay (receive/transmit)

Mobiles must be assigned an unused distinct channel (time slot/frequency pair) as if their multihop link was a direct GSM link to BTS.

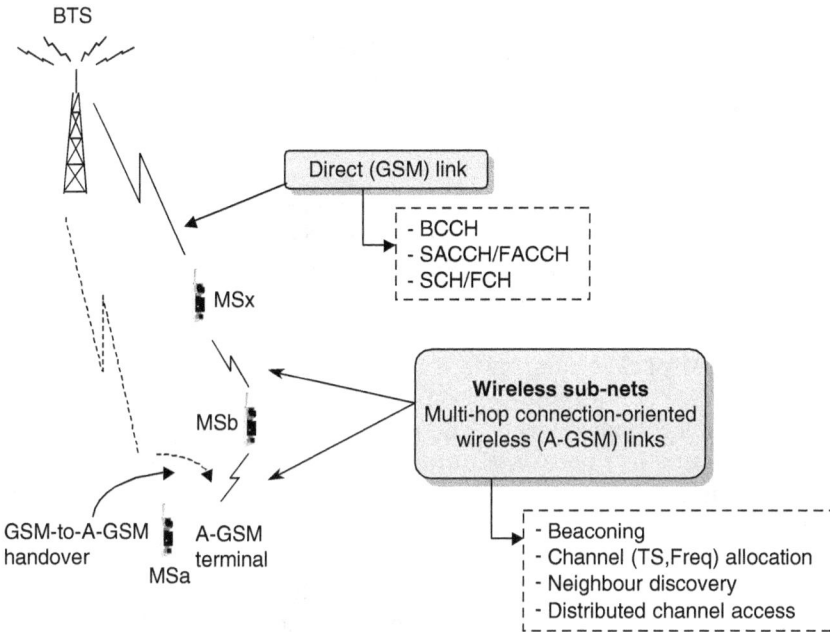

Figure 4.63 A-GSM call setup scenario.

This process involves a lot of commitment (in terms of monitoring) and coordination (in terms of synchronization) from the multihop mobiles. Clearly, one can see that the number of channels required for the establishment of the end-to-end (MSa to BTS) signaling and data paths are equal to the number of wireless hops (physical links) involved in the path. Thus, a single transmission along a multihop path is virtually treated as a distinct GSM transmission, where separate resources are allocated to transmitting nodes.

Transmitters should emit at minimum power levels to minimize *interhop interference*. It is a very difficult task to suppress interference from two asynchronous stations (e.g., a mobile relay and its associated base station). In the case of a downlink connection, for instance, a signal from a base station to some relay station and another signal from this relay station to another mobile terminal may severely interfere with each other. The impact and level of interhop interference varies based on the transmit power or the location of each station.

The handover phase proceeds as described in "A-GSM Call Rerouting (Handover)," Sec. 4.6.3. Upon successful connection, MSb acknowledges MSa the channel availability.

Moreover, some means of local resource management must exist so that nodes along the multihop path (i.e., MSa, MSb, MSc) are aware of the exact channel characteristics, such as the time slot number, timing advance, and transmitted power levels of their immediate neighbors. These become known during the channel access procedure in the A-GSM interface (see the following text).

Based on these rules, the call forwarding process in A-GSM is structured as follows:

Given that the channel assignment procedure is successful and a data link is established between MSa-MSb, MSa *encapsulates* its data blocks [IMEP-ID] and sends them to MSb. Similarly, forwarding at each mutihop relay involves encapsulation before transmission and decapsulation on reception of data.

An A-GSM packet comprises a header and a payload section. The packet header contains physical layer information such as transmit powers and local noise levels. The packet payload encapsulates either data segments, in the case of data (voice) volume, or signaling in the case of a signaling information transfer. A common downside of encapsulation schemes is the extra overhead added on a per-slot basis; thus reducing the effective net bit rate. The latter translates to a higher number of equivalent GSM time slots for a certain volume of data transfer in A-GSM. In practice, this means that instead of 116 data bits per TDMA frame per user, the effective data load per TDMA frame in A-GSM would be 116 bits minus the encapsulated overhead. This is illustrated in Fig. 4.62 where for, say, x encapsulation bits required per normal burst, $116 \times x$ bits will be the nominal payload in an A-GSM normal burst. Encapsulation in A-GSM is covered in more detailed in "A-GSM Protocol Layering," Sec. 4.6.3.

A-GSM protocol layering.   The ad hoc segment of an A-GSM communication system has fundamentally different characteristics from its GSM counterparts. This explains why modifications of the existing management modules are required. The A-GSM protocol architecture used for the exchange of signaling messages pertaining to mobility, radio resource, and connection management functions, comprises two broad segments—the GSM radio access together with the fixed network part, and the A-GSM radio access part. Practically, A-GSM layers should inherit the semantics and roles of their peer GSM layers.

The prototype protocol architecture of an A-GSM communication system is illustrated in Fig. 4.64.

Link layer protocol.   The data link layer over the radio link (connecting two mobiles) is based on the GSM LAPDm protocol [GSM 04.08], termed A-LAPDm, which is designed for operation within the constraints and functional requirements of an A-GSM protocol. In particular, the link access protocol for the D channel should be enhanced to support the following procedures:

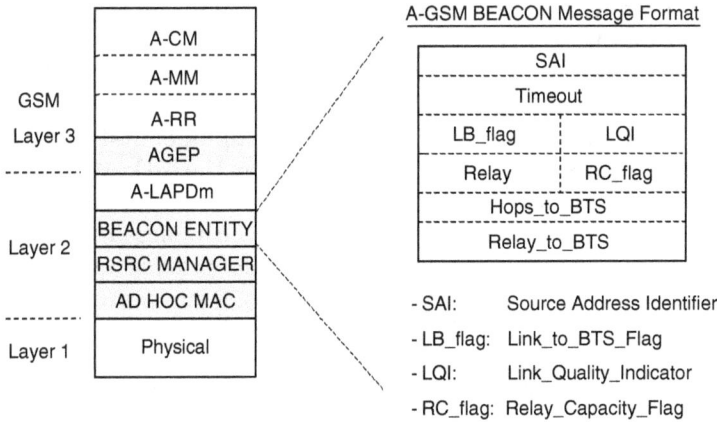

Figure 4.64   Proposed A-GSM protocol stack.

Beaconing.   Beaconing is a mechanism used to indicate mobile activity in an A-GSM network. During operation, a mobile *may* offer connectivity by broadcasting local beacon messages as follows. At periodic intervals, nodes generate a broadcast message, called *beacon* message. The message fields are illustrated in Fig. 4.64.

The LINK_TO_BTS flag is set to state *on* if the sender of the beacon has a direct link to BTS. If LINK_TO_BTS is set to *off*, the ID of the relay node through which the sender of the beacon can reach the BTS (RELAY_TO_BTS) along with the number of hops (HOPS_TO_BTS) is provided. If, however, the RELAY field is set to −1, it means that the sender has neither a direct link nor a multihop connection to BTS and the message is silently discarded. The significance and use of the RELAY_CAPACITY flag is illustrated later in this section.

Measuring the signal strength of the surrounding MTs, through the beaconing process, raises one major concern. "When can MTs perform this beaconing process?" In GSM, MTs measure the characteristics of the neighbor cells during the interval between the transmission of an uplink burst and the reception of a downlink burst. The uplink direction is derived from the downlink one by a delay of three burst periods (BPs). However, this approach presents fundamental problems if it is to be applied to the A-GSM beaconing. These problems arise from the fact that the beaconing process requires the MTs to be synchronized during the measurement period. On one hand, different nodes have different transmission/reception patterns, thus making it difficult to achieve synchronization of nodes between the uplink and downlink bursts, on the other hand the intervals during the transmission of an uplink burst and the reception of a downlink burst are of various lengths, depending on the dedicated channel type.

These factors prevent this approach from being a solid solution to the problem of beaconing transmission timing.

Within each of these cycles, 24 slots are used for the TCH, one slot for the corresponding SACCH, and one slot where nothing is sent. These 26 small intervals between the transmission of an uplink burst and the reception of a downlink burst could then be used by the A-GSM MTs in order to perform signal strength measurements on their surrounding MTs through the transmission and reception of beacon messages.

A second solution could be to lower the TCH rate and use one of the 24 slots to perform the measurements on the surrounding MTs. Alternatively, the idle frame could be shared for both GSM and A-GSM measurements although this scheme may have some implications on the strength measurements of the surrounding BTSs.

Resource manager. As capacity is a scarce resource in wireless environments, the protocol should ensure that the relaying of calls does not degrade the performance of the relay nodes. To this avail, the resource manager entity is responsible for coordinating the allocation of resources of a relay node. An A-GSM-to-GSM connection routed through a mobile, uses the resources of the mobile. These resources include link bandwidth, buffer space, and processing time. Thus, a certain number of connection requests can be established in parallel via a relay mobile. When an MT receives a request to set up a connection, the resource manager will execute a function called the *connection admission control* (CAC) to define whether it can accept the connection or not. If the connection is accepted, the resources are reserved.

Furthermore, the resource manager is responsible for informing the beacon entity when no resources are available for relaying. The lack of resources is in turn indicated in the beacon messages, using the RELAY_CAPACITY flag, so that nodes that receive the message and currently do not have a connection through the sender, silently discard the message.

Assuming that MTs are capable of relaying multiple calls, a protocol parameter then is the relaying (or forwarding) capacity per node; that is, the number of calls that an MT can simultaneously relay.

An alternative for relaying for a mobile radio with no resources would be to declare itself as "busy" and instead of broadcasting beacons with the RELAY_CAPACITY flag set to *off*. Following this approach the mobile radio will refrain from sending beacons until sufficient resources for forwarding are regained. By doing so, the neighbor mobiles of this radio would not unnecessarily be interrupted from their schedule to receive and process these otherwise "useless" beacons. Therefore, neighbor mobiles would avoid extra processing overhead, as the beacons from this radio do not have any significance in the A-GSM system because this radio cannot be of any help for the salvage of calls in case relay is needed from one of its neighbors.

However, a mobile should not defer continuously transmitting beacons even when the number of calls that are currently relaying exceeds its forwarding capacity threshold. That is because its neighbor nodes should constantly assess their link to this node so that when resources are found and the mobile node is again able to relay calls, its neighbor nodes would have a good quality indicator of their link to this radio over a long period of time.

Layer 3 protocols. The layer 3 is divided into three sublayers [GSM 04.08]—*radio resource management* (RR), *mobility management* (MM), and *connection management* (CM). Connection management is further subdivided into three protocol entities—*call control* (CC), *supplementary services* (SS), and SMS [GSM92]. The protocol architecture formed by these sublayers is shown in Fig. 4.64. The A-GSM Encapsulation protocol resides in between the layer-3 and layer-2 plane.

In the following sections we cover thoroughly the handling of calls at the network layer.

A-GSM encapsulation protocol (AGEP). Protocol encapsulation is a simple and easy to implement technique for passing arbitrary information through network entities. In this context, the AGEP platform is designed to transparently support different user terminal standards through a proprietary A-GSM-specific interface.

Besides, AGEP can be used to improve the overall network performance by reducing the number of network control packet broadcasts through the encapsulation and aggregation of multiple A-GSM-related control packets (e.g., routing protocol packets, acknowledgments, link status sensing packets, "network-level" address resolution, and the like) into larger AGEP messages. The AGEP could also provide an architecture for MANET router identification, interface identification, and addressing [IMEP-ID]. The AGEP runs at the network layer (Fig. 4.64) and is in fact an adjunct to whichever network protocol is using it.

Usage of the AGEP seems to be desirable because per-message, multiple access delay in contention-based schemes such as the IEEE 802.11 standard [GSM 04.06] is significant, and thus favors the use of fewer, larger messages. It also may be useful in reservation-based, time-slotted access schemes where smaller packets must be aggregated into appropriately sized network layer packets for transmission in a given time slot. Another purpose of AGEP concerns the commonality of certain functionality in many network-level control algorithms. Many algorithms intended for use in an A-GSM will require a common functionality such as link status sensing, security authentication with adjacent routers, one-hop neighbor broadcast (or multicast) reliability of control packets, and the like. This common functionality can be extracted from these individual protocols and put into a unified, generic protocol useful to all.

A-GSM radio resource management (A-RR). The radio resource management sublayer essentially handles the administration of the frequencies and channels. The A-RR management sublayer terminates at the BSS. This involves the A-RR module of the MS communicating with the A-RR module of the BSC. The general objective of the A-RR module is to set up, maintain, and take down A-RR connections, which enable point-to-point communication between the MS and the network. This also includes *cell selection* in idle mode and handover (GSM and A-GSM) procedures. Furthermore, the A-RR is responsible for monitoring BCCH and CCCH on the downlink when A-RR connections are active.

The following functions are realized in the A-RR module:

- Providing resource updates to the resource manager in A-LAPDm, as part of the beaconing process

- Monitoring of BCCH and PCH (readout of system information and paging messages)

- RACH administration. MSs send their requests for connections and replies to paging announcements to the BSS

- Requests for and assignment of data and signaling channels

- Periodic measurement of channel quality (quality monitoring)

- Transmitter power control and synchronization of the MS

- Handover always initiated by the network

- Synchronization of encryption and decryption on the data channel

To this end, the A-RR entity indicates to upper layers the unavailability of a BCCH/CCCH and the mode change (i.e., from GSM/A-GSM to A-GSM/GSM, respectively) when decided by the A-RR entity. An A-RR-connection includes a physical point-to-point bidirectional connection on the main DCCH. The upper layer can require the establishment/release of an A-RR connection.

When an A-RR-connection is established, A-RR procedures provide the following services:

- Indication of temporary unavailability of transmission

- Indication of loss of A-RR-connection

- Automatic mobile relay or cell selection and handover to maintain the A-RR-connection

- Allocation/release of an additional channel (for the Lm + Lm configuration)

The A-RR sublayer provides several services to the mobility management sublayer. These services are needed to setup and take down signaling connections and to transmit signaling messages.

A-GSM mobility management (A-MM).   The mobility management sublayer encompasses all the tasks resulting from mobility. The A-MM sublayer terminates at the MSC. Hence, the A-MM activities are exclusively performed in coordination between the MS and MSC, and they include

- TMSI assignment
- Localization of the MS
- Location updating of the MS; parts that are sometimes known as roaming functions
- Identification of the MS (IMSI, IMEI)
- Authentication of the MS
- IMSI attach and detach procedures (e.g., at insertion or removal of SIM)
- Ensuring confidentiality of subscriber identity

The A-MM sublayer provides registration services for higher layers. Registration involves the IMSI attach and detach procedures which are used by the MS to report state changes such as power-up or power-down, or SIM card removal or insertion.

The location tracking of MTs as well as the handling of location information of user is done in a similar manner as in GSM. Location information is maintained and used by the network to locate the user for call routing purposes. The network registers the user's location in user's HLR, which is associated with an MSC located in the PLMN, to which the user is subscribed. As change in the location of mobiles are detected (from the last information recorded by them), they each report the new location to the BSS which routes it to the VLR, of the MSC to which it is connected. The VLR , in turn, sends the location information to the user's HLR, where it is also recorded. In the mean time, the HLR directs the old VLR to delete the old visiting location of the mobile from its database, and also sends a copy of the user's service profile to the new VLR. The procedure for keeping the network informed of where the mobile is roaming is referred to as *location updating* (LU).

For as long as the mobile relay belongs to the same cell (or location area) of the mobile station, location updating in A-GSM should not be triggered either when performing GSM-to-A-GSM or A-GSM-to-A-GSM handover. In the special case where an MT switches on for the first time and sets up its signaling through multihop relaying, the LU procedure should be started as the MS is unknown in the VLR as a response to the A-MM-connection establishment request.

Periodic updating (PU) may be used to notify periodically the availability of the MS to the network. PU is performed by using the LU procedure.

The A-MM sublayer offers its services to CC, SS, and SMS entities. This is essentially a connection to the network side over which these units can communicate.

A-GSM connection management (A-CM). The connection management sublayer consists of three entities—call control, supplementary services, and short message service. *Call control* handles all tasks related to setting up, maintaining, and taking down calls. The services of call control encompass:

- Establishment of normal calls (MS-originating and MS-terminating)
- Establishment of emergency calls (only MS-originating)
- Termination of calls
- Call-related supplementary services

Signaling between the different entities in the fixed part of the network, such as between the HLR and VLR, is accomplished through the *mobile application part* (MAP). MAP is built on top of the *transaction capabilities application part* (TCAP, the top layer of signaling system number 7 [GSM92]).

A-GSM RR-connection transfer phase. While a mobile is engaged in the GSM RR-connected mode (TCH or SDCCH), the functions of channel measurement and power control serve to maintain and optimize the radio channel. Both have to be done until the current base can successfully hand over the current connection to the next base station. In the RR-connected mode the SACCH is used in the signaling layer for the transmission of measurement results from the MT to the network. A SACCH block contains measurement results about reception characteristics (power control and timing advance) from the current cell and from neighbor cells. (The measurement results are obtained as specified in Ref. [GSM 05.08]).

For GSM-to-A-GSM communication transfer purposes, the SACCH structure is extended to declare the case when an MT intends to switch to the relaying mode. This is done with an extra added flag, in the currently unused space of SACCH called *connection A-GSM* (CAG) flag. The free bits in the SACCH block are illustrated in Fig. 4.65; that is, the 6 to 8 bits in the first octet and the 7 to 8 in the second octet. Bits from either free block can be used to accommodate the new flag. Figure 4.65 depicts the format of an A-GSM SACCH block, which contains 21 octets of A-LAPDm data and a kind of a protocol header which carries the current power level, the value of the timing advance, and the CAG handover flag that indicates when a mobile terminal switches to the A-GSM connection mode.

By default this flag is set to the state *off*. When the MT decides to switch to the relay mode, it sets the flag to state *on* thus notifying the BTS its intention to handover its ongoing connection(s) to a mobile relay.

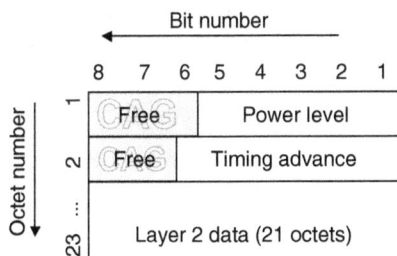

Figure 4.65 Format of an A-GSM SACCH block.

Note that when the CAG flag is *on* the path loss and timing advance fields serve no purpose and therefore could be set to a default value. An GSM-to-A-GSM handover procedure then follows, which is described later in "A-GSM call rerouting (handover)" in Sec. 4.6.3.

Measurement report.   A measurement report pertains to a given measurement period, that is to say the period during which the measurements were done. The duration of the measurement period is always equal to the periodicity of message transmission on the SACCH (i.e., 480 ms on the TACH/F, and around 471 ms for the TACH/8). From the reception of these measurements current mean values are calculated. At first, these measurement data are supplied to the transmitter power control to adapt the power of MS and BS to a new situation if necessary. Thereafter, the measurement data and the result of the PC activity are supplied to the handover process, which can then decode whether an HP is necessary or not. The measurement loop is left (at BSS) only when the connection is terminated.

A-GSM channel assignment.   An intracell change of channel can be requested by the upper layers for changing the channel type, or decided by the RR-sublayer, e.g., for an internal handover, as illustrated in "A-GSM call rerouting (handover)" in Sec. 4.6.3. This change is performed through the dedicated channel assignment procedure. The purpose of the dedicated channel assignment procedure is to completely modify the physical channel configuration of the MS staying in the same cell. The channel assignment procedure happens only in the RR-connected mode. This procedure cannot be used in the idle mode, where in this case the immediate assignment procedure is used.

The channel assignment procedure includes

- The suspension of normal operation including A-RR management features[7]

---

[7]Note that in GSM the RR management connections are not suspended during the channel assignment procedure [GSM92].

- The disconnection of the main signaling link, and of the other data links via local end release (layer 2), and the disconnection of TCHs if any
- The deactivation of previously-assigned channels (layer 1)
- The activation of the new channels and their connection if need be
- The triggering of the establishment of the data link connections

In contrast to the GSM channel assignment procedure, which is triggered and initiated by the network, in A-GSM this is triggered (but not initiated) by the MT itself. This is because until the channel assignment phase is triggered, the network side is only aware of the fact that an A-GSM handover is about to take place, but, however, knows nothing about the multihop path that is to be established by the ad hoc segment. To this end, since the channel assignment procedure is a network initiated process, the MT first sends a channel assignment-specific control message through the multihop path. The purpose if this signaling is to help the network side to identify either the identity of the MTs that compose the multihop path (source routing) or the A-GSM GW relay (next-hop routing).

On the reception of this message, the network initiates the channel assignment procedure by sending an encapsulated ASSIGNMENT COMMAND message to the MS on the main signaling link. The ASSIGNMENT COMMAND message contains the description of the configuration, including for the Lm + Lm + ACCHs configuration, the exact SACCHs to be used and a power command. Unlike GSM, however, this power command in the A-GSM system shall not affect the power decisions made by the MTs through the multihop path, as distributed power control is performed throughout the ad hoc segment instead.

On receiving this command, the MS initiates a local release of link-layer connections, disconnects the physical channels, commands the switching to the assigned channel and initiates the establishment of lower layer connections (including the activation of the channels, their connection, and the establishment of the data links).

After the main signaling is successfully established, the MS returns an ASSIGNMENT COMPLETE message to the network on the main DCCH.

A-GSM integrated dual-mode terminals.   A number of issues arise from the integration of GSM and ad hoc mobile networks. First, in such a system with dissimilar network- and air-interfaces, the issues of integration and/or interworking as well as the level of integration affect almost all the signaling transfer, radio resource, mobility, security, and communication management procedures. Special care has to be taken to design optimally a call routing or rerouting (handover) algorithm, as it

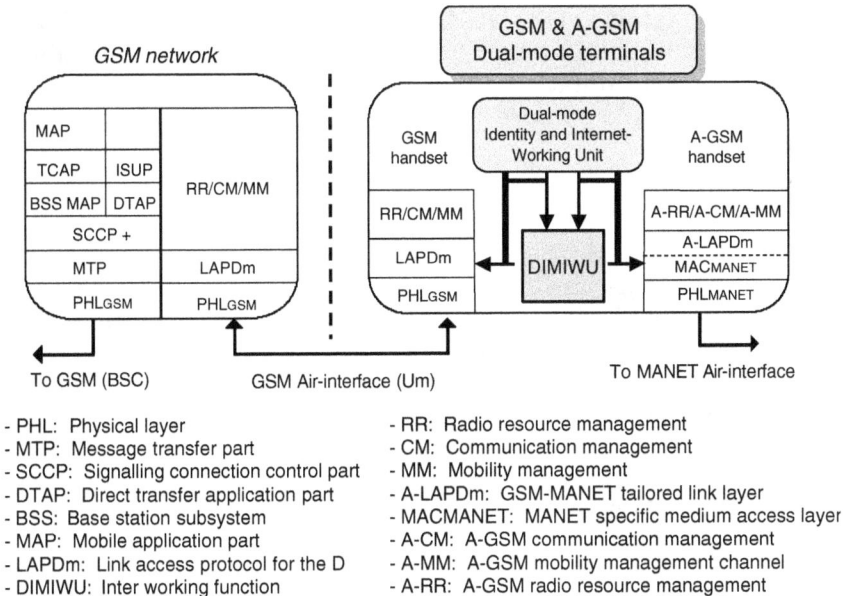

Figure 4.66   GSM-MANET dual mode terminal.

- PHL: Physical layer
- MTP: Message transfer part
- SCCP: Signalling connection control part
- DTAP: Direct transfer application part
- BSS: Base station subsystem
- MAP: Mobile application part
- LAPDm: Link access protocol for the D
- DIMIWU: Inter working function

- RR: Radio resource management
- CM: Communication management
- MM: Mobility management
- A-LAPDm: GSM-MANET tailored link layer
- MACMANET: MANET specific medium access layer
- A-CM: A-GSM communication management
- A-MM: A-GSM mobility management channel
- A-RR: A-GSM radio resource management

could significantly affect the performance of the system and thereby the quality of the offered services in the combined system. The capability and complexity of a dual mode handset will also impact the design of the signaling load, delay, required level of modification, implementation complexity, and cost as well as operators requirements.

As far as handover in the interworking system is concerned, a dual-mode handset, that is capable of inter-GSM/MANET handover, must function as two handsets with a direct interface (Fig. 4.66). While one A-GSM MT serves an active call either in the GSM or the MANET system, the other system is on the lookout for information indicating the presence of an alternative system. This allows the handset to access the alternative system in case handover needs to be performed. Thus, a very attractive feature of this type of terminal would be service mobility between modes by means of initiating a service in one mode then handing over the service to a more suitable system in terms of system capability and application.

Some interworking functionality would be required in the dual-mode handset to translate the information coming from two different systems and make some decision based on a predefined algorithm.

Dual mode identity and Internet-working unit (DIMUW).   This is a nonstandard, specially designed unit, responsible for providing access to the GSM/A-GSM network. It performs all the necessary user terminal protocol adaptations to the GSM/A-GSM protocol platform. DIMIWU also includes all physical layer functionalities such as channel coding, modulation, demodulation, and the radio frequency parts. All the supported terminals share the same access scheme and protocol stacks.

A-GSM call rerouting (handover)

Call handover phases.   During a call session, the system has to ensure the continuity of its provision when the mobile moves across different base stations, of different cells, as well as across different mobile ad hoc vicinities within the same cell (or even different cells). This feature is generally achieved by the provision of call rerouting or, else, *handover.* The rerouting process involves three successive phases, which are discussed as follows:

- Radio measurements
- Initiation and trigger
- Handover control

Measurements.   A mobile terminal participating in an A-GSM system has to perform three types of radio link measurements—one for the link to BTS, one to its neighbor base stations, and one to each of its neighbor nodes. The method for performing as well as for reporting the measurements is discussed later in this section.

Handover initiation and trigger.   Based on the signal measurement, a handover decision is made based on both absolute and relative signal strength measurements, and in particular, the measurements taken at the mobile rather than at the base stations (received signal strength at base stations may not be reliable for handover decisions, especially in systems employing power control). Additional parameters such as the bit error rate (BER) or the carrier-to-interference ratio (C/I) could be used as alarm condition indicators, in order to increase the efficiency of the handover mechanism.

A-GSM handover (GSM-to-A-GSM and A-GSM-to-A-GSM) occurs when a high probability exists that the call will be lost or the quality of the ongoing connection will be seriously degraded if the current link to BTS (direct or multihop) is not changed. Whereas in GSM this is interpreted as changing the serving cell, in A-GSM—depending on the type of handover—this is interpreted as changing the serving relay (A-GSM-to-A-GSM case) or the serving base station (GSM-to-A-GSM case).

The A-GSM handover may be triggered because of any one of the following:

1. Serving BTS failure

2. Change of the base station, frequency band, or time slot due to signal-quality degradation and interference—in order to provide a service with better QoS—but no neighbor BTS accessible with the adequate signal quality

3. Change of base station, frequency band, or time slot due to the mobility of the user in a multicell environment—e.g., lack of continuous coverage could initiate the handover procedure—where the user with a call in progress, crosses the border of the coverage area, but similar to the first two cases, the neighbor BTSs either not accessible or their communication link quality is below acceptable thresholds

Let us assume a mobile which is moving toward its serving BTS and is also crossing a number of relay mobiles along its way. As shown in Fig. 4.67, the received signal level in the A-GSM reception range shows a variation of about 95 dB whereas, within the same distance, the variation in the GSM microcell is less than 15 dB.

Many variations in the handover algorithm are possible whereas the choice of each method would depend on the characteristics of the system.

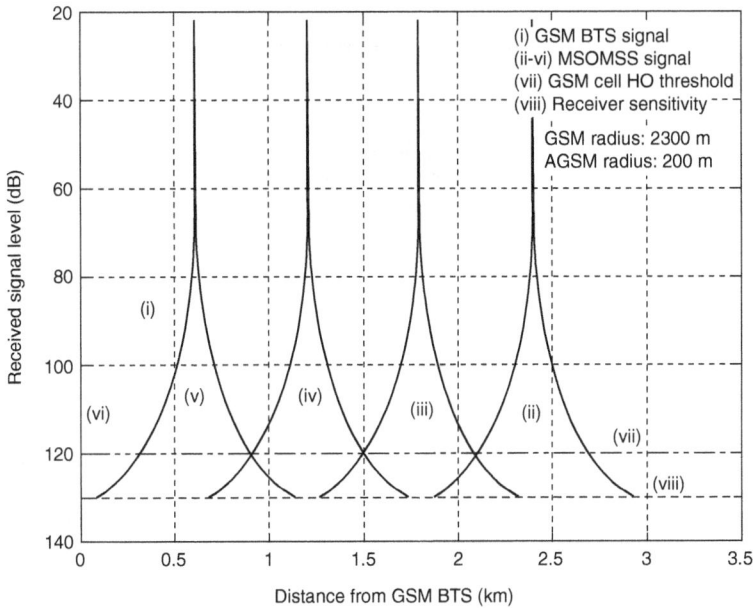

Figure 4.67   Received signal levels from GSM BTS and A-GSM relays [AGGE01].

A recent study [GHAHE99] on efficient handover techniques between micro- and picocell systems has shown that the signal level received in the dual-mode terminal is dependent on the location of the picocell base station relative to the position of the microcell base station. The study concludes that the hysterisis level may not be suitable for handover initiation criteria in an integrated micro-/picocell system.

In A-GSM one could envision the coverage of beacon transmissions from a single user as forming a virtual picocell such that the sender of the beacon is the base-station of the picocell whereas the radius of the picocell equals to the transmitting range of the node. The A-GSM terrain can thus be viewed as a platform that comprises a microcell and many virtual mobile picocells whose radius is changing due to the transmitting range adjustments during the beaconing process. Since prior knowledge of the crossover signal strength between two base stations is required to define the hysterisis margin, a hysterisis-based handover is not appropriate in an A-GSM integrated system. Hence, in a GSM-A-GSM system a threshold-based handover could be applicable.

A threshold-based handover algorithm is proposed for used in A-GSM [AGGE01]:

$$\{r_{\text{GSM}}(x) < T_{\text{GSM}}\} \text{ or } \{r_{\text{GSM}}(x) \text{ fails}\} \text{ and } \{\text{A-GSM relay exists}\} \text{ and}$$
$$\{\text{relay}(x) > T_{\text{A-GSM}}\} \text{ and } \{r_{\text{GSM\_NEIGHBOR}}(x) < T_{\text{GSM}}\}$$

where r_{GSM} and $r_{\text{A-GSM}}$ are the average signal levels received from the serving BTS and the neighbor mobile relay, respectively. T_{GSM} and $T_{\text{A-GSM}}$ are the GSM and A-GSM handover threshold levels, respectively. Therefore handover is initiated when r_{GSM} falls below the threshold level, there exists a neighbor relay with its signal ($r_{\text{A-GSM}}$) being the strongest among the mobile relay ranking list and higher than the A-GSM handover threshold level, and also, there is no GSM coverage available to perform GSM-to-GSM cell handover (to demonstrate the applicability of wireless multihop relaying and its feasibility for A-GSM, let us assume that the latter is not an option).

The proposed criteria for different A-GSM handover cases could be modeled as:

- GSM-to-A-GSM handover is performed if the following conditions are fulfilled:
  1. The averaged signal level of the serving BS falls below a threshold T_{GSM} (dBm)
  2. (*Optional*) No averaged signal level of any other BS is greater than that of the serving BS by a hysterisis of *h (dB)*

3. No averaged signal level of any other BS is greater than the threshold T_{GSM} (dBm)

4. The averaged signal level of a neighbor MT is greater than a threshold $T_{A\text{-}GSM}$ (dBm)

- A-GSM-to-A-GSM handover is performed if the following conditions are simultaneously fulfilled:

  1. No averaged signal level of any BS is greater than the threshold T_{GSM} (dBm)

  2. The averaged signal level of the serving MT falls below the threshold $T_{A\text{-}GSM}$ (dBm)

  3. The averaged signal level of a neighbor MT is greater than a threshold $T_{A\text{-}GSM}$ (dBm)

- A-GSM-to-GSM handover is performed if the average signal level of a BTS is greater than the threshold T_{GSM} (dBm)

The different handover decision algorithms can be seen in Figs. 4.68 and 4.69.

Handover control. When a mobile moves away from its current base station, it must quickly become affiliated with a new base station so as not to interrupt any calls in progress. Selecting the new base station and determining when to initiate this handoff depends on the perceived signal quality (e.g., power, bit error rate) measured at the mobile endpoint and the nearby base stations. Call handoff may be initiated by the mobile endpoint, by the base station, by the mobile switching center, or by a combination of these, and it may occur within a service area or between adjacent service areas. Handoffs between cells under control of the same base station are called internal handoffs, while handoffs between cells controlled by separate base stations are called external handoffs. The following are the principal types of external handoffs.

- *Mobile-controlled handoff.* The mobile endpoint constantly monitors the quality of the signal from its current base station and from other base stations in its vicinity. It chooses as its new base station the one producing the best signal, with some hysteresis built into the selection process in order to prevent frequent handoffs when a mobile endpoint crosses back and forth between two cells. This technique is used in both DECT [DECT] and WACS [BELLCORE93].

- *Network-controlled handoff.* The endpoint's current base station constantly monitors the quality of the signal from the mobile endpoint. When the signal quality falls below a specified threshold, that base station sends a handoff request to the MSC. The MSC then asks other base stations in the endpoint's vicinity to monitor the quality of the signal from the mobile endpoint, and these base stations respond with

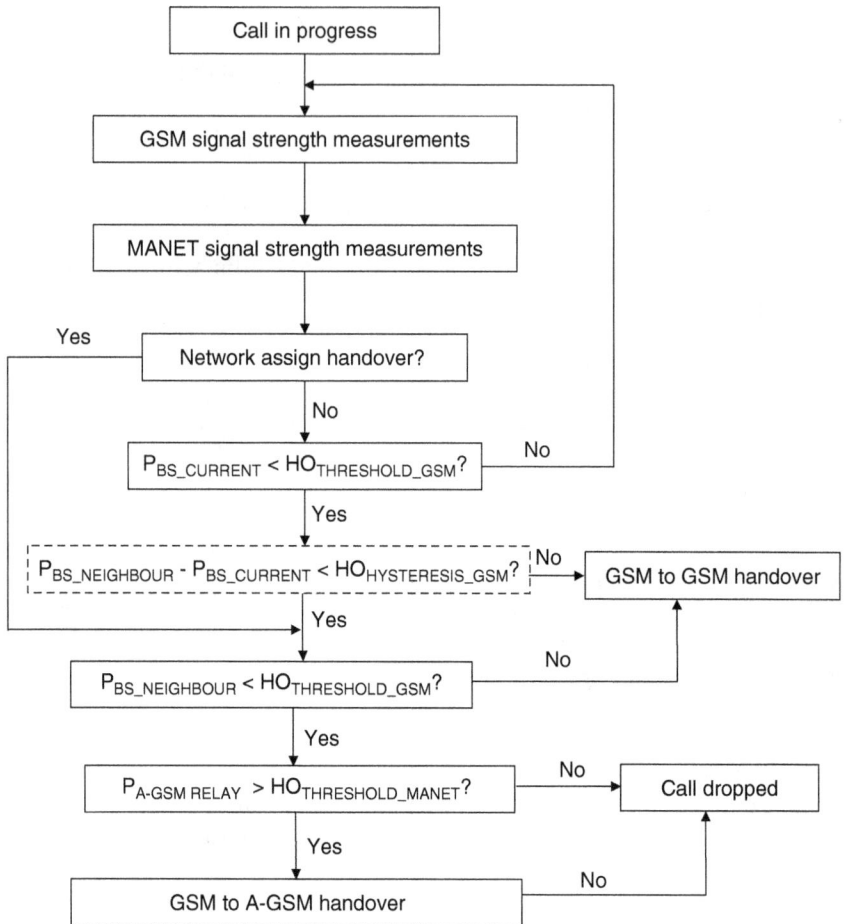

Figure 4.68 GSM-to-GSM and GSM-to-A-GSM handover algorithm.

measurements of signal quality. From among the base stations with sufficiently high signal quality, the MSC selects a new base station. This technique is used in AMPS [YOUNG79].

- *Mobile-assisted handoff.* The mobile endpoint's current base station asks it to constantly monitor the quality of the signals received from a specified set of neighboring base stations. These measurements are returned to the base station which in turn delivers them to the MSC. The base station may also provide the MSC with its own measurements of the quality of the signal from the mobile endpoint, as in a network-controlled handoff. Using all these measurements, the MSC determines when to initiate the handoff and which base station will become the endpoint's new base station. This technique results in

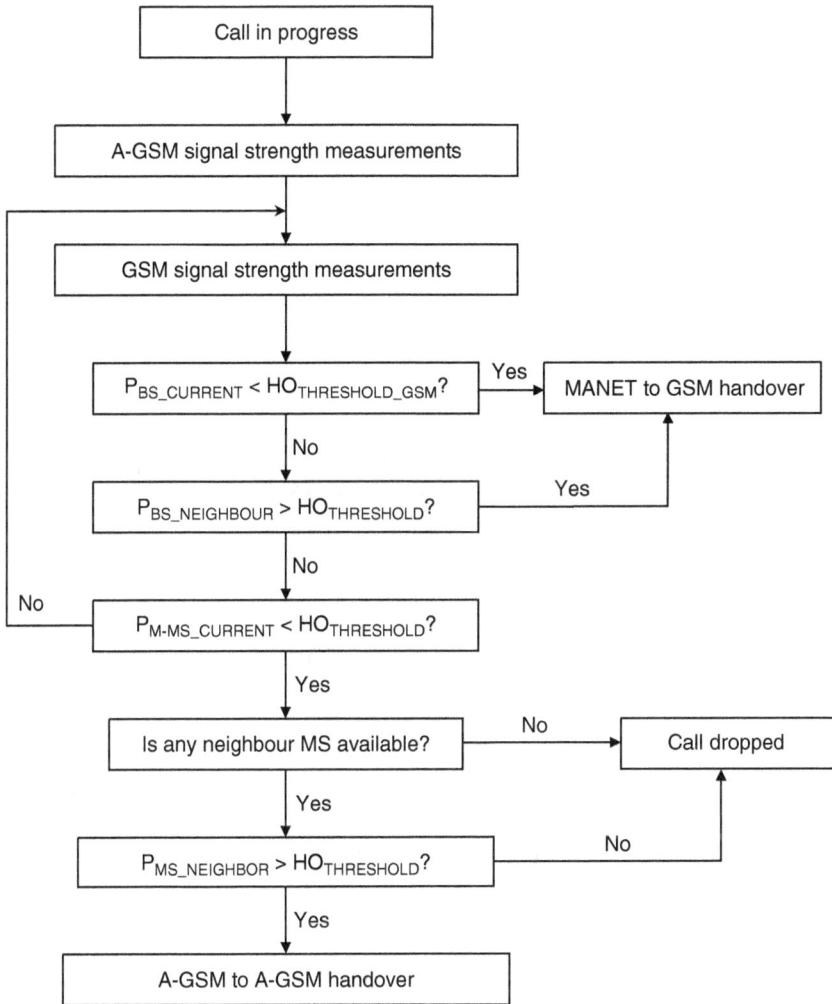

Figure 4.69   MANET-to-GSM and MANET-to-MANET handover algorithm.

lower delays than those encountered in a network-controlled handoff. In Ref. [POLLINI95], enhancements to the basic IS-41 [ITU91] handoff procedure are described, which provide sequential and/or single-copy delivery guarantees to support data services in a cellular network with an IEEE 802.6 MAN infrastructure.

Although the basic type of handover protocol followed in the present GSM systems is similar to the mobile-assisted handover (MAHO) [GSM92] in an A-GSM system, a hybrid handover scheme is applied. More specifically, for the case of A-GSM-to-GSM and GSM-to-GSM handover, the handover type is MAHO. For all other types of handover, including GSM-to-A-GSM and A-GSM-to-A-GSM, the *mobile-controlled handover* (MCHO) is applied. In an MCHO scheme, the MT itself makes both radio link measurements and handover decisions. In terms of the handover process, the network will be under the instructions of the relay MTs. As the handover decision is made totally by the MT, the handover can be initiated very fast. In addition, since the handover decision is made without assistance from the network, handover-related radio link signaling is relatively low. On the other hand, allowing an MT to make the handover decision may not be eventually a reliable practice if the radio link characteristics of the multihop uplink and downlink are uncorrelated.

Furthermore, two cases need to be distinguished with regard to the participation of network components in the handover, depending on whether the signaling sequences of a handover execution also involve an MSC. Since the *resource reservation* (RR) module of the network resides in the BSC, the BSS can perform the handover without the participation of the MSC. Such handovers occur between cells that are controlled by the same BSC; these types of handover are known as *internal* handovers. An internal handover can be performed independently by the BSS; the MSC is only informed about the successful execution of internal handovers. All other types of handovers require participation of at least one MSC, or their BSSMAP and MAP parts, respectively. These types of handover are known as *external* handover.

In the following text, the internal handover is further analyzed. The basic signaling structure for a GSM-to-A-GSM *internal* handover is presented in Fig. 4.70.

In a GSM-to-A-GSM handover protocol, an MT requests a handover from its BSS by sending an encapsulated HANDOVER REQUIRED message through its relay MT and starts the timer T3124, as specified in the GSM specification[8] [GSM 3.02].

The A-GSM HANDOVER REQUIRED message (Fig. 4.70) contains the following elements:

- Message type
- Cause (handover type)
- Cell identifier list

---

[8]Note however that all the timers specified by the GSM standards must be modified accordingly to account for the additional delay imposed by the multihop relaying. This implies that A-GSM timers may differ from the respective GSM times. Timer resolution might prove to be a problem though for some switching complex manufacturers hoping to reuse existing products.

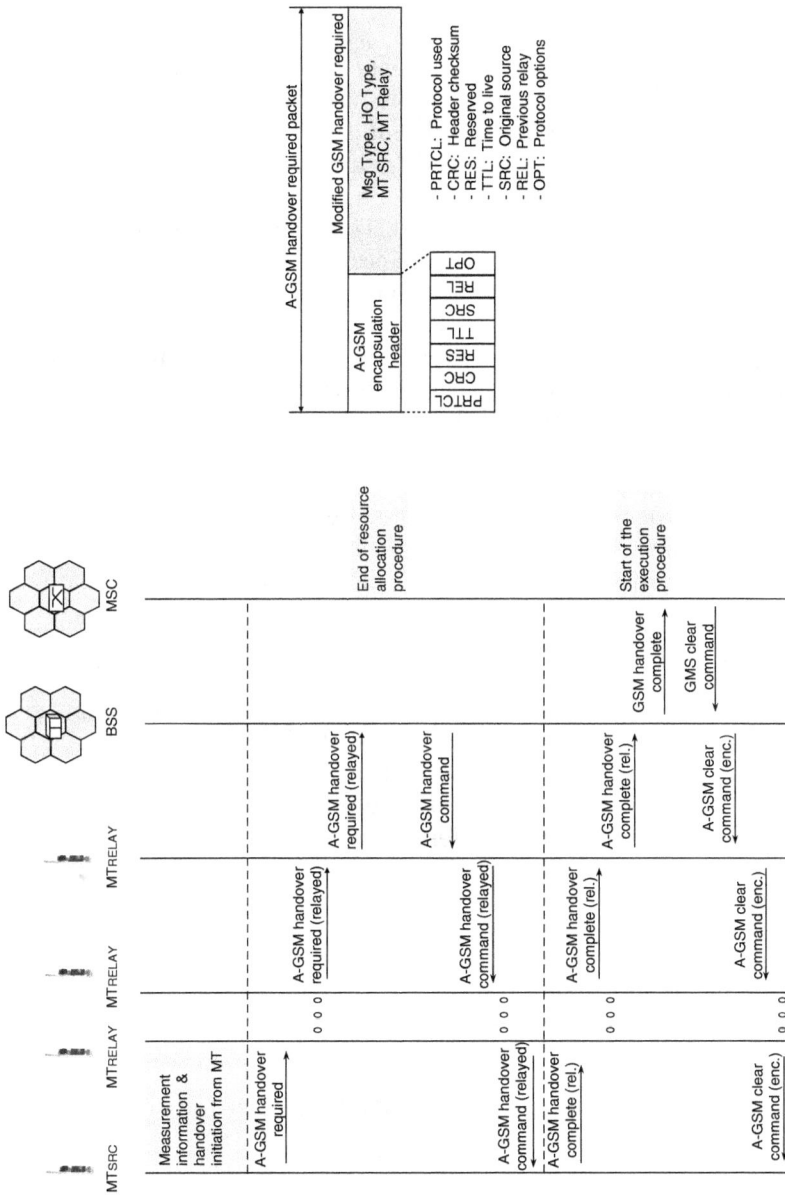

Figure 4.70 GSM-to-A-GSM internal handover signaling.

389

- Details of the resource that is required
- MT relay(s)

The recipient MTs along the multihop path try first to see if direct communication with BTS can be established. If so, the message is sent directly to BTS. Otherwise, the HANDOVER REQUIRED message is forwarded from each relay MT along the multihop path to the BTS. The forwarding process continues until either a relay MT is found to have a direct link to BTS or an MT has neither a relay nor a direct link to BTS. In the latter case, the packet is silently discarded and the handover phase resumes with failure. If failure occurs, the node that reports the handover failure could either send an error message to the initiator of the handover or not take any action. In the former case, the node that initiated the handover, upon reception of the error message, may initiate a new handover phase by sending a HANDOVER REQUIRED message through a different mobile, while in the latter case the initiator of the handover will eventually react to the failure on timeout of the time T3124.

If, however, the relaying process has successfully forwarded the encapsulated message to BSS, the message is forwarded to BSC which decapsulates it and reads the contents of the GSM-compatible HANDOVER REQUIRED message.

On receipt of this message the BSC shall choose a suitable idle radio resource. If radio resources are available, this will be reflected back to the MS in an A-GSM HANDOVER COMMAND within its layer 3 information element, which is in fact the RR-Layer3 A-GSM HANDOVER COMMAND. The timer T3103 is then started (again this timer is modified to A-GSM protocol semantics). Information about the appropriate new channels and a handover reference number chosen by the new BSS are contained in the A-GSM HANDOVER COMMAND.

In case the handover cannot be established, BSS informs the MS by sending an A-GSM HANDOVER REQUIRED REJECT (A-HANDOVER REQUIRED REJECT) message. The A-GSM HANDOVER REQUIRED shall be repeated by the MS periodically until:

- An A-GSM HANDOVER COMMAND is received from the MS
- An A-GSM HANDOVER REQUIRED REJECT is received
- The transaction ends, e.g., call clearing

The sending of the A-GSM HANDOVER COMMAND by the BSS to the MS ends the handover resource allocation procedure. The handover execution procedure can now proceed.

On receipt of the A-GSM HANDOVER COMMAND message, the MT initiates the release of link-level connections, disconnects the physical channels, commands the switching to the assigned channels and initiates the establishment of lower-layer connections (this includes the activation of the channels, their connection and the establishment of the data links). After lower-layer connections are successfully established, the MT returns a HANDOVER COMPLETE message. The sending of this message on the MT side and its receipt on the network side allow the resuming of the transmission of signaling layer messages other than those of RF management. When receiving this message, the network stops the A-GSM timer T3103 and releases the old channels. The BSS shall then take all necessary action to allow the MS to access the radio resource that the BSS has chosen. Since the new A-RR traffic connection is essentially an A-RR connection, the BSS shall then switch to the A-GSM mode.

Now, on the MT side, if timer T3124 times out or if lower layer failures occurs on the new channel before the HANDOVER COMPLETE message has been sent, the MT shall then deactivate the old channels and disconnect the associated TCHs, if any. On the network side, if the timer T3103 elapses before the HANDOVER COMPLETE message is received on the new channels, the old channels are released and all contexts related to the connections with that MT are cleared. Finally, BSC informs the MSC at the completion of the process.

Provision of handover. The connection transference scheme described earlier belongs to the general family of handover types, called "soft" handover. A soft handover maintains the current connection through the original link until the target link is firmly established. The MT keeps both streams simultaneously. It should also be noted that it is not necessary to drop the old link even after the successful establishment of the new link. This scheme forces the MT to work on two carriers in a time division fashion whereas, as previously mentioned, suitability to a mobile-controlled handover (MCHO) execution is evident—the MT can activate the new frequency in the new BS on the basis of its own quality perception, avoiding the measurement transmission on the radio interface. The very high performance is achieved at the cost of a higher load on the fixed network, due to the doubling of the various paths corresponding to a single call.

Although the soft handover is considered as a more suitable connection transfer mechanism for the A-GSM network platform, the main inefficiency of this handover scheme is its low reliability. If the signal level received from the original link is seriously shadowed before a new connection is available, the call has to be terminated. This is because handover-related signaling messages cannot be exchanged *on-time* through the initial link any more, so that the new signaling path can be established before the old one is destroyed. Clearly, this corresponds to the case where the time T3124 expires, as we illustrated earlier.

Therefore, the connection transference scheme might instead turn out to be a "hard" handover type. In a hard handover, the old connection is released before the new one is established; this is a break-before-make handover. Hard handover has two stages—in the first stage the call is detached from the original link; in the second stage the call is attached to the newly established link.

This scheme produces a short interruption of service and the handover has to be performed in a nonseamless mode. A significant advantage lies in the fact that the MT works on a single carrier at a time; however, being a network-controlled handover also implies that all data needed for handover decisions are to be sent to the network through the radio interface. This may then produce significant load on both the radio interface as well as the fixed network segment.

A-GSM relay selection process.   Within the vicinity of an A-GSM node, defined by its neighbor list, the mobile should select one of several relays from its neighbor nodes, which it will use to route its call through toward the BTS. The beacon packet structure is designed to enable nodes to listen to neighbor broadcasts and derive the required connectivity information. Clearly, a selection criterion should exist so as to determine which neighbor is the "best" relay for a multihop connection.

At periodic intervals, nodes perform a relay selection process in order to select the appropriate relay node from among the neighbors that are currently stored in their neighbor tables. Node selection is a two-step process: first the link qualities associated with each neighbor are averaged in order to determine the average *direct* link quality of each of the neighbors in the table, and second a minimum-hop route calculation is performed to determine the neighbor through which the BTS can be reached with the minimum number of hops. Due to node mobility and channel fading, the averaging procedure has to be performed periodically to detect and update any change in the physical links.

The *link quality value* (LQV) provides a measure of the propagation conditions in the communication channel whereas the number of hops provides a measure of the end-to-end delay. The use of link quality information together with the number of hops forms the basis for route selection. As a consequence, the *path quality value* (PQV) for a particular route from a mobile to its base station is a function of the LQVs of all the links that make up that route. The neighbor with the lowest PQV to BTS is the one to be selected as the default relay node until the next neighbor selection process.

Also included in the neighbor database are secondary outgoing links, which have higher PQV than the first one and are also sorted according to their PQVs. Although the intent is that a call is normally forwarded on the primary outgoing link, it is better not to rely on a single route. Because of the highly mobile nature of a packet radio network, the route with the highest PQV may become unsatisfactory before the routing algorithm has time to respond. The maintenance of redundant (secondary) routing information allows the protocol to react quickly to such degradation by forwarding the call through the secondary routes.

4.6.4 The future of ad hoc networking

Imagine the following scenarios—a wireless mesh of rooftop-mounted ad hoc routers; an ad hoc network of cars for instant traffic and other information; sensors and robots forming a multimedia network that allows remote visualization and control; multiple airborne routers (from tiny robots to blimps) automatically providing connectivity and capacity where needed (e.g., at a football game); an ad hoc network of spacecraft around and in transit between the Earth and Mars.

These may seem like science fiction, but are in fact ideas pursued seriously by the ad hoc research community. While only time can tell which of these imaginations will become real, the above offers a glimpse into both the technological potential and the evolving state of the art. We discuss in this section the forces at play that are likely to shape the future of ad hoc networking, and discuss the directions in which it may evolve.

To appreciate the role ad hoc networks are likely to play in the future, consider this: bandwidth-hungry applications and the laws of physics drive wireless architectures away from cellular toward ad hoc. This is because more capacity implies need for a higher communications bandwidth and better spatial spectral reuse. Higher bandwidth is found at higher frequencies, where the propagation is dismal. It is worth emphasizing here the basic difference in the fundamental goal of the conventional ad hoc networks and the multihop-augmented infrastructure-based networks—while the defining goal of ad hoc networks is the ability to function without any infrastructure, the goal in the latter types is the almost-ubiquitous provision of very high data rate coverage and throughput.

Further, mobile devices have to be power-thrifty. Propagation, spectral reuse, and energy issues support a shift away from a single long wireless link (as in cellular), to a mesh of short links (as in ad hoc networks). That this might be the wave of the future is attested to by burgeoning startups – e.g. Rooftop Communications (now part of Nokia), Mesh Networks, Radiant Networks – that use a "multihop mesh-based" architecture in place of conventional 3G architectures.

The other main impetus to ad hoc networks comes from the rapidly improving communications technologies. Wireless communication devices are getting smaller, cheaper, more sophisticated, and hence, more ubiquitous. Exploitation of these technologies for better ad hoc networking gives rise to new problems which point to new research. For instance, the use of smart antennas in ad hoc networking requires new medium access and neighbor discovery protocols. The ability to dynamically alter spread-spectrum codes, modulation schemes, and waveforms require corresponding innovations at the higher layers. Software radios, which signify an important change in radio architecture, offer more flexibility that is suitable for ad hoc networks.

How are ad hoc networks likely to evolve? It is likely that the nodes themselves will be smaller, cheaper, more capable and probably conformal, and come in all forms. Indoor ad hoc networks (perhaps based on Bluetooth, Wireless LAN or similar technologies) will probably be used to connect smart appliances to the Internet. Mesh-based last mile solutions will increase in popularity and may even be the dominant solution. Military ad hoc networks will have higher capacities and support multimedia applications, be more adaptive, stealthy, and evolve toward a system where all battlefield elements, mobile or stationary, are multimedia networked.

Finally, there is the utopian idea of a "global infosphere" where all network elements form a gigantic ad hoc wireless network using unlicensed spectrum, bypassing the existing infrastructure. While fascinating from a research viewpoint, the realization of this vision will depend not only on overcoming the capacity and other hurdles, but also the pragmatics of a "cooperative" network. Notwithstanding our predictions, however, like the Internet, which existed for more than 20 years before the World Wide Web came along, it may be a surprise "killer app" that shapes the future of ad hoc networking.

A

The Router Architecture

A.1 Overview of the Forwarding Process

In the network layer, the real work is the forwarding of datagrams. A key component in this forwarding process (see Fig. A.1) is the transfer of a datagram from a router's incoming link to an outgoing link; a process which is commonly known as *switching*. In the following text we elaborate on the switching functionality of a router and how the forwarding process is accomplished. Our coverage here is necessarily brief, as an entire course would be needed to cover router design in depth.

A high-level view of a generic router architecture is shown in Fig. A.2. Four components of a router can be identified.

- *Input ports.* The input port performs two functions—the physical-layer functionality (shown in light blue in Fig. A.2) of terminating an incoming physical link to a router, and a datagram lookup and forwarding function (shown in red) so that a datagram is forwarded into the switching fabric of the router. In practice, multiple input ports are often gathered together on a single *line card* within the router.

- *Switching fabric.* The switching fabric connects the router's input ports to its output ports.

- *Output ports.* An output port stores the datagrams that have been forwarded to it through the switching fabric and transmits them on the outgoing link.

- *Routing processor.* The routing processor executes the routing protocols, maintains the routing tables, and performs network management functions within the router.

This sequel highlights the specifics of each of these components.

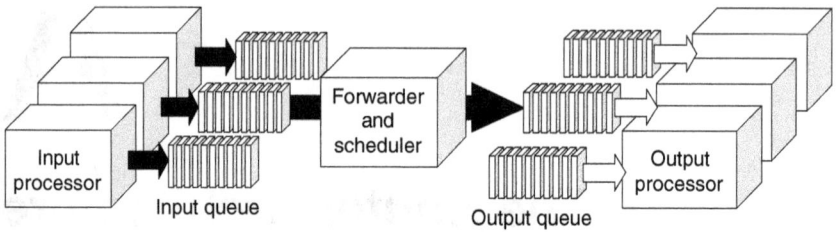

Figure A.1   Router queuing and scheduling: a block diagram.

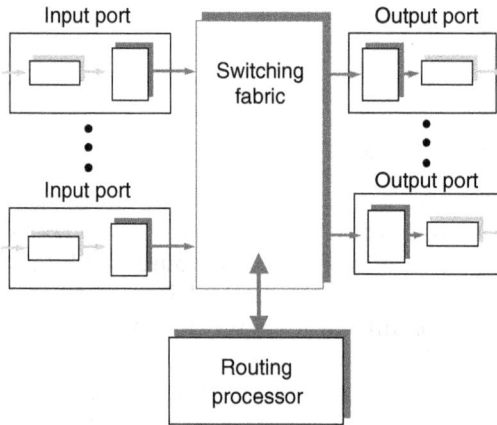

Figure A.2   Router architecture.

A.2   Switching Fabrics

The switching fabric is at the very heart of a router. It is through this switching that the datagrams are actually moved from an input port to an output port. Switching can be accomplished in a number of ways, as indicated in Fig. A.3.

A.2.1   Switching via memory

The simplest, earliest routers were often traditional computers, with switching between the input and output port being done under the direct control of the CPU (routing processor). Input and output ports functioned as traditional I/O devices in a traditional operating system. An input port with an arriving datagram first signalled the routing processor via an interrupt. The packet was then copied from the input port into the processor memory. The routing processor then extracted the destination address from the header, looked up the appropriate output port

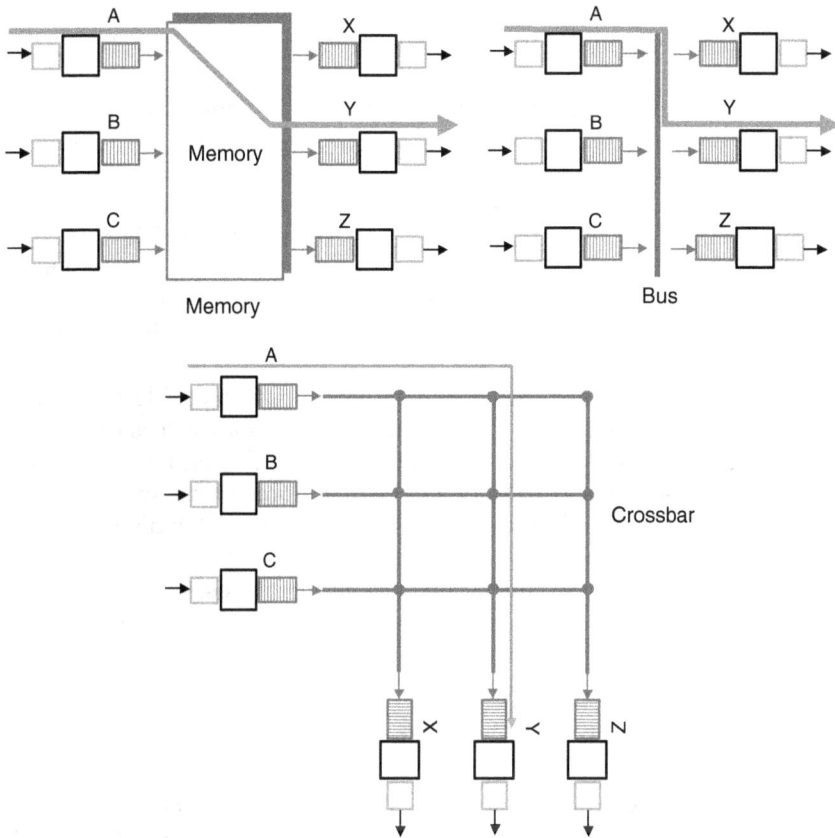

Figure A.3 Three switching techniques.

in the routing table, and copied the packet to the output port's buffers. Note that if the memory bandwidth is such that B packets per second can be written into, or read from, the memory, then the overall switch throughput (the total rate at which packets are transferred from input ports to output ports) must be less than B/2.

A number of modern routers also switch via the memory. A major difference from early routers, however, is that the lookup of the destination address and the storing (switching) of the packet in the appropriate memory location are performed by processors on the input line cards. In some ways, routers that switch via the memory look very much like shared memory multiprocessors, with the processors on a line card storing datagrams into the memory of the appropriate output port. Cisco's Catalyst 8500 series switches [CISCO98a] and Bay Networks Accelar 1200 Series routers switch packets via a shared memory.

A.2.2   Switching via a bus

In this approach, the input ports transfer a datagram directly to the output port over a shared bus, without having the routing processor to intervene (Note that when switching via the memory, the datagram must also cross the system bus going to/from the memory). Although the routing processor is not involved in the bus transfer, since the bus is shared, only one packet at a time can be transferred over the bus. A datagram arriving at an input port and finding the bus busy (with the transfer of another datagram) is blocked from passing through the switching fabric and queued at the input port. Because every packet must cross the single bus, the switching bandwidth of the router is limited to the bus speed.

Given that bus bandwidths of over 1 Gbps are possible in today's technology, switching via a bus is often sufficient for routers that operate in access and enterprise networks (e.g., local area and corporate networks). Bus-based switching has been adopted in a number of current router products, including Cisco 1900 [CISCO97b], which switches packets over a 1 Gbps *packet exchange bus*. 3Com's CoreBuilder 5000 systems [KAPOOR97] interconnect ports that reside on different switch modules over its *packet channel* data bus, with a bandwidth of 2 Gbps.

A.2.3   Switching via an interconnection network

One way to overcome the bandwidth limitation of a single, shared bus is to use a more sophisticated interconnection network, such as those that have been used in the past to interconnect processors in a multiprocessor computer architecture. A crossbar switch is an interconnection network consisting of 2N buses that connect N input ports to N output ports, as shown in Fig. A.3. A packet arriving at an input port travels along the horizontal bus attached to the input port until it intersects with the vertical bus leading to the desired output port. If the vertical bus leading to the output port is free, the packet is transferred to the output port. If the vertical bus is being used to transfer a packet from another input port to this same output port, the arriving packet is blocked and must be queued at the input port.

Delta and Omega switching fabrics have also been proposed as an interconnection network between input and output ports. See Ref. [TOBAGI90] for a survey of switch architectures. Cisco 12000 Family switches [CISCO98b] use an interconnection network, providing up to 60 Gbps through the switching fabric.

A.3   Where Does Queuing Occur?

Suppose that the input line speeds and output line speeds are all identical, and that there are n input ports and n output ports. If the switching fabric speed is at least n times as fast as the input line speed, then no queuing would occur at the input ports. This is because even in the worst case, when all n input lines are receiving packets, the switch will be able to transfer n packets from the input port to the output port in the time it takes each of the n input ports to (simultaneously) receive a *single* packet. But what can happen at the output ports? Let us suppose still that the switching fabric is at least n times as fast as the line speeds. In the worst case, the packets arriving at each of the n input ports will be destined to the *same* output port. Since the output port can only transmit a single packet in a unit of time (the packet transmission time), the n arriving packets will have to queue (wait) for transmission over the outgoing link. n more packets can then possibly arrive in the time it takes to transmit just one of the n packets that had previously been queued, and so on. Eventually, buffers can grow large enough to exhaust the memory space at the output port, in which case packets are dropped.

Output port queuing is illustrated in Fig. A.4. At time t, a packet has arrived at each of the incoming input ports, each destined for the uppermost outgoing port. Assuming identical line speeds and a switch operating at three times the line speed, one time unit later (i.e., the time needed to receive or send a packet), all three original packets have been transferred to the outgoing port and are queued awaiting transmission. In the next time unit, one of these three packets will have been transmitted over the outgoing link. In our example, two *new* packets have arrived at the incoming side of the switch; one of these packets is destined for this uppermost output port.

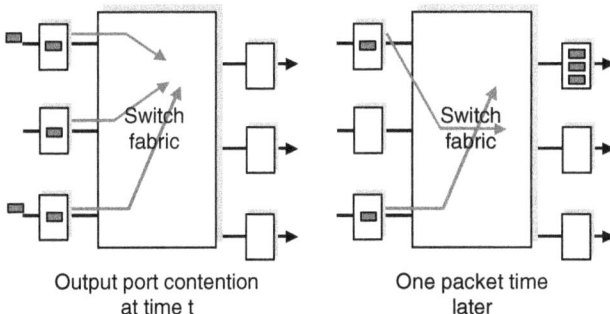

Output port contention   One packet time
    at time t   later

Figure A.4   Output port queuing.

A consequence of output port queuing is that a packet scheduler at the output port must choose one packet among those queued for transmission. This selection might be done on a simple basis such as *first-come-first-served* (FCFS) scheduling, or a more sophisticated scheduling discipline such as *weighted fair queuing* (WFQ), which shares the outgoing link "fairly" among the different end-to-end connections that have packets queued for transmission. Packet scheduling plays a crucial role in providing QoS guarantees. We covered this topic extensively in Chap. 3. Output port packet scheduling disciplines used in routers today are discussed in Ref. [CISCO97a].

If the switch fabric is not fast enough (relative to the input line speeds) to transfer *all* arriving packets through the fabric without delay, then packet queuing will also occur at the input ports, as packets must join input port queues to wait their turn to be transferred through the switching fabric to the output port. To illustrate an important consequence of this queuing, consider a crossbar switching fabric and suppose that (1) all link speeds are identical, (2) that one packet can be transferred from any one input port to a given output port in the same time it takes for the packet to be received on an input link, and (3) packets are moved from a given input queue to their desired output queue on an FCFS basis. Multiple packets can be transferred in parallel as long as their output ports are different. However, if two packets at the front of two input queues are destined to the same output queue, then one of the packets will be blocked at the input queue—the switching fabric can only transfer one packet to a given output port at a time.

Figure A.5 shows an example where two packets (red) at the front of their input queues are destined for the same upper right output port. Suppose that the switch fabric chooses to transfer the packet from the front of the upper left queue. In this case, the red packet in the lower left queue must wait. But not only must this red packet wait, but so too

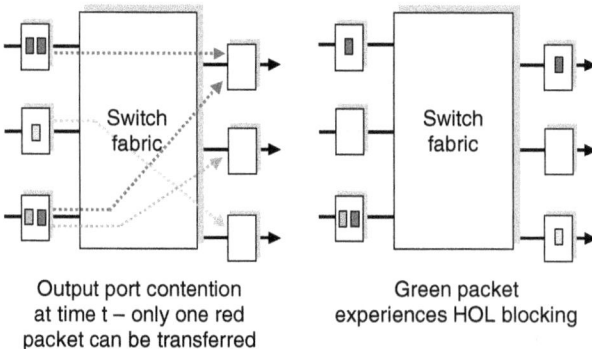

Output port contention at time t – only one red packet can be transferred

Green packet experiences HOL blocking

Figure A.5 HOL blocking at an input queued switch.

must the green packet that is queued behind that packet in the lower left queue, even though there is *no* contention for the middle right output port (the destination for the green packet). This phenomenon is known as *head-of-the-line* (HOL) blocking in an input-queued switch— a queued packet in an input queue must wait for transfer through the fabric (even though its output port is free) due to the blocking of another packet at the head-of-the-line. Reference [KAROL87] shows that due to the HOL blocking, the input queue will grow to unbounded length (informally, this is equivalent to saying that a significant packet loss will occur) as soon as the packet arrival rate on the input links reaches only 58 percent of their capacity. A number of solutions to HOL blocking are discussed in Ref. [McKEO97].

The Data Link Layer: Error Detection and Correction Techniques

B.1 The Link-Layer Protocol

Whereas the network layer has the end-to-end job of moving transport-layer segments from the source host to the destination host, a link-layer protocol has the job of moving a network-layer datagram over a *single link* in the path. The link-layer protocol defines the format of the packets exchanged between the nodes at the ends of the link, as well as the actions taken by these nodes when sending and receiving packets. Each link-layer frame typically encapsulates one network-layer datagram. Examples of link-layer protocols include Ethernet, token ring, FDDI, and PPP; in some contexts, ATM and frame relay can be considered link-layer protocols as well.

An important characteristic of the link layer is that a datagram may be handled by different link-layer protocols on the different links in the path. For example, a datagram may be handled by Ethernet on the first link, PPP on the last link, and frame relay on all intermediate links. It is also important to note that the services provided by the different link-layer protocols may differ from protocol to protocol. For example, a link-layer protocol may or may not provide reliable delivery.

Possible services that can be offered by a link-layer protocol include

- *Framing and link access.* Link-layer protocols encapsulate each network-layer datagram within a link-layer frame before transmission onto the link. A frame consists of a data field, in which the network-layer datagram is placed, and a number of header fields. (A frame may also include trailer fields; however, we will refer to both header and trailer fields as header fields.)

- *Reliable delivery.* If a link-layer protocol provides the reliable-delivery service, then it guarantees to move each network-layer datagram across the link without error. Similar to a transport-layer reliable-delivery service (such as TCP), a link-layer reliable-delivery service is achieved with acknowledgments and retransmissions. A link-layer reliable-delivery service is often used for links that are prone to high error rates, such as a wireless link, with the goal of correcting an error locally, on the link at which the error occurs, rather than forcing an end-to-end retransmission of the data by transport- or application-layer protocol.

  Link-layer reliable delivery is however considered to be an unnecessary overhead for low bit-error links, including fibber, coax, and many twisted-pair copper links.

- *Flow control.* The nodes on each side of a link have a limited amount of packet buffering capacity. This is a potential problem as a receiving node may receive frames at a rate faster than it can process the frames (over some time interval). Without flow control, the receiver's buffer can overflow and frames can get lost. A link-layer protocol provides then flow control functions in order to prevent the sending node on one side of a link from overwhelming the receiving node on the other side of the link.

- *Error detection.* Errors are introduced by signal attenuation and electromagnetic noise. Many link-layer protocols provide a mechanism for a node to detect the presence of one or more errors. This is achieved by having the transmitting node set error detection bits in the frame, and having the receiving node perform an error check. Error detection is a common service among link-layer protocols.

- *Error correction.* Error correction is similar to error detection, except that a receiver does not only detect whether errors are introduced in the frame but can also determine exactly where in the frame the errors have occurred and hence correct them. Some protocols (such as ATM) provide link-layer error correction for the packet header rather than for the entire packet. We cover error detection and correction in the following section.

- *Half-duplex and full-dulpex.* With full-duplex transmission, both nodes at the ends of a link may transmit packets at the same time. With half-duplex transmission, a node cannot transmit and receive at the same time.

B.2   Error Detection and Correction Techniques

We examine here a few of the simplest techniques that are used to detect and, in some cases, correct bit errors. Our goal here is to develop an intuitive feel for the capabilities that error detection and correction techniques provide, and to see how a few simple techniques are used in practice. A full treatment of the theory and implementation of this topic is itself the topic of many textbooks (e.g., [SCHWAR80]), and our treatment here is necessarily brief.

Figure B.1 illustrates the setting for the discussions that follow. At the sending node, data D, to be "protected" against bit errors, is augmented with error detection and correction bits, EDC. Typically, the data to be protected includes not only the datagram passed down from the network layer for transmission across the link, but also link-level addressing information, sequence numbers, and other fields in the data link frame header. Both D and EDC are sent to the receiving node in a link-level frame. At the receiving node, a sequence of bits, D' and EDC' are received. Note that D' and EDC' may differ from the original D and EDC as a result of in-transit bit flips.

The receiver's task is to determine whether or not D' is the same as the original D, given that it has only received D' and EDC'. Error detection and correction techniques allow the receiver to sometimes, *but not always*, detect that bit errors have occurred.

In the following, we examine three techniques for detecting errors in the transmitted data—parity checks (to illustrate the basic ideas behind

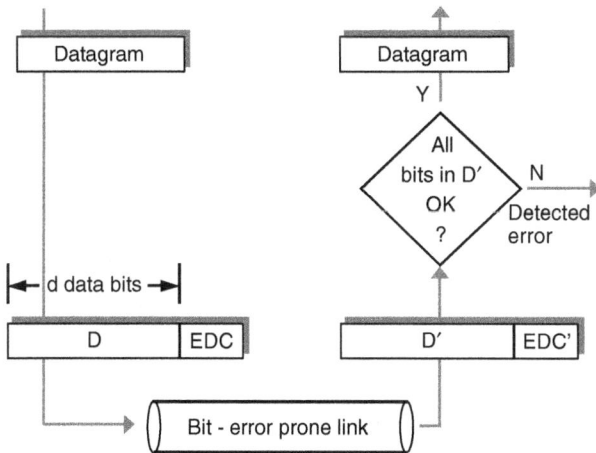

Figure B.1   Error detection and correction scenario.

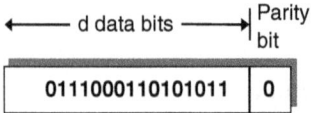

Figure B.2 One-bit even parity.

error detection and correction), checksumming methods (which are more typically employed in the transport layer), and cyclic redundancy checks (which are typically employed in the data link layer).

B.2.1 Parity checks

Perhaps the simplest form of error detection is the use of a single *parity bit*. Suppose that the information to be sent, D in Fig. B.1, has d bits. In an even parity scheme, the sender simply includes one additional bit and chooses its value such that the total number of 1s in the $d + 1$ bits (the original information plus a parity bit) is even. For odd parity schemes, the parity bit value is chosen such that there are an odd number of 1s. Figure B.2 illustrates an even parity scheme, with the single parity bit being stored in a separate field.

The receiver only needs to count the number of 1s in the received $d + 1$ bits. If an odd number of 1-valued bits are found with an even parity scheme, the receiver knows that at least one bit error has occurred. More precisely, it knows that some *odd* number of bit errors has occurred.

But what happens if an even number of bit errors occurs? If the probability of bit errors is small and errors can be assumed to occur independently from one bit to the next, the probability of multiple bit errors in a packet would be extremely small. In this case, a single parity bit might suffice. However, measurements have shown that rather than occurring independently, errors are often clustered together in "bursts". Under burst error conditions, the probability of undetected errors in a frame protected by single-bit-parity can approach 50 percent [SPRAG91]. Clearly, a more robust error detection scheme is needed (and, fortunately, is used in practice!). But before examining error detection schemes that are used in practice, let's consider a simple generalization of one-bit parity that will provide us with insight into error-correction techniques.

Figure B.3 shows a two-dimensional generalization of the single-bit parity scheme. Here, the d bits in D are divided into i rows and j columns. A parity value is computed for each row and for each column. The resulting $i + j + 1$ parity bits are the data-link frame's error-detection bits.

Suppose now that a single bit error occurs in the original d bits of information. With this *two-dimensional parity* scheme, the parity of both the column and the row containing the flipped bit will be in error.

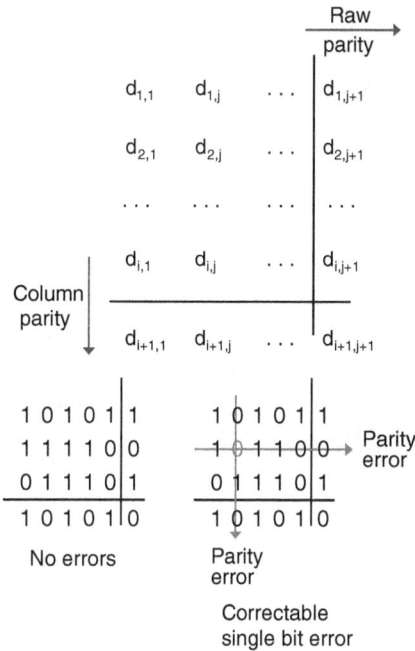

Figure B.3 Two-dimensional even parity.

The receiver can thus not only *detect* the fact that a single bit error has occurred, but can use the column and row indices of the column and row with parity errors to actually identify the bit that was corrupted and *correct* that error! Figure B.3 shows an example in which the 0-valued bit in position (1,1) is corrupted and switched to a 1—an error that is both detectable and correctable at the receiver.

The ability of the receiver to both detect and correct errors is known as *forward error correction* (FEC). These techniques are commonly used in audio storage and playback devices such as audio CDs. In a network setting, FEC techniques can be used by themselves to allow for immediate correction of errors at the receiver. This avoids having to wait the round-trip propagation delay needed for the sender to receive a negative acknowledgement packet and for the retransmitted packet to propagate back to the receiver—a potentially important advantage for real-time network applications [RUBENS98].

B.2.2 Cyclic redundancy check

An error detection technique used widely in today's computer networks is based on *cyclic redundancy check* (CRC) codes. CRC codes are also known as *polynomial codes*, since it is possible to view the bit string to be sent as

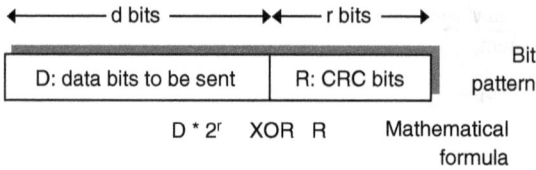

Figure B.4 CRC codes.

a polynomial whose coefficients are the 0 and 1 values in the bit string, with operations on the bit string interpreted as polynomial arithmetic.

CRC codes operate as follows. Consider that the sender (see Fig. B.4) wants to send the *d-bit* piece of data *D*. The sender will choose *r* additional bits *R* and append them to *D* such that the resulting $d + r$ bit pattern (interpreted as a binary number) is exactly divisible by an $(r + 1)$ bit pattern using modulo-2 arithmetic. Let us denote the $(r + 1)$ bit pattern as BP. The process of error checking with CRCs is thus simple— the receiver divides the $d + r$ received bits by BP. If the remainder is nonzero, the receiver knows that an error has occurred; otherwise the data is accepted as being correct.

All CRC calculations are done in modulo-2 arithmetic without carries in addition or borrows in subtraction. This means that addition and subtraction are identical, and both are equivalent to the bitwise exclusive-or (XOR) of the operands. Thus, for example,

$$0011 \text{ XOR } 0001 = 0010$$

$$1101 \text{ XOR } 1100 = 0001$$

Also, we similarly have

$$0011 - 0001 = 0010$$

$$1101 - 1100 = 0001$$

Multiplication and division are the same as in base 2 arithmetic, except that any required addition or subtraction is done without carries or borrows. As in regular binary arithmetic, multiplication by 2^k left shifts a bit pattern by k places. Thus, given *D* and *R*, the quantity $D \times 2r \text{ XOR } R$ yields the $d + r$ bit pattern shown in Fig. B.4.

Let us now turn to the crucial question of how the sender computes *R*. Recall that we want to find *R* such that there is an *n* such that

$$D \times 2r \text{ XOR } R = n\text{BP}$$

That is, we want to choose *R* such that BP divides into $D \times 2r\text{XOR } R$ without a remainder. If we exclusive-or (i.e., add modulo 2, without

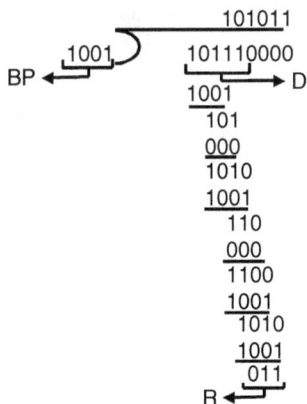

```
            101011
        ┌─────────────
  1001 )  101110000
BP ←──┘            → D
         1001
          101
          000
         1010
         1001
          110
          000
         1100
         1001
         1010
         1001                Figure B.5   An example CRC calculation.
          011
   R ←────┘
```

carry) R to both sides of the above equation, we get

$$D \times 2r = n\mathrm{BP}\ \mathrm{XOR}\ R$$

This equation basically tells that if we divide $D \times 2^r$ by BP, the value of the remainder is precisely R. In other words, we can calculate R as

$$R = \text{remainder}\left(D \times \frac{2r}{\mathrm{BP}} \right)$$

Figure B.5 illustrates this calculation for the case of $D = 101110$, $d = 6$ and BP $= 1001$, $r = 3$. The nine bits transmitted in this case are 101110011.

International standards have been defined for 8-, 12-, 16- and 32-bit generators G. An 8-bit CRC is used to protect the 5-byte header in ATM cells. The CRC-32 32-bit standard, which has been adopted in a number of link-level IEEE protocols, uses a generator of

$$G_{CRC\text{-}32} = 100000100110000010001110110110111$$

Each of the CRC standards can detect burst errors of less than $r + 1$ bits and any odd number of bit errors. Furthermore, under appropriate assumptions, a burst of length greater than $r + 1$ bits is detected with a probability of 1 to 0.5^r. An excellent introduction to this topic is provided in Ref. [SCHWAR80].

B.2.3   Checksumming methods

In checksumming techniques, the d bits of data in Fig. B.1 are treated as a sequence of k-bit integers. One simple checksumming method is to simply sum these k-bit integers and use the resulting sum as the error detection bits. The so-called *Internet checksum* [RFC 1071] is based on this approach—bytes of data are treated as 16-bit integers and their ones-complement sum forms the Internet checksum. A receiver calculates the checksum, which it calculates over the received data, and checks whether it matches the checksum carried in the received packet.

RFC1071 [RFC 1071] discusses the Internet checksum algorithm and its implementation in detail. In the TCP/IP protocols, the Internet checksum is computed over all fields (header and data fields included). In other protocols, such as XTP [STRAYER92], one checksum is computed over the header, while another checksum is computed over the entire packet.

The Global Positioning System: Introduction and Services

Global Positioning System

The *global positioning system* (GPS) is a worldwide satellite-based radio navigation system tool to calculate a device's physical location [KAPLAN96]. The system provides accurate, continuous, and worldwide three-dimensional position and velocity information to GPS receivers. The satellite constellation consists of 24 satellites arranged in six orbital planes with four satellites per plane. Satellites transmit navigation messages periodically. Each navigation message contains the satellite's orbit element, clock, and status. After receiving the navigation messages, a GPS receiver can determine its position and roaming velocity. To determine the receiver's longitude and latitude, at least three satellites are required. To determine also the altitude, another satellite is needed.

To calculate the position of a user, GPS utilizes the concept of one-way time of arrival (TOA) ranging. The results of TOA ranging measurements are applied to the *triangulation method* and then the location information with some error can be calculated. The errors are due to the fact that the signal is deteriorated by ionospheric and tropospheric effects, noisy channel, and clock inaccuracy during its travel from the satellite to the receiver. More satellites can increase the positioning accuracy. The positioning accuracy of GPS ranges to about a few tens of meters. GPS receivers can be used almost anywhere near the surface of the Earth.

Basically, GPS provides two services—the *standard positioning service* (SPS) and the *precise positioning service* (PPS). The SPS is designated for the public and provides a predictable accuracy of at least 100 m (95 percent) in the horizontal plane and 156 m (95 percent) in the vertical plane.

Since SPS and PPS do not provide exact location information, the *differential GPS* (DGPS) system has been introduced to provide more accurate location information. In the basic form of DGPS, a reference station with a precisely-known location is used. The reference station also performs GPS signal calculation. By comparing the result of location information obtained by the GPS signal with the known location information, the reference station can produce error-correction information. The error-correction information is broadcast by the reference station and used for error correction of DGPS receivers, which can hear signals both from the satellites and the reference station. Some DGPSs can provide exact location information with no more than a 1-m error.

D

Analysis of the Relative Distance Microdiscovery Mechanism

D.1 Overview

In the following sections an analytical model for the RDM mechanism is introduced and the network route searching cost is derived. An assumption made is that the average transmitting power level of nodes is similar for all nodes and known by the system.[1]

D.1.1 The performance analysis model: the concept of virtual wireless ring (VWR) and terminal mobility

Consider a wireless ad hoc network system with N mobile users. Let us assume that a mobile terminal powers up at time $t = 0$. At every discrete time $t(t \geq 0)$ after the initiation, a mobile terminal may either move or stay at its position.

Assume $X^t, Y^t \in \mathfrak{R}^M$ are the coordinates of the position where a mobile terminal is located at discrete time t. The first step upon position prediction initiation is to find the discrete positions (here called movement spots or, simply, spots) the mobile can possibly be after a period of time t. The spot where mobile is located at $t = 0$ is referred to as the *reference spot*.

---

[1]In a real-packet radio network where power control is applied to adjust transmission power levels, this information can be shared among devices through a separate signaling channel on the data-link layer as in Ref. [SURE98].

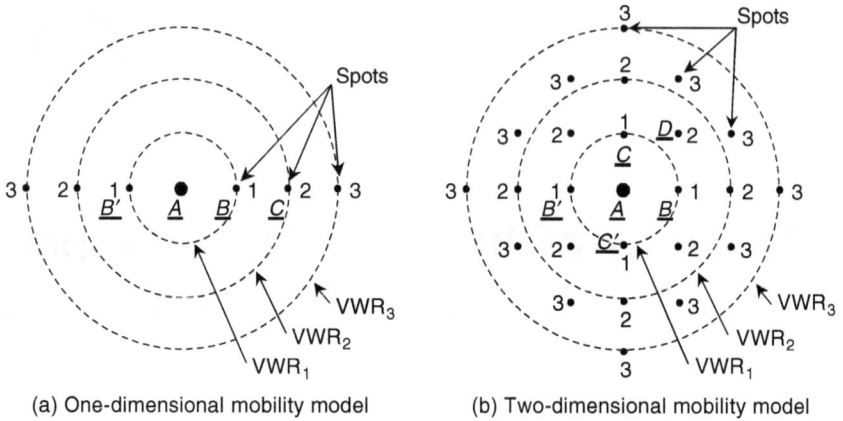

(a) One-dimensional mobility model   (b) Two-dimensional mobility model

Figure D.1   One and two-dimensional mobility model.

Here, the concept of a *virtual wireless ring* (VWR) is introduced, such that the MANET coverage area is divided into a number of virtual wireless rings each of which is centered at the source of the route discovery. A VWR_t is a circular area with radius equal to the maximum distance that the user may have traveled from its reference spot during time t ($t \geq 0$). Figure D.1 illustrates the partitioning of a residing area and the formation of virtual wireless rings for the one- and two-dimensional mobility models.

Each VWR comprises a number of spots whose distribution for the one- and two-dimensional mobility models is depicted in Fig. D.1(a) and Fig. D.1 (b), respectively. In the same figure, let us assume that at time $t = 0$ a mobile resides at spot A. Then, for the one-dimensional mobility model, after time $t = 1$ unit, the node may be located in any of the spots A, B, or B′, assuming that the mobile moves across the direction of the x axis. For the two-dimensional case, again for time $t = 1$ unit, the mobile may reside in any of the spots A, B, B′, C, or C′, assuming that the motion of the mobile is restricted to the direction of the x and y axes only. The innermost ring corresponds to the initiation time ($t = 0$). It consists of only one spot, which will be referred to as the *reference virtual wireless ring* (RVWR). Clearly, the spot of the RVWR is the reference spot. For a given RVWR, let us assume S_t ($t \geq 0$) to be the group of spots that reside within the area of the tth ring (VWR_t) at time t.

Each group of spots is labelled according to its position in the ring configuration area such that the group of spots i refers to the spots that reside in VWR_i. Let us further assume S_t to be the number of spots in

VWR$_t$ such that

$$S_t = \begin{cases} 2 \times t + 1 & \text{for one} - \text{dimensional model,} \quad t = 0,1,2,... \\ \sum_{i=0}^{t}(4 \times i)+1 & \text{for two} - \text{dimensional model,} \quad t = 0,1,2,... \end{cases}$$

It is evident therefore that, after time t, the terminal will be located somewhere within an area centred at the reference spot S_0 and with *maximum* radius equal to $t \times ST$, where ST, called the spot step, is the minimum distance the mobile can traverse during one discrete time unit. It is evident that the maximum distance results when the terminal has moved in the *same* direction for the time t. The spot step is not fixed but varies based on the velocity of the terminal; the higher the velocity the larger the spot step. For example, let us assume that one time unit equals 1 s of real time. For a given average velocity $|v| = 30$ m/s, the distance between two neighbor spots is 30 m; this is the spot step. In Fig. D.1(a), the distance of the reference spot (A) to each one of the spots B and C, is $|v| \times 1$ and $|v| \times 2$, respectively. The distance to the spot D, however, is not $|v| \times 2$, albeit it resides in VWR$_2$, but instead is given through other methods such as the Euclidean formula.

D.1.2   The analytical model

Assume, without loss of generality, that a mobile terminal was initially at the spot S_0 with coordinates (0,0) and after time t is located at spot S_t with coordinates (x, y). To model the movements of the mobile users in the system, it is assumed here that the time is slotted, and a user can make at most one move during a slot. It is obvious that movements are restricted to the immediate neighboring spots. (Note that the two-dimensional model is described herein, as this involves more complicated analysis than the one-dimensional model. A similar approach can then be followed for the analysis of the one-dimensional model.)

The Markovian model.   In the two-dimensional model, during each slot a user can be in one of the following five states—(1) stationary state (S), (2) right-move state (R), (3) left-move state (L), (4) up-move state (UP), and (5) down-move state (DN).

Let $\mathfrak{I} = \{S,R,L,UP,DN\}$ be the state space and $\mathfrak{I}(t)$ be the state during slot t. Assume that a user is in position $k = (x, y)$ at the beginning of slot t. The movement of the user during that slot depends on the state $\mathfrak{I}(t)$ as follows: if the user is in state S, it then remains in position k, if the user is in state R then it moves to position $(x+1, y)$, if the user is in state L then it moves to position $(x-1, y)$, if the user is in state UP then it

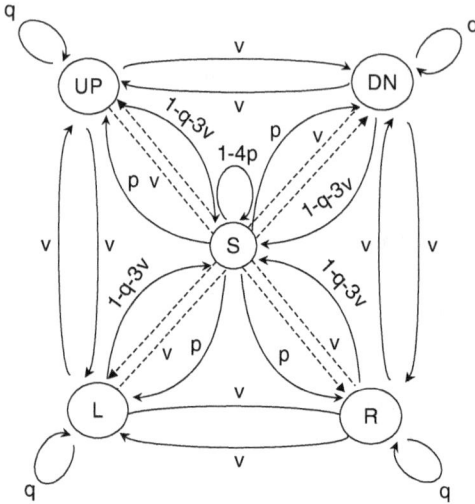

Figure D.2 State diagram of the two dimensional Markov walk model.

moves to position $(x, y - 1)$, and if the user is in state DN then it moves to position $(x, y + 1)$.

Let us assume that the sequence of states $\{\Im(t),\ t = 0,1,2,...\}$ is a Markov chain with transition probabilities (see Fig. D.2) $p_{i,j}$ for each state $i, j \in \Im(t)$:

$$p_{i,j} = \begin{cases} p & \text{for } i = S,\ j \neq i \\ q & \text{for } i \neq S, j \neq S, i = j \\ v & \text{for } i \neq j, j \neq S \qquad 0 < [p, q, v] < 1 \\ 1 - q - 3v & \text{for } i \neq S,\ j = S \\ 1 - 4p & \text{for } i = j = S \end{cases}$$

Given for example that a terminal is in stationary state Yt at time t, its location (spot) at time $t + 1$ is formally given as follows:

$$\Psi_{t+1}(x, y) = \begin{cases} \Psi_t(x, y), & \text{with probability } 1 - 4p \\ \Psi_t(x + 1, y), & \text{with probability } p \\ \Psi_t(x - 1, y), & \text{with probability } p \qquad 0 < p \leq 1/4 \\ \Psi_t(x, y + 1), & \text{with probability } p \\ \Psi_t(x, y - 1), & \text{with probability } p \end{cases}$$

In the above equation, the first case (i.e., $\psi_i(x, y)$) results when the terminal stays idle for a duration of one time slot, while the next four cases result when the terminal makes a movement to the left, right, up, and down, respectively.

Steady-state probabilities.   In this section the steady state probabilities for each spot of the VWR coverage area for the two-dimensional model are derived. Let $D(t)$ be the distance between the spot in which the user is located at time t and the reference spot. In the Markovian model it is important to distinguish between the state in which the user is and its previous spot. As described above, if the user's position at time t is at distance d from the reference spot, then during the next time slot the user's new distance (for the two dimensional Markov mobility model, as described above) could be d, $d - ST$, $d + ST$, or $\sqrt{d^2 + ST^2}$.

Clearly, $\{(D(t), \Im(t), t = 0,1,2,...\}$ is a Markov chain. Let $\Phi_t^s(x, y) = \lim_{t \to \infty}$ Prob$[X(t) = x, Y(t) = y,$ and $s \in \Im(t) \mid t = 0,1,2,...]$ be the stationary probability distribution of this Markov chain. The balance equations for these probabilities are

$$\Phi_t^S(x, y) = (1 - 4p)\Phi_{t-1}^S(x, y) + (1 - q - 3v)\Phi_{t-1}^R(x - 1, y)$$

$$+ (1 - q - 3v)\Phi_{t-1}^L(x + 1, y) + (1 - q - 3v)\Phi_{t-1}^U(x, y + 1)$$

$$+ (1 - q - 3v)\Phi_{t-1}^D(x, y - 1)$$

$$\Phi_t^L(x, y) = p\Phi_{t-1}^S(x, y) + v\Phi_{t-1}^S(x - 1, y) + q\Phi_{t-1}^L(x + 1, y) + v\Phi_{t-1}^U(x, y + 1)$$

$$+ v\Phi_{t-1}^D(x, y - 1)$$

$$\Phi_t^R(x, y) = p\Phi_{t-1}^S(x, y) + q\Phi_{t-1}^R(x - 1, y) + v\Phi_{t-1}^L(x + 1, y) + v\Phi_{t-1}^U(x, y + 1)$$

$$+ v\Phi_{t-1}^D(x, y - 1)$$

$$\Phi_t^U(x, y) = p\Phi_{t-1}^S(x, y) + v\Phi_{t-1}^R(x - 1, y) + v\Phi_{t-1}^L(x + 1, y) + q\Phi_{t-1}^U(x, y + 1)$$

$$+ v\Phi_{t-1}^D(x, y - 1)$$

$$\Phi_t^D(x, y) = p\Phi_{t-1}^S(x, y) + v\Phi_{t-1}^R(x - 1, y) + v\Phi_{t-1}^U(x + 1, y) + v\Phi_{t-1}^S(x, y + 1)$$

$$+ q\Phi_{t-1}^D(x, y - 1)$$

with initial values

$$\Phi_0^S(x, y) = \Phi_0^R(x, y) = \Phi_0^L(x, y) = \Phi_0^U(x, y) = \Phi_0^D(x, y) = \begin{cases} 1 & \text{if } x = y = 0 \\ 0 & \text{otherwise} \end{cases}$$

The probabilities $\text{Prob}[X(t) = x, Y(t) = y \mid t = 0,1,2,...]$ are

$$\Phi_t(x, y) = \text{Prob}[X(t) = x, Y(t) = y \mid t = 0,1,2,...] = \tau(S) \times \Phi_t^S(x, y) + \tau(R)$$
$$\times \Phi_t^S(x, y) + \tau(L) \times \Phi_t^L(x, y) + \tau(U) \times \Phi_t^U(x, y) + \tau(D) \times \Phi_t^D(x, y)$$

where $\tau(\vartheta)$ is the stationary probability of being in state ϑ, i.e.,

$$\tau(S) = \frac{1 - q - 3v}{1 + 4p - q - 3v}$$

$$\tau(R) = \tau(L) = \tau(U) = \tau(D) = \frac{p}{1 + 4p - q - 3v}$$

D.2   The RDE Algorithm

As mentioned earlier, route discovery in RDMAR relies for its operation on the relative distance estimation (RDE) algorithm, whose primary objective is to estimate the relative distance between two nodes to achieve query localization. As such, the RDE algorithm presents a virtual topology to the routing algorithm and accepts feedback from the routing algorithm in order to adjust that virtual routing topology and make routing estimations. This section presents a methodology to compute the expected relative distance (d_t) between two nodes at time t, if d_0 is their distance at time t_0.

Formally, in order to calculate an estimate of the relative distance between two nodes, the stochastic model of the RDE algorithm needs the following information: the previous relative distance of the two mobiles (if known) and the time elapsed since this information was last received. If no prior history is available for their previous relative distance, then the new RD is set to some maximum value; this results in broadcasting into the entire network area, a mechanism similar to flooding.

It is obvious that since both terminals have identical moving probabilities, the steady state probabilities of the spots in the coverage area of each node are identical. That is, the source and destination nodes have identical spot configurations. In addition, it is assumed that mobiles have similar transmitting power levels as well as speed and mobility patterns. Therefore, the two circles (or, according to Fig. 2.11, RDM areas) centred at the source and destination nodes, respectively, have similar characteristics.

Two cases are distinguished here—when the two virtual areas overlap and when they do not. The former case results when the product of the average nodal velocity and the elapsed time is greater than half the previous RD while the latter case results when the product is smaller than half the previous RD. Let S^{SRC} and S^{DST} denote the spot space for source and destination nodes, respectively. Based on this model, an iterative algorithm (Fig. D.3) that calculates the convolution of the two spot configurations is used to calculate the probability that the terminals could be at relative distance d_t after time t.

The probability of the two terminals being in relative distance d_t after time t is formally given by

$$P(d_t) = \sum_{\substack{SRC\_x \in S^{SRC} \\ SRC\_y \in S^{SRC}}} \sum_{\substack{DST\_x \in S^{DST} \\ DST\_y \in S^{DST}}} \{\Phi(SRC\_x, SRC\_y)$$

$$\times \Phi(DST\_x, DST\_y) \mid ED(SRC\_x, SRC\_y, DST\_x, DST\_y) = d_t\}$$

See Fig. D.3 for details on the complete RDE algorithm.

Moreover, note that for an RDM area of radius RD, the number of virtual areas can be regulated through adjustments in each node's transmitter power. Subject to the local propagation conditions and receiver sensitivity, the transmission power determines the set of VWRs.

Finally, it is evident that the RDE algorithm relies on the up-to-date information that is maintained in the RDE database for the accuracy of its estimations; the more frequent the database update, the more precise the RDE information will be. To increase the rate of updating the RDE information (namely, TLU and RD fields), a node on reception of a packet, control or data, proceeds and updates these two fields for the previous and source nodes of the packet.

D.3   Communication Cost

A key factor for the design of an efficient MANET routing protocol is the length or range of the route discovery or else, according to RDM, the number of VWRs required to locate the end terminal. This important term here is referred to as $E(A, B)$, which is the expected number of forwarding VWRs between two nodes A, B.

Assume that the cost for broadcasting in a VWR is Y. This cost accounts for the wireless bandwidth utilization to forward a broadcast message and the computational requirements in order to process the message. Obviously, the larger the number of estimated VWRs ($E(A, B)$) the higher the number of nodes "polled" and thus, the higher the overhead induced. This observation implies that the wider the RDM area the higher the cost incurred to broadcast in this area. This comes from the fact that a higher number of mobile terminals reside in wider rings rather than in smaller ones.

```
if (Do >= 2L) {
  offset = Do − 2L;
  /* Initialize array of distance probabilities */
  D = 0;
  P[D] = −1.0;
  /* Calculate the convolution of the two spot configurations (i.e., SRC & DST) */
  for(Src_x = 0; Src_x <= 2L; Src_x++){
    for(Src_y = 0; Src_y <= 2L; Src_y++){
      for(Dst_x = 0; Dst_x <= 2L; Dst_x++){
        for(Dst_y = 0; Dst_y <= 2L; Dst_y++){
          RD = EucledianDistance(Src_x, Dst_x+2L-offset, Src_y, Dst_y);
          if( RD ≠ P[i] , ∀ i <= D ) {
            D++;
            P(D) = Φ(Src_x, Src_y)* Φ(Dst_x,Dst_y);
          }
          else  P(i) += Φ(Src_x, Src_y)* Φ(Dst_x,Dst_y);
        }
      }
    }
  }
}
```

```
if (Do < 2L) {
  offset = 2L − Do;
  /* Initialize array of distance probabilities */
  D = 0;
  P[D] = −1.0;
  /* Calculate the convolution of the two spot configuration (i.e., SRC & DST) */
  for(Src_x = 0; Src_x <= 2L; Src_x++){
    for(Src_y = 0; Src_y <= 2L; Src_y++){
      for(Dst_x = 0; Dst_x <= 2L; Dst_x++){
        for(Dst_y = 0; Dst_y <= 2L; Dst_y++){
          RD = EucledianDistance(Src_x, Dst_x+2L-offset, Src_y, Dst_y);
          if(RD ≠ P[i] , ∀ i <= D) {
            D++;
            P(D) = Φ(Src_x, Src_y)* Φ(Dst_x, Dst_y);
          }
          else  P(i) += Φ(Src_x, Src_y)* Φ(Dst_x, Dst_y);
        }
      }
    }
  }
}
```

Figure D.3 RDE algorithm for a two-dimensional random walk model for nonoverlapped (left) and overlapped (right) virtual areas.

However, an important observation is that the control overhead induced by a single route discovery does not thoroughly depend on the number of VWRs in which the broadcast propagates, but also on the number of mobiles that reside in the RDM area. Two RDM areas with the same number of VWRs but different mobile population per VWR result in different volumes of control overhead. That is, the area with the higher population contributes to a larger amount of control messages. In a highly dynamic scenario, such as a MANET, it has been proven extremely difficult to estimate the control-signalling load per VWR as the number of nodes change dynamically and constantly. To simplify the analysis, let us assume that the system is in an equilibrium state such that for each VWR, the number of nodes entering a VWR equals the number of nodes departing from this VWR.

In other words, each VWR is assumed to comprise an even distribution of mobiles, which then results in two RDM areas (of the same size) to induce the same amount of control overhead. The benefit of assuming uniform nodal distribution per VWR is that the total generated network control overhead becomes merely a function of the number of VWRs, which in turn is a function of the average transmission range (Tx). More sophisticated stochastic models that account for different populations per VWR could also leverage the design analysis.

To realize the impact of mobile population in VWRs, the number of nodes in each VWR is then counted. It can now be concluded that the network cost of a single query search is a function of the number of nodes per VWR as well as of the number of VWRs covered by the propagation of the query thread.

Given that the newly calculated relative distance of the mobiles is RD_{AB}, the cost for terminal searching when a hit occurs is defined as $C(RD_{AB}) = Y \times E(A, B), = Y \times (RD_{AB} + 1)$. "1" is added to allow some degree of searching redundancy during the route discovery. Searching redundancy should be allowed, as there are situations where the smallest path cannot be discovered if the range of discovery is limited to the exact distance between the source and destination nodes of the discovery process. As illustrated in Ref. [AGGE99], even though the distance estimator (RDE) may produce a correct estimated relative distance for route discovery, yet the rapid and unpredictable topological changes, as a result of mobility, during the route discovery phase may have an immediate effect on the actual distance between the source and destination nodes, which may lead to discovery failure. In such situations, path redundancy is a viable solution for accommodating these fast network changes during the discovery process.

From the equations above, the total network cost per route discovery, C_{RDMAR_HIT}, when a hit occurs, is $C_{RDMAR\_HIT} = C(RD_{AB}) + N_{RP} \times E(A, B)$, where N_{RP} is the total number of route reply messages generated from the destination of the discovery as a response to the same query thread. That is, the cost of a single route discovery phase is the total terminal searching cost as well as the total cost spent for the propagation of replies. Note here that the parameter N_{RP} is controllable in RDMAR, since the route discovery process is end-to-end, and can be set as a protocol parameter such that the destination of the discovery sends N_{RP} replies per route query thread. However, N_{RP} is totally uncontrollable in query quenching schemes as any node with routing information for the destination of the discovery can respond, thus N_{RP} can be as high as N_{net}-1 messages, where N_{net} is the node population of the system.

However, the equations above present the searching cost as a function of the average number of VWRs (i.e., $E(S, D)$), only. To have a better insight into the searching cost, one should take into account the number of nodes per VWR as well as their transmission powers and the success probability of *RDM*. In this sequel, we derive the formulas that relate all these factors with the cost of a single discovery attempt.

Let us assume that a network with N_{net} mobile nodes is partitioned into M_{max} virtual areas and that $M = E(A, B) \leq M_{max}$ is the estimated RD between two nodes located at A and B respectively. Assuming that the N_{net} nodes are distributed evenly within each VWR, one can easily find that there are N_{net}/M_{max} nodes per VWR or $(N_{net}/M_{max}) \times M$ nodes in *RDM* area. Obviously, if $M = M_{max}$ then flooding-based route searching is used.

In the following text, the communication cost is derived. This is the initiated and relayed routing control messaging induced by the route discovery algorithm of a flooding-based query quenching scheme and the *RDMAR* protocol. Throughout the analysis, it is assumed that no network partitions occur. For RDMAR, the cost incurred by the route request propagation during a single *RDM* attempt C_{rdisc_rdm} is:

$$C_{RDisc\_RDM} = \frac{M \times N_{net} - M_{max}}{M_{max}}$$

Because each VWR has to forward one route reply, the number of replies is $(M - 1)$ or, if the destination sends more than one reply, say N_{RP}, this becomes $(M - 1) \times N_{RP}$. So, the total amount of communication overhead generated by RDM, C_{RDM}, is

$$C_{RDMAR} = p \times [C_{Rdisc\_RDM} + (M - 1) \times N_{RP}] + (1 - p)$$
$$\times [C_{Rdisc\_RDM} + C_{FLOODING} + (M - 1) \times N_{RP}]$$

$$= C_{Rdisc\_RDM} + (M - 1) \times N_{RP} + (1 - p) \times C_{FLOODING}$$

where p is the probability that the RDM attempt is successful and C_{RDM} is the cost induced from a single invocation of the RDM algorithm. The network cost induced from the route discovery in RDMAR, therefore, depends on the probability that correct RD estimates are derived from RDE, thus leading to discovery success. In case RDM fails, the source of discovery may reinitiate an RDM with an increased RD value or use flooding-based searching instead. The analysis above uses flooding which is the worst-case scenario. Assuming no network partitions, it is evident that an RDM miss may result in an overall searching overhead that will be greater than the cost of using flooding, as well as to an even higher connection set-up delay.

For flooding-based query quenching schemes, two cases are distinguished —either nodes that receive a route request to forward have a valid route to generate a route reply, or there are no such nodes and the destination sends a route reply instead. If for a single query thread, N_{IN_RP} number of nodes use their routing tables to send replies, then the cost incurred from a single query thread will be:

$$C_{FB\_QQ} = q \times \left[C_{FLOODING} + \sum_{i=1}^{N_{IN\_RP}} Length(i) \right] + (1-q) \times [C_{FLOODING}$$

$$+ (M-1) \times N_{RP}]$$

$$= C_{FLOODING} + q \times \left[\sum_{i=1}^{N_{IN\_RP}} Length(i) \right] + (1-q) \times (M-1) \times N_{RP}$$

where the term $Length(i)$ expresses the length of the route from the source of the route discovery to the node that generates the route reply, q expresses the reliability of the returned path, whereas for both RDMAR and flooding-based query quenching schemes, $C_{FLOODING} = N_{net}$ messages. The last equation shows clearly that in flooding-based query quenching schemes, a single query thread causes two types of flooding; one caused by the propagation of the query thread (i.e., $C_{FLOODING}$) and the other due to the uncontrolled generation of route reply messages (i.e., $q \times [\sum_{i=1}^{N_{IN\_RP}} Length(i)]$).

Furthermore, the propagation of replies is proportional to the number of nodes that generate. replies as a response to a single query thread as well as to the length of the route (i.e., $Length(i)$) from the source of the route discovery to the nodes that generate the replies. As the route length increases, the distance between the source of the query as well as the number of nodes that generate replies increases; thus the cost induced from the propagation of replies increases likewise. In addition, the correctness of a route returned to a query from a node other than the destination of the query, depends on q, which expresses the reliability of the path between the node that returned the route to the destination of the query. Clearly, q depends on the stability of the underlying topology. If the topology is stable enough (i.e., low-mobility case) q is expected to be high and the offered routes to be more reliable, but also the number of replies is high. On the other hand, in a high-mobility scenario where the underlying routing topology changes more frequently, the q factor is expected to be low, thus causing more frequent network failures.

Assuming $N_{RP} = 1$ for simplicity it can be shown that, for a network area of size [max_x × max_y] and an average transmission range, T_x, C_{RDMAR} is always smaller than C_{FB_QQ} for

$$M \leq \left\lceil \frac{\sqrt{\text{max\_x}^2 + \text{max\_y}^2}}{T_x} \right\rceil \left(\frac{1 + pN_{net} + qnk + q}{N_{net} + q \left(\left\lceil \frac{\sqrt{\text{max\_x}^2 + \text{max\_y}^2}}{T_x} \right\rceil \right)} \right)$$

(D.1)

where $N_{IN\_RP} = n$ and Length(i) = k, for each $i = 1,2,3,...,N_{net}$.

Equation (D.1) is true as the optimal number of VWRs is system dependent and varies with T_x. Specifically, M is inversely proportional to T_x since higher T_x values result in the decrease of M. This is because the number of nodes per VWR increases and therefore the probability of locating a user increases. In addition, it is clear from Eq. (D.1) that for n equal to zero, which is the case in RDMAR, the optimum number of VWRs depends on the success probability p of RDM.

Table D.1 summarizes the simulation results from Ref. [AGGE02] and the results predicted in Eq. (D.1) on the VWR distribution with respect to T_x. As illustrated, the simulation results closely match the predicted ones.

TABLE D.1   VWR Distribution

Tx(m)	Simulation (99 Percentile)	Predicted (Eq.1)
150	<9	8.69
200	<7	6.03
250	<5	5

Appendix

E

Spead Spectrum Code Sequences

E.1 Pseudo-Random Noise Codes

The importance of the code sequence choice for the spread spectrum communications is emphasized here. The type of code used, its length as well as its chip rate set bounds on the capability of the system itself.

In direct-sequence systems, as every data symbol is combined with a complete PNcode, the DS processing gain is equal to the code-length. To be usable, a PNcode must follow a number of constraints such as

- The codes must have a sharp (1-chip wide) autocorrelation peak to enable code-synchronization.

- The codes must have a low cross-correlation value; the lower this cross-correlation, the more the users can be allowed in the system [GOL67, HOFFM71, ROEFS77, GLA94]. This holds true for both full-code correlation and partial-code correlation.

- The codes should be "balanced"—the difference between ones and zeros in the code may only be 1. This last requirement stands for good spectral density properties (equally spreading the energy over the whole frequency-band).

Codes can be broadly divided into two classes—orthogonal codes and nonorthogonal codes. Common examples for each class are the Walsh-Hadamard sequences [BEAU75] for the first class, and the shift-register sequences, M-sequences, Gold-codes, and Kasami-codes [GOL67, ROEFS77, SARWA80] for the second class.

E.2   Walsh Hadamard codes

Walsh-sequences have the advantage of being orthogonal; in this way we could get rid of any multiaccess interference. There are, however, a number of drawbacks:

- The codes do not have a single, narrow autocorrelation peak.
- The spreading is not over the whole bandwidth; instead the energy is spread over a number of discrete frequency-components.

Although the full-sequence cross-correlation is identically zero, this does not hold for a partial-sequence cross-correlation function. The consequence is that the advantage of using orthogonal codes is lost. Orthogonality is also affected by channel properties like multipath. In practical systems, equalization is applied to recover the original signal.

These drawbacks make Walsh-sequences unsuitable for noncellular systems. Systems in which Walsh-sequences are applied are, for instance, multicarrier CDMA [YEE94] and the cellular CDMA system IS-95 [QUALC92]. Both systems are based on a cellular concept that all users (and so all interferers) are synchronized with each other. Multicarrier CDMA uses another way of spreading while IS-95 uses a combination of a Walsh-sequence and a shift-register sequence to enable synchronization.

E.2.1   Shift-register sequences

Shift-register sequences are not orthogonal, but they do have a narrow autocorrelation peak. The name already makes clear that the codes are created using a shift-register with feedback-taps. By using a single shift-register, maximum length sequences (M-sequences) can be obtained. Such sequences can be created by applying a single shift-register with a number of specially selected feedback-taps. If the shift-register size is n then the length of the code is equal to $2^n - 1$. The number of possible codes is dependent on the number of possible sets of feedback-taps that produce an M-sequence.

These sequences have a number of special properties, such as

- M-sequences are balanced—the number of ones exceeds the number of zeros by only 1.
- The spectrum of an M-sequence has a sync^2-envelope.
- The shift-and-add property can be formulated as follows:

$$T^k u = T^t u T^j u$$

here u is an M-sequence, by combining two shifts of this sequence (relative shifts i and j) we obtain again the same M-sequence, yet with another relative shift.

- The autocorrelation function is two-valued:

$$R_u(\tau) = \begin{cases} N & \tau = kN \\ -1 & \tau \neq kN \end{cases} \tag{E.1}$$

where k is an integer value, and τ is the relative shift.

- There is no general formula for the cross-correlation of two M-sequences, only some rules can be formulated [ROEFS77].

- A so called "preferred pair" is a combination of M-sequences for which the cross-correlation only shows three different values— $-1, -2^{\lfloor(n+2)/2\rfloor}$, and $2^{\lfloor(n+2)/2\rfloor}$. There do not exist preferred pairs for shift-registers with a length equal to $4k$ where k is an integer.

Combining two M-sequences which form a "preferred pair" leads to a so-called Gold-code. By giving one of the codes a delay with respect to the other code, we can get different sequences. The number of sequences that are available is $2^n + 1$ (the two M-sequences alone, and a combination with $2^n - 1$ different shift positions). The maximum full-code cross-correlation has a value of $2^{\lfloor(n+2)/2\rfloor} + 1$.

If we combine a Gold-code with a decimated version of one of the 2 M-sequences that form the Gold-code we obtain a *Kasami-code* from the large set. Such a code can then be formulated as follows:

$$C = u \times T^k \times v \times T^m \times w$$

here u and v are M-sequences of length, $N_{DS} = 2^n - 1$ (n is even), which form a preferred pair. w is an M-sequence resulting after decimation the v-code with a value $2^{n/2} + 1$. T denotes a delay of one chip, k is the offset of the v-code with respect to the u-code and m is the offset of the w-code with respect to the u-code.

Kasami-codes have the same correlation properties as Gold-codes, the difference lies in the number of codes that can be created. For a large set of Kasami-codes this number is equal to $2^{n/2}(2^n + 1)$. Choosing n equal to, for instance, 6 leaves us 520 possible codes. It is important to have a large code-set—the number of available codes determines the number of different code addresses that can be created. Also a large code-set enables us to select those codes which show good cross-correlation characteristics.

Some families of such coding methods include the *maximal sequences*—maximal codes that can be generated by a given shift register or a delay element of given length. In binary shift resgister sequence generators, the maximum length sequence is 2^{n-1} chips where n is the number of stages in the shift register. A shift register generator consists of a shift register in conjunction with the appropriate logic, which feeds back a logical combination of the state of two or more of its stages to its input. The output, and its contents of its n stages at any clock time, is a function of the outputs of the stages fed back at the preceding sample time. Some codes can be of length 7 to $[2^{36} - 1]$ chips.

Properties of maximal code sequences are:

1. Number of ones in a sequence equals the number of zeros within one chip, e.g., for a 1023 chip (i.e., $2^{10}-1$), there are 512 ones and 511 zeros.

2. Statistical distribution of ones and zeros is well defined and always the same. Relative positions of the run vary from code sequence, but the number of each run does not.

3. Autocorrelation of a maximal linear code is such that for all values of the phase shift, the correlation values are −1 (except for 0 phase shift where it has a value which is positive).

4. A modulo-2 addition of a maximal linear code with a phase-shifted replica of itself results in another replica with a different phase from the original two phases.

5. Every possible state of a given n-stage generator exists at sometime during the generation of a given code cycle. Each state exists for one and only one clock interval (the only exception is that the all-zero state does not and cannot be allowed to occur).

The above properties are very useful in communication, and so a further discussion and explanation of each one is needed.

From (1), it is important that there are equal number of ones and zeros because the DC component in a code or modulated signal can then be neglected. However, it should be noted there is always one less zeros then ones. That is,

Number of ones = $2^{n/2}$

Number of zeros = $2^{n/2} - 1$ where n = no. of stages in generator.

When modulating a carrier with a code sequence, the one–zero balance can limit the degree of carrier suppression obtainable since carrier suppression is dependant on the symmetry of the modulating signal. Therefore, the longer the code sequence, the less effect on the carrier balance.

Run length distribution (2) is similar to the above point. This is obvious when considering that if there is an equal one-zero ratio, then the run length of zeros approximates the run length of ones.

Autocorrelation (3) refers to the degree of correspondence between a code and its phase shifted replica. Autocorrelation plots show the number of agreements (A) minus disagreements (D) for the overall length of the two codes being compared, as the codes assume every phase shift of interest. This can be done using a pair of code generators or a computer simulation.

The combinatorial property (4) of maximal linear code is particularly interesting. This property, which allows the generation of any desired code phase (e.g., a delay of up to 2^{n-1} chips), is valuable any time a different phase code is required. One use is in operating multiple correlators to reduce the synchronization time. Another use is where the phase-shifted code is treated like another sequence, with a normal output and the shift- and add-generated output for separate communication links (possible due to autocorrelation).

Possibly the most valuable linear addition property is that the addition of two m sequences (length r) results in a sequence of length r but not maximal. However, the composite sequence is different for each combination of delay between the two sequences. Hence it can give a large number of code sequences. A pair of sequence generators of length r, can generate r nonmaximal linear codes. It should be noted that since the linear maximal codes are also predictable by anyone who knows the current code state, future operation can be anticipated.

The number of states possible (5) for a set of n elements, each capable of r uncorrelated states is r^n. These states are n-tuples that may be used to control a processor such as a frequency synthesizer. This pseudorandom ordering of n-tuples is typical of all maximal linear sequences, ordering dependant on the feedback use.

Jet propulsion laboratory (JPL) ranging codes are constructed by the modulo-2 addition of two or more maximal linear sequences whose lengths are not necessarily equal. The advantage of this technique is

1. Very long codes useful for unambiguous ranging over long ranges.

2. These long codes are generated by a relatively small number of shift register stages.

3. Synchronization of a receiver can be accomplished by separate operations on the component codes, which can greatly reduce the time required for synchronization.

Error detection and correction codes (EDAC) are mandatory for use in frequency-hopping systems in order to overcome the high rates of error induced by partial band jamming. The usefulness of these codes has a threshold that must be exceeded before satisfactory performance is achieved.

In direct-sequence systems, EDAC may not be advisable because of the effect it has on the code, increasing the apparent data transmission rate, and maybe also the jamming threshold. Some demodulators can operate detecting errors with approximately the same accuracy as an EDAC, so it may not be worthwhile to include a complex coding/decoding scheme in the system.

However, it should be noted that error detection and correction has found application in some spread spectrum systems.

Glossary

A/D	Analog/digital
CDMA	Code division multiple access
3GPP	3rd generation partnership project
P2	3rd generation partnership project 2
8PSK	8 phase shift keying
ATC	ATM transfer capability
ATM	Asynchronous transfer mode
AAL2	ATM adaptation layer type 2
AAL5	ATM adaptation layer type 5
ABR	Available bit rate
ABT	ATM block transfer
ACELP	Algebraic code excitation linear prediction
ACIR	Adjacent channel interference ratio
ACLR	Adjacent channel leakage ratio
ACTS	Advanced Communications Technologies and Services
AICH	Acquisition indication channel
ALCAP	Access link control application part
AM	Acknowledged mode
AMD	Acknowledged mode data
AMR	Adaptive multirate
ANSI	American National Standards Institute
AODV	Ad hoc on-demand distance vector
API	Application programmable interface
APN	Access point name
ARIB	Association of radio industries and businesses
ARQ	Automatic repeat request
ASC	Access service class
ASN 1	Abstract syntax notation one
ATM	Asynchronous transfer mode
AWGN	Additive white Gaussian noise
BB SS7	Broadband signaling system no. 7

BCCH	Broadcast channel
BCFE	Broadcast control functional entity
BCH	Broadcast channel
BCH	Broadcast channel
BER	Bit error ratio
BLER	Block error rate
BMC	Broadcast/multicast control protocol
BNF	Backus-Naur form
BoD	Bandwidth on demand
BPSK	Binary phase shift keying
BS	Base station
BSC	Base station controller
BSI	British Standards Institution
BSS	Base station subsystem
BSS	Base station system
BSSGP	Base station system GPRS protocol
BTS	Base transceiver station
C-RNTI	Cell-RNTI, radio network temporary identity
C-NBAP	Common NBAP
C/I	Carrier to interference
CA-ICH	Channel assignment indication channel
CB	Cell broadcast
CBC	Cell broadcast center
CBR	Constant bit rate
CBS	Cell broadcast service
CCCH	Common control channel
CCH	Common transport channel
CCH	Control channel
CCITT	Comite Consultatif International des Telegraphes et des Telephones
CDF	Cumulative distribution function
CD-ICH	Collision detection indication channel
CDMA	Code division multiple access
CDR	Call data record
CDV	Cell delay variation
CER	Cell error rate
CFN	Connection frame number
CGF	Charging gateway functionality

CIR	Carrier to interference ratio
CLR	Cell loss ratio
CM	Continuous media
CM	Connection management
CMR	Cell misinsertion rate
CN	Core network
CODIT	Code division test bed
CoS	Class of service
CPCH	Common packet channel
CPICH	Common pilot channel
CRC	Cyclic redundancy check
CRNC	Controlling RNC
CS	Circuit switched
CSGR	Clusterhead gateway switch routing
CSICH	CPCH status indication channel
CTCH	Common traffic channel
CTD	Cell transfer delay
CWTS	China wireless telecommunications standard group
DBR	Deterministic bit rate
DCA	Dynamic channel allocation
DCCH	Dedicated control channel
DCFE	Dedicated control functional entity
DCH	Dedicated channel
DECT	Digital enhanced cordless technology
DECT	Digital enhanced cordless telephone
DF	Decision feedback
Diff-Serv	Differentiated services
DL	Downlink
DlA	Digital/analog
D-NBAP	Dedicated NBAP
DPCCH	Dedicated physical control channel
DPDCH	Dedicated physical data channel
DPE	Distributed processing environment
DRNC	Drift RNC
DRX	Discontinuous reception
DS-CDMA	Direct spread code division multiple access
DSCH	Downlink shared channel

DSDV	Destination-sequenced distance-vector
DSP	Data service profile
DSR	Dynamic source routing
DTCH	Dedicated traffic channel
DTX	Discontinuous transmission
EDGE	Enhanced data rates for GSM evolution
EDR	Event data record
EFR	Enhance full rate speech codec
EIRP	Equivalent isotropic radiated power
EP	Elementary procedure
ETSI	European Telecommunications Standards Institute
FACH	Forward access channel
FBI	Feedback information
FDD	Frequency division duplex
FDMA	Frequency division multiple access
FER	Frame error rate
FTP	File transfer protocol
FP	Frame protocol
FPLMTS	Future public land mobile telecommunications system
FRAMES	Future radio wideband multiple access system
FSR	Fisheye state routing
FTP	File transfer protocol
GFR	Guaranteed frame rate
GGSN	Gateway GPRS support node
GMSC	Gateway MSC
GPRS	General packet radio system
GPS	Global positioning system
GSIC	Groupwise serial interference cancellation
GSM	Global system for mobile communications
GTP	GPRS tunnel protocol
GTP-U	User plane part of GPRS tunnelling protocol
HLR	Home location register
HO	Handover
HSCSD	High-speed circuit-switched data
HTTP	Hyper text transfer protocol
IBT	Intrinsic burst tolerance
IC	Interference cancellation

ICMP	Internet control message protocol
ID	Identity
IDL	Interface description language
IETF	Internet Engineering Task Force
IMSI	International mobile subscriber identity
IMT-2000	International Mobile Telecommunications 2000.
IN	Intelligent network
Int-Serv	Integrated services
IP	Internet protocol
IPI	Interpath interference
IPPM	IP performance metrics
IPv6	IP version 6
IRC	Interference rejection combining
ISDN	Integrated services digital network
ISI	Intersymbol interference
ISO	International Standard Organisation
ITU	International Telecommunications Union
L2	Layer 2
LAI	Location area identity
LAN	local area network
LAP	Link access procedure
LBNL	Lawrence Berkeley National Laboratory
LCS	Location services
LLC	Logical link control
LP	Low pass
MA	Midamble
MAC	Medium access control
MAl	Multiple access interference
MAP	Maximum a posteriori
MCU	Multipoint control unit
ME	Mobile equipment
MF	Matched filter
MLSD	Maximum likelihood sequence detection
MM	Mobility management
MMSE	Minimum mean square error
MPEG	Motion picture experts group
MR-ACELP	Multirate ACELP

MS	Mobile station
MSC	Mobile switching centre
MSC/VLR	Mobile switching centre/visitor location register
MT	Mobile terminal
MTP-b	Message transfer part (broadband)
MUD	Multiuser detection
NAS	Non access stratum
NBAP	Node B application part
NRT	Nonreal time
O&M	Operation and maintenance
OOMA	Opportunity driven multiple access
OVSF	Orthogonal variable spreading factor
PAD	Padding
PC	Power control
PCCC	Parallel concatenated convolutional coder
PCCCR	Physical common control channel
PCCH	Paging channel
PCCPCR	Primary common control physical channel
PCH	Paging channel
PCPCH	Physical common packet channel
PCS	Persona communication systems,
PER	Packed encoding rules
PI	Page indicator
PIC	Parallel interference cancellation
PICR	Paging indicator channel
PLMN	Public land mobile network
PNFE	Paging and notification control function entity
PDC	Personal digital cellular
PDCP	Packet data converge protocol
PDN	Public data network
PDP	Packet data protocol
PDP context	Packet data protocol context
PDSCR	Physical downlink shared channel
PDU	Protocol data unit
PRACR	Physical random access channel
PHS	Personal handy phone system
PHY	Physical layer

PS	Packet switched
PSCH	Physical shared channel
PSTN	Public switched telephone network
P-TMSI	Packet-TMSI
PU	Payload unit
PVC	Predefined virtual connection
QoS	Quality of service
QPSK	Quadrature phase shift keying
RAB	Radio access bearer
RACH	Random access channel
Radius	Remote access dial-in user server
RAI	Routing area identity
RAN	Radio access network
RANAP	RAN application part
RB	Radio bearer
RDMAR	Relative distance microdiscovery ad hoc routing
RF	Radio frequency
RLC	Radio link control
RNC	Radio network controller
RNS	Radio network subsystem
RNSAP	RNS application part
RNTI	Radio network temporary identity
RRC	Radio resource control
RRM	Radio resource management
RSSI	Received signal strength indicator
RSVP	Resource reservation protocol
RT	Real time
RTCP	Real time transport control protocol
RTP	Real time protocol
RTSP	Real time streaming protocol
RU	Resource unit
SAAL- NNI	Signaling ATM adaptation layer for network to network interfaces
SAP	Session announcement protocol
SCCP	Signaling connection control part
SCCPCH	Secondary common control physical channel
SCH	Synchronisation channel
SCTP	Simple control transmission protocol

SDO	Space division duplex
SF	Spreading factor
SFN	System frame number
SGSN	Serving GPRS Support Node
SHO	Soft handover
SIB	System information block
SIC	Successive interference cancellation
SINR	Signal-to-noise
SIO	Silence indicator
SIP	Session initiation protocol
SIR	Signal to interference ratio
SLA	Service level agreement
SM	Session management
SMS	Short message service
SN	Sequence number
SNDCP	Subnetwork dependent convergence protocol
SNR	Signal-to-noise ratio
SDP	Session description protocol
SDU	Service data unit
SRB	Signaling radio bearer
SRNC	Serving RNC
SRNS	Serving RNS
SRP	Secure routing protocol
SS7	Signaling system no. 7
SSCF	Service specific coordination function
SSCOP	Service specific connection oriented protocol
STD	Switched transmit diversity
STTD	Space time transmit diversity
TCH	Traffic channel
TCP	Transport control protocol
TCTF	Target channel type field
TD/CD MA	Time division COMA, combined TDMA and COMA
TDD	Time division duplex
TDMA	Time division multiple access
TE	Terminal equipment
TF	Transport format
TFCI	Transport format combination indicator

TFCS	Transport format combination set
TFI	Transport format indicator
TIC	Telecommunication technology commission
TMSI	Temporary mobile subscriber identity
TORA	Temporally ordered routing algorithm
TPC	Transmission power control
TR	Transparent mode
TS	Technical specification
TSTD	Time switched transmit diversity
TTA	Telecommunications technology association
TxAA	Transmit adaptive antennas
UDP	User datagram protocol
UE	User equipment
UL	Uplink
UM	Unacknowledged mode
UMTS	Universal mobile telecommunications system.
URA	UTRAN registration area
URL	Universal resource locator
U-RNTI	UTRAN RNTI
USCH	Uplink shared channel
USIM	UMTS subscriber identity module
UTRA	Universal terrestrial radio access
UTRAN	UMTS terrestrial radio access network
VAD	Voice activation detection
VlR	Visitor location register
VoIP	Voice over IP
VPN	Virtual private network
WAP	Wireless application protocol
WARC	World administrative radio conference
WCDMA-	Wideband code division multiple access
WLL	Wireless local loop
WRP	Wireless routing protocol
WWW	World Wide Web
ZF	Zero forcing
ZRP	Zone routing protocol

References

[3GPP] Third-generation Partnership Project, http://www.3gpp.org/

[3GSM01] 3GSM^{TM} World Focus 2001 Magazine

[3GSM02] 3GSM^{TM} World Focus 2002 Magazine

[ABRAMS70] N. Abramson, "The Aloha system," *AFIPS Conf. Proc.*, Vol. 37, 1970 Fall Joint Computer Conference, AFIPS Press, Montvale, N.J., 1970, pp. 281–285.

[ABRAMS85] N. Abramson, "Development of the Alohanet," *IEEE Transactions on Information Theory*, Vol. IT-31, No. 3, March 1985, pp. 119–123.

[ADACHI95] F Adachi et al., "Multimedia mobile radio access based on coherent DS-CDMA," *Proc. 2nd International workshop on Mobile Multimedia Commun.*, A23, Bristol University, UK, April 1995.

[ADCON] Adcon Telemetetry. http://www.adcon.com.

[ADLER00] M. Adler and C. Scheideler, "Efficient communication strategies for ad hoc wireless networks," *Theory of Comp. Sys.*, Vol. 33, No. 5–6, pp. 337–391, 1432–4350, Sep-Dec 2000.

[AFEK96] Y. Afek, M. Cohen, E. Haalman, and Y. Mansour, "Dynamic bandwidth allocation," *Proc. IEEE INFOCOM*, Vol. 2, pp. 880–887, March 1996.

[AGGE01] G. Aggelou, R.Tafazolli, "On the Relaying Capability of Next Generation GSM Cellular Network," *IEEE Personal Communications*, Vol. 8, No. 1, pp. 6–13, February 2001 (Special Issue on Advances in Mobile Ad Hoc Networking).

[AGGE02] G. Aggelou, R. Tafazolli, "Determining the Optimal Configuration for the Relative Distance Microdiscovery Ad Hoc Routing Protocol," *IEEE Transactions on Vehicular Technology*, Vol. 51, No. 2, pp. 354–370, March 2003.

[AGGE04] G. Aggelou, "On the Performance Analysis of the Minimum Blocking and Bandwidth Reallocation Channel Assignment (MBCA/BRCA) Methods for Quality-of-Service Routing Support in Mobile Multimedia Ad Hoc networks," *IEEE Transactions on Vehicular Technology*, Vol. 53, No. 3, pp. 770–782, May 2004.

[ALWAN96] A. Alwan, R. Bagrodia, N. Bambos, M. Gerla, L. Klenrock, J. Short, and J. Villasenor, "Adaptive Mobile Multimedia Networks," *IEEE Personal Communications*, April 1996, pp. 34–51.

[ANDER95] Andermo (ed.), "UMTS Code Division Testbed (CODIT)," *CODIT Final Review Report*, September 1995.

[ANGIN98] O. Angin, A. T. Campbell, M. E. Kounavis, and R. R.-F. Liao, "The mobiware toolkit: Programmable support for adaptive mobile networking," *IEEE Personal Communications Magazine*, August 1998 (Special Issue on Adaptive Mobile Systems).

[ARIB] ARIB FPLMTS Study Committee, "Report on FPLMTS Radio Transmission Technology SPECIAL GROUP, (Round 2 Activity Report)," Draft v.E.1.1, January 1997.

[AS-ID] J. Heinanen, F. Baker, W. Weiss, and J. Wroclawski, "Assured Forwarding PHB Group," *Internet Draft <draft-ietf-diffserv-af-04.txt>*, January 1999.

[BAHL01] P. Bahl, R. Wattenhofer, L. Li, and Y. Wang, "Distributed Topology Control for Power Efficient Operation in Multihop Wireless Ad Hoc Networks," *Proc. 20th Annual Joint Conf of IEEE Computer and Communications Societies (INFOCOM 2001)*, Vol. 3, pp. 1388–1397, Anchorage, April 2001.

[BAIER94] A. Baier. U. -C. Fiebig, W. Granzow, W. Koch, P. Teder, and J. Thielecke, "Design Study for a CDMA-Based Third Generation Mobile Radio System," *IEEE Journal on Selected Areas in Communications*, Vol. 12, No. 4, pp 733– 743, May 1994.

[BALAKR98] H. Balakrishnan, "Challenges in Reliable Data Transport Over Heterogeneous Wireless Networks," Ph.D. Thesis, Department of Computer Science, University of California, Berkeley, 1998.

[BANERJ02] S. Banerjee and A. Misra, "Minimum energy paths for reliable communication in multi-hop wireless networks," *Proceedings of ACM Mobihoc 2002*, Vol. 1,
pp. 146–156, Laussane, Switzerland,
June 2002.

[BARRET01] C. L. Barret, M. Drozda, M. V. Marathe, "A Comparative Experimental
Study of Media Access Protocols for Wireless Radio Networks," Research Report LA-
UR-01-2879, Los Alamos National Laboratory, Los Alamos, NM, 2001.

[BASAGN98] S. Basagni, I. Chlamatac, v. Syrotiuk, and B. Woodward, "A Distance
Routing Effect Algorithm for Mobility (DREAM)," *Proceedings of the 4th Annual
ACM/IEEE International Conference on Mobile Computing and Networking (MobiCom)
'98*, Dallas, TX, Oct. 25–30, 1998, pp. 76–84.

[BASAGN98] S. Basagni, I. Chlamtac, and V.R. Syrotiuk, "Geographic Messaging in
Wireless Ad Hoc Networks," *Proc. 4th Annual ACM/IEEE Inti. Conf on Mobile
Computing and Networking (MobiCom '98)*, Dallas, Oct. 25–30, 1998, pp. 76–84.

[BEAU75] K. G. Beauchamp, *Walsh Functions and their Applications*," London: Acadamic
Press, 1975.

[BELLCORE93] Bellcore, Generic criteria for version 0.1 wireless access communications
systems (WACS) and supplement, Bellcore Technical Reference TR-INS-001313 1 (1993).

[BENJIE01] B. Chen, K. Jamieson, H. Balakrishnan, and R. Morris, "Span: An energy-
efficient coordination algorithm for topology maintenance in ad hoc wireless networks,"
ACM Wireless Networks Journal, Vol. 8, No. 5, pp. 481–494, September 2002.

[BERTS92] D. Bertsekas, R. Gallager, *Data Networks*, 2d ed., Prentice Hall, New Jersey,
1992.

[BHAGW95] P. Bhagwat, "A framework for integrating Mobile Hosts within the Internet,"
PhD thesis, University of Maryland, December 1995.

[BHARGH94] V. Bharghavan, A. Demers, S. Shenker, and L. Zhang, "MACAW: A Media
Access Protocol for Wireless LANs," *Proc. 1994 SIGCOMM Conference*, London, UK,
pp. 212–225, 1994.

[BIRK95] Y. Birk and Y. Nachman, "Using direction and elapsed-time information to
reduce the wireless cost of locating mobile units in cellular networks," *Wireless Networks*,
Vol. 1, No. 4, pp. 403–412, 1995.

[BISDIK01] C. Bisdikian, An Overview of the Bluetooth Wireless Technology, *IEEE
Communication Magazine*, Vol. 39, No. 12, pp. 86–94, December 2001.

[BLAZEV01] L. Blazevic, S. Giordano, J. -Y. Le Boudec, "Self-Organizing Routing," *IEEE
Communication Magazine*, Vol. 39, No. 1, pp. 118–124, January 2001.

[BLUETH-SPEC] Specification of the Bluetooth System, Version 1.0B, December 1999.

[BOSE99] P. Bose, P. Morin, I. Stojmenovic, and J. Urrutia, "Routing with Guaranteed
Delivery in Ad Hoc Wireless Networks," *Proc. Jrd ACM Inti. Workshop on Discrete
Algorithms and Methods for Mobile Computing and Communications (DIAL M '99)*,
Seattle, Aug. 20, 1999, pp. 48–55.

[BRAY00] J. Bray, and F. C. Sturman, "Bluetooth: Connect Without Cables," Prentice Hall,
ISBN 0130898406, 2000.

[BROCH98] J. Broch, D. A. Maltz, D. B. Johnson, Y. C. Hu, and J. Jetcheva, "A
Performance Comparison of Multi-Hop Wireless Ad Hoc Network Routing Protocols,"
*Proc. of the Fourth Annual ACM/IEEE International Conference on Mobile Computing
and Networking (MOBICOM '98)*, Dallas, TX, Oct. 25–30,1998, pp. 85–97.

[BROCH98] J. Broch, D. B. Johnson, and D. A. Maltz, The Dynamic Source Routing
Protocol for Mobile Ad Hoc Networks, Internet-Draft, draft-ietf-manet-dsr-10.txt, July
2004, Work in progress.

[BRUNO01] R.Bruno, M.Conti, and E.Georgi, "WLAN Technologies for Mobile Ad Hoc
Networks," *Proceedings of the 34th Annual Hawaii International Conference on System
Sciences (HICSS-34)*, Vol. 9, p. 9003,
Maui, January 3–6, 2001.

[BRUNO02] Bruno R., M. Conti, E. Gregory, "Traffic Integration in personal, local and
geographical wireless networks," *Handbook of Wireless Networks and Mobile
Computing*, Wiley, ISBN 0-471-41902-8, 2002.

[CAPKUN01] S. Capkun, M. Hamdi, and J.P. Hubaux, "GPS-free Positioning in Mobile
Ad Hoc Networks," *Cluster Computing Journal*, Vol. 5, No. 2, April 2002.

[CBRP-ID] M. Jiang, J. Li, and Y. C. Tay, "Cluster Based Routing Protocol (CBRP) Functional specification," Internet draft, draft-ietf-manet-cbrp-spec-00.txt, August 1998. Work in progress.

[CDMAWEB] CDMA online, http://www.cdmaonline.com/interactive/workshops/terms1/1035.htm.

[CDPD] CDPD Industry Input Coordinator, *Cellular Digital Packet Data System Specification*, Preliminary Release V. 0.9, Kirkland, WA:, April 30, 1993.

[CHAMBER02] B.A. Chambers, "The Grid Roofnet: a Rooftop Ad Hoc Wireless Network," Master's Thesis, June 2002 (available on http://www.pdos.lcs.mit.edu/grid/pubs.html)

[CHANG00] J. Chang and L. Tassiulas, "Energy Conserving Routing Wireless Ad Hoc Networks," *Proc. 19th Annual Joint Conf of IEEE Computer and Communications Societies (INFOCOM 2000)*, Vol. 1,2000, pp. 22–31.

[CHATTER02] M. Chatterjee, S. K. Das, and D. Turgut, "WCA: A weighted clustering algorithm for mobile ad hoc networks," *Journal of Cluster Computing (Special Issue on Mobile Ad hoc Networks)*, Vol. 5, pp. 193–204, April 2002.

[CHAUD99] P. Chaudhury, W. Mohr, and S. Onoe, "The 3GPP Proposal for IMT-2000," *IEEE Communications Magazine*, Vol. 37. No. 12, pp. 72 – 81, December 1999.

[CHEN97] T. Chen , M. Gerla, and J. T. Tsai , "QoS Routing Performance in a Multi-Hop Wireless Network," *Proceedings of IEEE ICUPC '97*, 1997.

[CHEN97] T. Chen, J. Tsai, and M. Gerla, "QoS Routing Performance in Multihop, Multimedia, Wireless Networks," *Proceedings of ICUPC '97*, Vol. 2, pp. 557–561, 1997.

[CHEN99] S. Chen, and K. Nahrstedt, "Distributed quality-of -service routing in ad hoc networks," *IEEE Journal on Selected Areas in Communications*, Vol. 17, pp. 1488–1505,1999.

[CHEN99] S. Chen, Routing Support For Providing Guaranteed End-To-End Quality-Of-Service, thesis, University of Illinois, Urbana-Champaign, 1999.

[CHENG89] C. Cheng, R. Reley, S.P.R Kumar, and J. J. Garcia-Luna-Aceves, "A loop-free extended Bellman-Ford routing protocol without bouncing effect," *ACM Computer Communications Review*, Vol. 19, No. 4, pp. 224–236, 1989.

[CIDO95] I. Cidon, R. Rom, and Y. Shavit, "Analysis of one way reservation algorithms," *Fourteenth Annual Joint Conference of the IEEE Computer and Communication Societies*, pp. 1256 – 1263, Boston, MA, 1995.

[CISCO97a] Cisco Systems, "Queue Management," http://www.cisco.com/warp/public/614/quemg_wp.htm, 1997.

[CISCO97b] Cisco Systems, "Next Generation Clear Channel Architecture for Catalyst 1900/2820 Ethernet Switches," http://www.cisco.com/warp/public/729/c1928/nwgen_wp.htm, 1997.

[CISCO98a] Cisco Systems "Catalyst 8500 Campus Switch Router Architecture," http://www.cisco.com/warp/public/729/c8500/csr/8510_wp.htm, 1998.

[CISCO98b] Cisco Systems, "Cisco 12000 Series Gigabit Switch Routers," http://www.cisco.com/warp/public/733/12000/12000_ov.htm, 1998.

[CLARK92] D. Clark, S. Shenker, and L. Zhang, "Supporting Real-Time Applications in an Integrated Services Packet Network: Architecture and Mechanism," *ACM SIG-COMM 92*, August 1992.

[CNDN] The Cellular-News Daily Newsletter, http://www.cellular-news.com/story/10501.shtml.

[CONTI02] M. Conti and S. Giordano, "Mobile Ad Hoc Networking," *Cluster Computing Journal*, Vol. 5, No. 2, April 2002.

[CONTI97] M. Conti, E. Gregori, and L. Lenzini, *Metropolitan Area Networks*, Springer Limited Series on Telecommunication Networks and Computer Systems, New York, November 1997.

[CORS00] M.S. Corson, J.P. Maker, and J.H. Cernicione, "Internet-based Mobile Ad Hoc Networking," *IEEE Internet Computing*, Vol.3, No. 4, pp. 63–70, July-August 1999.

[CORS95] M. S. Corson and A. Ephremides, "A Distributed Routing Algorithm for Mobile Wireless Networks," *ACM/Baltzer Wireless Networks Journal*, Vol. 1, No. 1, pp. 61–81, February 1995.

[CORS97] M.S. Corson, "Issues in Supporting Quality of Service in Mobile Ad Hoc Networks," *IFIP 5th Into Workshop on Quality of Service - (IWQOS '97)*, May 1997.

[COUTO02] D. S. J. De Couto, D. Aguayo, B. A. Chambers, and R. Morris, "Effects of loss rate on ad hoc wireless routing," *Tech. Rep. MIT-LCS-TR-836, MIT Laboratory for Computer Science*, March 2002.

[COUTO98] D. S. J. De Couto and R. Morris, "Location Proxies and Intermediate Node Forwarding for Practical Geographic Forwarding," *MIT Laboratory for Computer Science technical report MIT-LCS-TR-824*, June 2001, Dallas, Oct. 25–30, 1998.

[COX92] D. Cox, "Wireless network access for personal communications," *IEEE Communications*, December 1992, pp. 96–115.

[COX95] D. C. Cox, "An Evolution Toward Three Large Groups of Applications and Services - Wireless Personal Communications: What Is It?," *IEEE Pers. Commun.*, April 1995, pp. 20–35.

[CROWCR95] J. Crowcroft, and Z. Wang, A Rough Comparison of the IETF and ATM Service Models, *IEEE Network*, November 1995, pp. 12–16.

[DAM01] K. V. Dam, et al., "From PAN to BAN: Why Body Area Networks," *Proceedings of the WWRF meeting*, Helsinki, Finland, May 10–11, 2001.

[DAS00] S.R. Das, R. Castaneda, and J. Yan, "Simulation Based Performance Evaluation of Mobile, Ad Hoc Network Routing Protocols," *ACM/Baltzer Mobile Networks and Applications Journal*, July 2000, pp. 179–189.

[DAS00a] S.R. Das, C.E. Perkins, and E. Royer, "Performance Comparison of Two On-demand Routing Protocols for Ad Hoc Networks," *Proc. 19th Annual Joint Conf of IEEE Computer and Communications Societies (INFOCOM 2000)*, Tel Aviv, March 2000, pp. 3–12.

[DAS98] S.R. Das, R. Castaneda, J. Yan, and R. Sengupta, "Comparative Performance Evaluation of Routing Protocols for Mobile Ad Hoc Networks," *Proc. IEEE 7th Inti. Conf on Computer Communication and Networks (I3CN)*, Lafayette, LA, October 1998, pp. 153–161.

[DECT] ETSI, Digital European Cordless Telephone common interface, version 05.03 (May 1991).

[DEME89] A. Demers, S. Keshav, and S. Shenker, "Analysis and Simulation of a Fair Queueing Algorithm," *ACM SIGCOMM 89*, Vol. 19, No. 4, pp. 2—12, August 19–22, 1989.

[DIFFSERV-ID] The IETF Differentiated Services Working Group homepage, http://www.ietf.org/html.charters/diffserv-charter.html.

[DIXIT01] S. Dixit et al., "Resource Management and Quality of Service in Third Generation Wireless Networks," *IEEE Communications Magazine*, February 2001, pp. 125–133.

[DUBE97] R. Dube , C. D. Rais, K.- Y. Wang, and S. K. Tripathi, "Signal Stability Based Adaptive Routing (SSA) for Ad-Hoc Networks," *IEEE Personal Communications*, February 1997, pp. 36–45.

[DWYE98] D. Dwyer, S. Ha, J. Li, and V. Bharghavan, "An adaptive transport protocol for multimedia communications," *In Proc of IEEE Conference on Multimedia Computing Systems*, Durham, North Carolina, August 1998.

[ESSEL00] N. Esseling, H.S. Vandra, and B. Walke, A Forwarding Concept for HiperLAN/2, Proceedings of European Wireless 2000, pp. 13–18, Dresden, Germany, September 2000.

[ESSEL02] N. Esseling, E. Weiss, A. Krämling, and W. Zirwas: A Multi Hop Concept for HiperLAN/2: Capacity and Interference. In Proc. European Wireless 2002, Vol. 1, pp. 1–7, Florence, Italy, February 2002.

[ETSI-HLAN] European Telecommunications Standards Institute, ETSI HIPERLAN/l standard, Oct. 2000. Available at http://www.etsi.org/technicalactiv/hiperlan l.htm.

[ETSI-QoS] ETSI Technical Report, "Telecommunications and Internet Protocol Harmonization Over Networks (TIPHON); General Aspects of Quality of Service (QoS)," TR 101 329 V1.2.5 (1998-10).

[ETSI-UMTS] ETSI ETR 271 Objectives and Overview of UMTS.

[ETSI-UMTSa] ETSI DTR/SMG-O5-040 1 Overall Requirements on the radio interface(s) of UMTS.

[ETSI-UMTSb] ETSI Press Release, SMG Tdoc 40/98, "Agreement Reached on Radio Interface for Third Generation Mobile System, UMTS," Paris, France, January 1998.

[EUGENE98] T. S. Eugene Ng, I. Stoica, H. Zhang, "Packet Fair Queueing Algorithms for Wireless Networks with Location-Dependent Errors," *Proceedings of INFOCOM '98*, 1998.

[EX-ID] V. Jacobson, K. Nichols, K. Poduri, "An Expedited Forwarding PHB," Internet Draft *<draft-ietf-diffserv-phb-ef-02.txt>* Feb. 1999.

[FALAH01] S. Falahati, N.C. Ericsson, A. Ahlén, and A. Svensson, "Hybrid type-II ARQ/AMS and Scheduling using Channel Prediction on Fading Channels," *Nordic Radio Symposium 2001*, 2001.

[FEENEY01] L. M. Feeney and M. Nilsson, "Investigating the energy consumption of a wireless network interface in an ad hoc networking environment," *INFOCOM 2001*, April 2001.

[FRAGOU98] C. Fragouli, V. Sivaraman, and M. Srivastava, "Controlled Multimedia Wireless Link Sharing via Enhanced Class-based Queueinf with Channel-state-dependent Packet Scheduling," *Proceedings of INFOCOM'98*, 1998.

[FREEBER01] J. Freebersyser and B. Leiner, "A DoD Perspective on Mobile Ad Hoc Networks", in Ad Hoc Networking, ed. C.E. Perkins, New York, Addison-Wesley, 2001, pp. 29-51.

[FRODIG01] M. Frodigh et al., "Future-Generation Wireless Networks," *IEEE Personal Communications*, October 2001, pp.10–17.

[GALI00] S. Galli, K.D. Wong, B.J. Koshy, and M. Barton, "Bluetooth Technology: Link Performance and Networking Issues," *Proc. European Wireless 2000*, Dresden, Germany, September 2000.

[GERLA95] M. Gerla and J.T.-C. Tsai, "Multicluster, mobile, multimedia radio network," *ACM-Baltzer Journal of Wireless Networks*, Vo l. 1, No. 3, pp. 255–265, 1995.

[GIBS95] G.A. Gibson, D. Stodolsky, F.W. Chang, W.V. Courtright II, C.G. Demetriou, E. Ginting, M. Holland, L. Neal Q. Ma, R.H. Patterson, R. Youssef J. Su, and Jim Zelenka, "The Scotch Parallel Storage Systems," *Proceedings of the IEEE CompCon Conference*, San Francisco, CA, March 5–8, 1995.

[GIBS97] G.A. Gibson, D.F. Nagle, K. Amiri, F.W. Chang, E.M. Feinberg, H. Gobioff, C. Lee, B. Ozceri, E. Riedel, D. Rochberg, and J. Zelenka, "File Server Scaling with Network-Attached Secure Disks," *ACM SIGMETRICS '97*, pp. 272–284, Seattle, Washington, June 15–18 1997.

[GIBSON96] J.D. Gibson, *The Mobile Communications Handbook*, 2d ed., ISBN: 0849385970, Boca Raton, Fla., J.D. Gibson (ed.), CRC Press, Boca Raton, Fl, 1996.

[GILH91] K. Gilhousen, I. M. Jacobs, et al. "On the capacity of a cellular cdma system," *IEEE Trans. Veh. Tech.*, Vol. 40, pp. 303–312, 1991.

[GIORD00] S. Giordano, "Mobile Ad Hoc Networks," *Handbook of Wireless Networks and Mobile Computing*, Ivan Stojmenovic (ed.), Wiley, New York, 2000.

[GLA94] J. P. F. Glas, "On multiple access interference in a DS/FFH spread spectrum communication system," "*Proceedings of the Third IEEE International Symposium on Spread Spectrum Techniques and Applications*," pp. 3–2, Oulu, Finland, July 1994.

[GOL67] S. W. Golomb, *Shift Register Sequences*, Holden-Day, San Francisco, CA, 1967.

[GOLDB95] L. Goldberg, "Wireless LANs: Mobile computing's second wave," *Electronic Design*, Vol. 43, pp. 55–72, June 26, 1995.

[GOMEZ99] J. Gomez, A.T. Campbell, and H. Morikawa, "The Havana Framework: Supporting Application and Channel Dependent QOS in Wireless Network," *Proceedings of the 7th International Conference on Network Protocols*, 1999.

[GOOD97] J. Cai and D. Goodman, "General Packet Radio Service in GSM," *IEEE Communications Magazine*, October 1997.

[GOYA95] P. Goyal and H.M. Vin, "Generalized Guaranteed Rate Scheduling Algorithms: A Framework. Technical Report" *Technical Report TR95–30*," Department of Computer Science, UT Austin, TX, 1995.

[GROSS95] M. Grossglauser, S. Keshav, and D. Tse, "RCBR: A simple efficient service for multiple time-scale traffic," *Proc. ACM SIGCOM*, pp. 219–230, 1995.

[GSM 03.02] GSM-Rec. 03.02, Network Architecture.

[GSM 04.06] GSM-Rec. 04–06, MS-BSS Data Link Layer Specification.

[GSM 04.08] GSM Rec. 04.08, Mobile Radio Interface Layer 3 Specification.

[GSM 05.01] GSM Rec. 05.01, Physical Layer on the Radio Path.

[GSM 05.08] GSM Rec. 05.08, Radio Sub-system Link Control.

[GSM92] M. Mouly and M.B. Pautet, *The GSM system for mobile communications*, France, 1992.

[GUPTA00] P. Gupta and P. R. Kumar, "The capacity of wireless networks," *IEEE Transactions on Information Theory*, Vol. IT-46, pp. 388–404, 2000.

[HA98] S. Ha, K-W. Lee, and V. Bharghavan, "Performance Evaluation of Scheduling and Resource Reservation Algorithms in an Integrated Packet Services Network Environment," *Proceedings of IEEE ISCC '98* (The Third IEEE Symposium on Computers and Communications), Athens, Greece, July 1998.

[HAART99] J.C. Haartsen and S. Zurbes, "Bluetooth Voice and Data Performance in 802.11 DS WLAN Environment," *Technical Report Ericsson*, May 1999.

[HAAS97] Z. J. Haas, "A new routing protocol for the reconfigurable wireless networks," *IEEE ICUPC'97*, October 1997.

[HAAS98] Z. J. Haas and J. Deng, "Dual Busy Tone Multiple Access (DBTMA): A New Medium Access Control for Packet Radio Networks," *IEEE International Conference on Universal Personal Communications*, October 1998.

[HAAS98a] Z. J. Haas and S. Tabrizi, "On some challenges and design choices in ad hoc communications," *IEEE MILCOM '98*, Bedford, MA, October 1998.

[HAAS99] Z. J. Haas and B. Liang, "Ad-hoc mobility management with uniform quorum systems," *IEEE/ACM Trans. on Networking*, Vol. 7, pp. 228–240,1999.

[HABETHA02] J. Habetha, B. Walke: Fuzzy rulebased mobility and load management for self-organizing wireless networks. Journal of Wireless Information Networks, Special Issue on Mobile Ad Hoc Networks (MANETs): Standards, Research, Applications, Vol. 9, No. 2, pp. 119–140, April 2002

[HAFID97] A. Hafid, "Providing a Scalable Video-on-Demand System using Future Reservation of Resources and Multicast Communications," *Proc. 5th International Workshop on Quality of Service (IWQOS'97)*, Columbia University, New York, pp. 277–288., 1997.

[HALLS95] G. Halls, "HIPERLAN - the 20 Mbit/s radio LAN," *IEE Electronics Division Colloqium on Radio Lans and Mans*, (IEE Colloqium (Digest), No. 071, Stevenage, England), pp. 1/1 – 1/8, 1995.

[HENKEL02] D. Henkel, N. Houchaime, N. Locatelli and S. Singh, "The Impact of Emerging WLANs on Incumbent Cellular Providers in the U.S.," M.Sc. thesis, University of Colorado, Boulder, 2002.

[HJELM00] B. Hjelm, "The Rough Road to IMT-2000 RTT standard," *Proceedings of the 52nd IEEE Conference on Vehicular Technology*, Vol. 4, pp. 1493–1497, 2000.

[HLAN-SPEC] ETSI TC-RES, *Radio Equipment and Systems (RES); HIgh Performance Radio Local Area Network (HIPERLAN); Functional Specification*, ETSI, 06921 Sophia Antipolis, Cedex , France, July 1995. Draft prETS 300 652.

[HLAN-SPEC2] ETSI Normalization Committee, BRAN, "HIPERLAN Type 2; Physical (PHY) Layer," doc. RTS0023003-R2, Feb. 2001.

[HOFFM71] G. H. de Visme, *Binary Sequences*, The English Universities Press, London, 1971.

[HOLMA00] H. Holma and A. Toskala, *WCDMA for UMTS: Radio Access for Third Generation Mobile Communications*," Wiley, ISBN: 0471720518, June 2000.

[HSU98] Y. C. Hsu, T. C. Tsai, and Y. D. Lin, "QoS routing in multihop packet radio environment," *Proc. of IEEE ISCC'98*, 1998.

[HUBER00] J. F. Huber, D. Weiler, and H. Brand, "UMTS, the Mobile Multimedia Vision for IMT-2000: A Focus on Standardization," *IEEE Communications Magazine*, Vol. 38, pp. 129–136, September 2000.

[HUMBL91] P.A. Humblet, "Another adaptive shortest-path algorithm," *Trans. Commun.*, Vol. 39, pp. 995–1003, 1991.

[IARP-ID] Z J. Haas, M.R. Pearlman, and P Samar, The Intrazone Routing Protocol (IARP) for Ad Hoc Network, draft-ietf-manet-zone-iarp-01.txt, IETF MANET working group, Dec. 2001.

[IEEE-PERS98] IEEE Personal Communications Magazine (several articles on antenna concepts), Vol. 5, No.1, February 1998, pp. 10–48.

[IEEE-PERS99] IEEE Communications Magazine (several articles on software defined radio), Vol. 37, No. 2, February 1999, pp. 82–112.

[IEEE-STDRD] IEEE 802 LAN/MAN Standards Committee. "Wireless LAN medium access control (MAC) and physical layer (PHY) specifications," *IEEE Standard 802.11,* 1999 ed, 1999.

[IERP-ID] Z. J. Haas, M.R. Pearlman, and P. Samar, The Interzone Routing Protocol (IERP) for Ad Hoc Networks, draft-ietf-manet-zone-ierp-01.txt, IETF MANET working group, Dec. 2001.

[IMEP-ID] M. S. Corson and V. D. Park. An Internet MANET Encapsulation Protocol (IMEP) Specification. Internet-Draft, draft-ietf-manet-imep-spec-02.txt, August 1999, Work in progress.

[IMT-2000] IMT-2000, http://www.itu.int/imt.

[IMT-ITU] Special Issue, IMT-2000: Standards Efforts of the ITU, *IEEE Personal Communications,* Vol. 4, No. 4, August 1997.

[INRIA02] "Optimizing Route Discovery in Reactive Protocols for Ad Hoc Networks," *Research Report,* RR-4509, 2002, Inria, France.

[IPFLOOD-ID] C. E. Perkins, E. M. Royer, and S. R. Das, IP flooding in ad hoc networks. Internet draft (draft-ietf-manet-bcast-00.txt), Nov 2001. Work in Progress.

[IPV6-ID] S. Deering and R. Hinden, "Internet Protocol, Version 6 (IPv6) Specification," IETF Internet Draft, draft-ietf-ipngwg-ipv6-spec-v2-01.txt, November 21 1997.

[ISSA97] V. Issarny, C. Bidan, F. Leleu, and T. Saridakis, "Towards Specifying QoS-Enabling Software Architectures," *Proc. 5th International Workshop on Quality of Service (IWQOS'97),* Columbia University, New York, pp. 363–366., 1997.

[ITU2002] ITU-R WP8F: Preliminary Draft New Recommendation (PDNR): Vision framework and overall objectives of the future development of IMT-2000 and of systems beyond IMT-2000. IMT-VIS (020304), 7th Meeting, Queenstown, New Zealand, February 7 to March 5, 2002.

[ITU91] Telecommunications Industry Assocation, Cellular radiotelecommunication inter-system operation, TIA/EIA IS-41B, (1991).

[IWATA99] A. Iwata, C-C. Chiang, G. Pei, M. Gerla, and T-W. Chen, "Scalable Routing Strategies for Ad Hoc Wireless Networks," *IEEE JSAC,* August 1999, Vol. 17, No. 8, pp. 1369-79.

[JACOBS96] J.M. Jacobsmeyer, "Congestion Relief on Power-controlled CDMA networks," *IEEE Journal on Selected Areas in Communications,* Vol. 18, No. 9, pp. 1758–1761, 1996.

[JACQUET00] P. Jacquet and L. Viennot, "Overhead in Mobile Ad-hoc Network Protocols," *INRIA Research Report RR3965,* INRIA, Rocquencourt, France, 2000.

[JAFFE82] J. Jaffe and M. Moss, "A Responsive Distributed Routing Algorithm for Computer Networks," *IEEE Transactions on Communications,* Vol. COM-30, No. 7, July 1982, pp. 1758–1762.

[JAIN01] R. Jain, A. Puri, and R. Sengupta, "Geographical Routing Using Partial Information for Wireless Ad Hoc Networks," *Personal Communications,* February 2001, pp. 48–57.

[JAMALI01] A. Jamalipour and T. Tung, "The Role of Satellites in Global IT: Trends and Implications," *IEEE Personal Communications,* June 2001, pp. 5–11.

[JAMI95] S. Jamin, P. Danzig, S. Shenker, and L. Zhang, "A Measurement-based Admission Control Algorithm for Integrated Services Packet Networks," *Proceedings of SIGCOMM'95,* pp. 2–13, Boston, MA, September 1995.

[JANNOT00] J. Li, J. Jannotti, D.SJ. De Couto, D.R. Karger, and R. Morris, "A Scalable Location Service for Geographic Ad Hoc Routing," *Proc. 6th Annual ACM/IEEE Intlo Conf on Mobile Computing and Networking (MobiCom '00),* Boston, August 2000, pp. 120–130.

[JINYA01] J. Li, C. Blake, D. S.J. De Couto, H. I. Lee, and R. Morris, "Capacity of Ad Hoc wireless networks," *The seventh annual international conference on Mobile computing and networking (MOBICOM 2001),* pp.61–69, Rome, Italy, 2001.

[JOHANS99] P. Johansson, T. Larsson, N. Hedman, B. Mielczarek, and M. Degermark, "Scenario-Based Performance Analysis of Routing Protocols for Mobile Ad-hoc Networks," *Proc. 5th Annual ACM/IEEE International Conference on Mobile Computing and Networking (MobiCom '99),* Seattle, WA, August 15–19, pp. 195–206, 1999.

[JOHN94] D. B. Johnson, "Routing in Ad Hoc Networks of Mobile Hosts," *Proceedings of the IEEE Workshop on Mobile Computing Systems and Applications (WMCSA)*, IEEE Computer Society, Santa Cruz, CA, pp. 158–163, December 1994.

[JOHNS95] D. B. Johnson, "Scalable support for transparent mobile host internetworking," *Wireless Networks*, Vol. 1, No. 3, pp. 311–322, 1995.

[JOHNS96] D. B. Johnson and D. A. Maltz, "Dynamic Source Routing in Ad-Hoc Wireless Networks," Mobile Computing, T. Imielinski and H. Korth, eds., Kluwer Academic Publishers, pp. 153–181, 1996.

[JUBIN87] J. Jubin and J. D. Tornow, "The DARPA Packet Radio Network Protocols," *Proceedings of the IEEE*, Vol. 75, No. 1, pp. 21–32, January 1987.

[JUNG94] P. Jung, J.J. Blanz, M.M. Naβhan, and P. W. Baier, "Simulation of the uplink of JD-CDMA mobile radio systems with coherent receiver antenna diversity," *Wireless Personal Communications*, Vol. 1, pp. 61–89, 1994.

[KAPLAN96] E.D. Kaplan, *Understanding the GPS: Principles and Applications*, Artech House, Norwood, MA, February 1996.

[KAPOOR97] H. Kapoor, "CoreBuilder 5000 SwitchModule Architecture," http://www.3com.com/technology/tech_net/white_papers/500645.html, 1997.

[KARN90] P. Karn, "MACA - a New Channel Access Method for Packet Radio," *ARRL/CRRL Amateur Radio 9th Computer Networking Conference*, pp.134–140, ARRL, 1990.

[KAROL87] M. Karol, M. Hluchyj, and A. Morgan, "Input Versus Output Queueing on a Space-Division Packet Switch," *IEEE Transactions on Communications*, Vol. COM-35, No. 12, pp. 1347–1356, December 1987.

[KARP00] B. Karp and H. T. Kung, *GPSR: Greedy Perimeter Stateless Routing for Wireless Networks*, MobiCom, Boston, MA, 2000, pp. 243–254.

[KARP01] B. N. Karp, "Geographic Routing for Wireless Networks," thesis, Harvard University, Cambridge, MA, 2001.

[KAWAD03] V. Kawadia and P. R. Kumar, "Clustering by Power Control in Ad-Hoc Networks," *Proceedings of Infocom*, 2003.

[KAWAD03] V. Kawadia and P. R. Kumar, "Power Control and Clustering in Ad Hoc Networks," *INFOCOM 2003*, San Francisco, March 30 - April 3, 2003.

[KETCHUM95] J. W. Ketchum, "Routing in cellular mobile radio communication networks," *Routing in Communication Networks*, M. Steenstrup, ed., Prentice Hall, Englewood Cliffs, NJ, 1995.

[KLEIN75] L. Kleinrock and F. A. Tobagi, "Packet Switching in Radio Channels: Part I - Carrier Sense Multiple-Access Modes and their Throughput-Delay Characteristics," *IEEE Transactions of Communications*, Vol. COM-23, No. 12, pp. 1417–1433, 1975.

[KLEIN75a] F. A. Tobagi and L. Kleinrock, "Packet Switching in Radio Channels: Part II - The Hidden Terminal Problem in Carrier Sense Multiple-Access Modes and the Busy-Tone Solution," *IEEE Transactions of Communications*, Vol. COM-23, No. 12, pp. 1417–1433, 1975.

[KLEIN93] A. Klein and P.W. Baier, "Linear unbiased data estimation in mobile radio systems applying CDMA," *IEEE Journal on Selected Areas in Communications*, Vol. SAC-11, pp. 1058–1066, 1993.

[KLEIN95] L. Kleinrock, "Nomadic Computing - An Opportunity," *ACM SIGCOMM, Computer Communications Review*, Vol. 25, No. 1, pp. 36–40, January 1995.

[KLEIN97] A. Klein, R. Pirhonen, J Sköld, and R Suoranta, "FRAMES Multiple Access Mode 1 - Wideband TDMA with and without Spreading" *Proceedings of PIMRC97*, pp. 37–41, Helsinki, September 1997.

[KLEINR97] L. Kleinrock and K. Faroukh, "Hierarchical routing for large networks," *Computer Networks*, Vol. 1, 1997.

[KNOC97] H. Knoche and H. de Meer, "Quantitative QoS-Mapping: A Unifying Approach," *Proc. 5th International Workshop on Quality of Service (IWQOS'97)*, Columbia University, New York, pp. 347–358., 1997.

[KO98] Y. Ko and N. Vaidya, "Location-aided routing (LAR) in mobile ad hoc networks," *Proc. 4th ACM/IEEE Mobile Computing and Networking*, Dallas, p. 66, 1998.

[KO99] Y.-B. Ko and N.H. Vaidya, Using Location Information in Wireless Ad Hoc Networks, IEEE Vehicular Technology Conf (VTC '99), May 1999.

[KRAVETS98] . R. Kravets and P. Krishnan, "Power Management Techniques for Mobile Communication," *ACM International Conference on Mobile Computing and Networking (MobiCom)*, Dallas, TX, 1998.

[KRISHN97] P. Krishna , N. H. Vaidya, M. Chatterjee , and D. K. Pradhan, "A cluster-based approach for routing in dynamic networks," *ACM SIGCOMM Computer Communication Review*, Vol. 27, No. 2, pp.49–64, April 1997.

[KUROSE] J.F. Kurose, M. Schwartz, and Y. Yemini, Multiple access protocols and time constraint communications, ACM Computing Surveys, Vol. 16, pp. 43–70., 1984.

[KUROSE02] J. F. Kurose, K. W. Ross *Computer Networking: A Top-Down Approach Featuring the Internet*, Pearson Addison Wesley, ISBN: 0201976994, New York, 2002.

[KWON99] T. J. Kwon and M. Gerla, "Clustering with power control," *IEEE MILCOM*, 1999.

[LAM80] S. Lam, A Carrier Sense Multiple Access Protocol for Local Networks," *Computer Networks*, Vol. 4, pp. 21–32, 1980.

[LAW03] C. Law, A.K. Mehta, and K.Y. Siu, A New Bluetooth Scatternet Formation Protocol, *ACM/Kluver Mobile Networks and Applications Journal*, Special Issue on Ad Hoc Networks, A. T. Campbell, M. Conti, and S. Giordano, eds., Vol. 8, No. 5, October 2003.

[LEE98] S. B. Lee and A. T. Campbell, "INSIGNIA: In-band Signaling Support for QoS in Mobile Ad Hoc Networks," *5th Int. Workshop on Mobile Multimedia Commo (MoMuc'98)*, Berlin, October 1998.

[LEIN87] B. M. Leiner, D. L. Nielson, and F. A. Tobagi, "Issues in packet radio network design," *Proceedings of IEEE*, Vol. 75, No. 1, pp. 6–20, January 1987.

[LI01] L. Li and J. Halpern, "Minimum-Energy Mobile Wireless Networks Revisited," *Proc. IEEE Int. Conf Communo (ICC '01)*, 2001.

[LI01] Q. Li , J. Aslam and D. Rus, "Online Power-aware Routing in Wireless Ad-hoc Networks," *Int'l Conf on Mobile Computing and Networking (MobiCom '2001)*, July 2001.

[LI01] Q. Li , J. Aslam , and D. Rus, "Online power-aware routing in wireless Ad-hoc net-works," *Proceedings of the 7th annual international conference on Mobile computing and networking*, pp.97–107, Rome, Italy, July 2001.

[LIAO01] W.-H. Liao, Y.-C. Tseng, and J.-P. Sheu, "GRID: a fully location-aware routing protocol for mobile ad hoc networks," *Telecommunication Systems*, Vol. 18, pp. 61–84,2001.

[LIN97] C. R. Lin and M. Gerla, "Adaptive clustering for mobile wireless networks," *IEEE Journal on Selected Areas in Communications*, Vol. 15, No. 7, September 1997.

[LIN97] C.R. Lin and M. Gerla, "Asynchronous Multimedia Multihop Wireless Networks," *Proceedings of IEEE INFOCOM '97*, Kobe, Japan, April 1997, pp. 118–125.

[LIN99] C. Lin and J. Liu, "QoS Routing in Ad Hoc Wireless Networks," *IEEE Journal on Selected Areas in Communications*, Vol. 17, No. 8, pp.1426–1438, August 1999.

[LIN99] X. Lin and I. Stojmenovic, "GEDIR: Loop-Free Location Based Routing in Wireless Networks," *Proco IASTED International Conference on Parallel and Distributed Computing and Systems*, pp. 1025–1028, 1999.

[LU00] S. Lu, T. Nandagopal, and V. Bharghavan, "Design and Analysis of an Algorithm for Fair Service in Error Prone Channels," *ACM/Baltzer Wireless Networks Journal*, Vol. 6, No. 4, pp. 323–343, August 2000.

[LU96] S. Lu and V. Bharghavan, "Adaptive Resource Management Algorithms for Indoor Mobile Computing Environments," *Proceedings of ACM SIGCOMM*, Stanford, CA, August 1996.

[LU97] S. Lu, K-W. Lee, and V. Bharghavan, "Adaptive Service in Mobile Computing Environments," *Proceedings of IFIP IWQoS*, New York, May 1997.

[LU97a] S. Lu, V. Bharghavan, and R. Srikant, "Fair Scheduling in Wireless Packet Networks," *Proceedings of ACM SIGCOMM*, Cannes, France, August 1997.

[LUI01] Chi-Jui H. et al., "Call admission control in the microcell/macrocell overlaying system," *IEEE Transactions on Vehicular Technology*, Vol. 50, pp. 992–1003, 2001.

[MADHOW95] U. Madhow, M.L. Honig, and K. Steiglitz, "Optimization of wireless resources for personal communications mobility tracking," *IEEE/ACM Trans. on Networking*, Vol. 3, No. 6, pp. 698–706, 1995.

[MALTZ99] D. Maltz, J. Broch, J. Jetcheva, and D.B. Johnson. "The Effects of On-Demand Behavior In Routing Protocols for Multi-Hop Wireless Ad Hoc Networks," *IEEE Journal on Selected Areas in Communications*, August 1999 (special issue on mobile and wireless networks).

[MATTHIA01] M. Gerlach, "WLAN-MAC Architectures," Study thesis on Ad Hoc Networking Architectures, May 10, 2001.

[MAUVE01] M. Mauve, J. Widner, and H. Hartenstein, "A Survey on Position-Based Routing in Mobile Ad-Hoc Networks," *IEEE Network*, Nov./Dec. 2001, pp. 30–39.

[McDONA00] A. B. McDonald, and T. Znati, "Predicting Node Proximity in Ad-Hoc Networks: A Least Overhead Adaptive Model for Electing Stable Routes," *MobiHoc 2000*, Boston, August 4, 2000.

[McDONA99] A. B. McDonald and T. Znati " A Mobility-Based Framework for Adaptive Clustering in Wireless Ad-Hoc Networks," *IEEE Journal on Selected Areas in Communication*, Vol. 17, No. 8, August 1999.

[McKEO97] N. McKeown, "A Fast Switched Backplane for a Gigabit Switched Router," *Business Communications Review*, Vol. 27. N0 12, 1998.

[McQUILL77] J. M. McQuillan and D. C. Walden, "The ARPA Network Design Decisions," *Computer Networks*, Vol. 1, No. 5, pp. 243–289, August, 1977.

[McQUILL80] J. M. McQuillan, I. Richer, and E. C. Rosen, "The New Routing Algorithm for ARPANET," *IEEE Trans. on Communications*, Vol. 28, No. 5, pp. 711–719, May, 1980.

[METCAL76] R. Metcalfe and D. Boggs, "Ethernet: Distributed packet switching for local computer networks," *Communications of the ACM*, Vol. 19, No. 7, pp. 395–404 1976.

[MIGU94] J. M. del Rosario and A. Choudhary, "High performance I/O for parallel computers: Problems and prospects," *IEEE Computer*, Vol. 27, No. 3, pp. 59–68, March 1994.

[MILLER00] B. A. Miller and C. Bisdikian, *Bluetooth Revealed*, Prentice Hall, New York, 2000.

[MOBCOM03] Mobile Communications International, Issue 102, June 2003.

[MOBILE-IP] IETF Mobile-IP Working Group, IPv4 mobility support, working draft (1995).

[MOGHE97] P. Moghe and A. Kalavade, "Terminal QoS of Adaptive Applications and its Analytical Computation," *Proc. 5th International Workshop on Quality of Service (IWQOS'97)*, Columbia University, NY, pp. 369–380, 1997.

[MOHAN94] S. Mohan and R. Jain, "Two user location strategies for personal communications services," *IEEE Personal Communications* (First Quarter 1994), pp. 42–50, 1994.

[MOHR02] W. Mohr, R. Lüder, and K.-H. Möhrmann, "Data rate estimates, range calculations and spectrum demand for new elements of systems beyond IMT-2000", *Proceedings of the 5th International Symposium on Wireless Personal Multimedia Communications (WPMC'02)*, pp. 37–46, Vol. 1, October 2002, Honolulu, Hawaii.

[MPLS-ID] E. Rosen, A. Viswanathan, and R. Callon, "Multiprotocol Label Switching Architecture," IETF Internet Draft, draft-ietf-mpls-arch-06.txt, August 1999.

[MPLS-WP] IP Traffic Engineering using MPLS Explicit Routing in Carrier Networks, Nortel Networks, White Paper, April 1999, http://www.nortelnetworks.com/products/library/collateral/55046.25-10-99.pdf.

[MRSVP] Talukdar A., Badrinath R. B., and Acharya A., "MRSVP: A Resource Reservation Protocol for Integrated Services Network with Mobile Hosts," *Wireless Networks*, Vol. 7, No. 1, pp. 5–19, 2001.

[MURHY96] S. Murthy and J. J. Garcia-Luna-Aceves, "An Efficient Routing Protocol for Wireless Networks," *ACM Mobile Networks and Applications Journal*, , pp. 183–197, October 1996 (Special Issue on Routing in Mobile Communication Networks).

[NA·HAN93] M.M. Naßhan, P. Jung, A. Steil, and P. W. Baier. "On the effects of quantization, non-linear amplification and band limitation in CDMA mobile radio systems using joint detection," *Proceedings of the Fifth Annual International Conference on Wireless Communications WIRELESS'93*, pp. 173–186, Calgary, Canada, 1993.

[NEUMAN97] R. Neumann, "Internet Routing Black Hole," *The Risks Digest: Forum on Risks to the Public in Computers and Related Systems*, Vol. 19, No. 12 (2-May-1997).

[NIEMEG01] I.G.M.M. Niemegeers et al., "Personal Area Networks," *Proceedings of the WWRF meeting*, Helsinki, Finland, May 10–11, 2001.

[NKULA98] E. Nikula, A. Toskala, E. Dahlman, L. Girard, and A. Klein, "FRAMES Multiple Access for UMTS and IMT-2000," *IEEE Personal Communications Magazine*, April 1998, pp. 16–24.

[NOY93] A. Bar-Noy and I. Kessler, "Tracking mobile users in wireless communications networks," *IEEE Trans. on Information Theory*, Vol. 39, No. 6, pp. 1877–1886, 1993.

[NOY95] A. Bar-Noy, I. Kessler, and M. Sidi, "Mobile users: to update or not to update?," *Wireless Networks*, Vol. 1, No. 2, pp. 175–186, 1995.

[OH97] S. Oh, H. Sugano, S. Shimojo, H. Miyahara, K. Fujikawa, T. Matsuura, and M. Arikawa, "A Dynamic QoS Adaptation for Networked Virtual Reality," *Proc. 5th International Workshop on Quality of Service (IWQOS'97)*, Columbia University, NY, pp. 397–400, 1997.

[OJANPE96] T. Ojanperä, K. Rikkinen, H. Häkkinen, K. Pehkonen, A. Hottinen, and J. Lilleberg, , "Design of a 3rd Generation Multirate CDMA Systems with Multiuser Detection, MUD-CDMA," *Proc. ISSSTA'96*, Mainz, Germany, September 1996, pp. 334–338, 1996.

[OJANPE97] T. Ojanperä, "Overview of Research Activities for Third Generation Mobile Communications," *Wireless Communications TDMA vs CDMA*, S. Glisic and P. A. Leppanen ,ed., pp 415–446, Kluwer Academic Publishers, 1997.

[OJANPE97] T. Ojanperä, A Klein, and P.O. Anderson, "FRAMES Multiple Access for UMTS," *IEE Colloquium on CDMA Techniques and Applications for Third Generation Mobile Systems*, London, May 1997.

[OJANPE98] T. Ojanperä and R. Prasad, "An Overview of Air Interface Multiple Access for IMT-2000/UMTS," *IEEE Communications Magazine*, Vol. 36, pp. 82–95, September 1998.

[OJANPER] T. Ojanperä and R. Prasad, An Overview of CDMA Evolution toward Wideband CDMA, http://www.comsoc.org/pubs/surveys/4q98issue/prasad.html.

[ONOE97] S. Onoe, K Ohno, K Yarnagata, and T Nakamura, "Wideband-CDMA Radio Control Techniques for Third Generation Mobile Communication Systems," *Proceedings of VTC97*, Vol.2, pp. 835–839, Phoenix, May 1997.

[OSLR-ID] T. Clausen et al., Optimized Link State Routing Protocol, IETF MANET Working Group Internet Draft, draft-ietf-MANET-olsrr-06.txt, Sep. 2001.

[OTT97] M. Ott, G. Michelitsch, D. Reininger, and G. Welling, "Adaptive QoS and in Multimedia Systems," *Proc. 5th International Workshop on Quality of Service (IWQOS'97)*, Columbia University, New York , pp. 393–396, 1997.

[OVESJO97] F. Ovesjö, E. Dahlman. T. Ojanperä, A Toskala, and A. Klein, "FRAMES Multiple Access Mode 2 - Wideband CDMA," *Proceedings of PIMRC97*, pp. 42–46, Helsinki, September 1997.

[PAHL98] K. Pahlavan, "Wireless Intraoffice Networks," *ACM Trans. Office Inf. Systems*, Vol. 6, No.3, pp. 227–302, July 1998.

[PAHLAV02] K. Pahlavan and P. Krishnamurthy, *Principles of Wireless Networks: A Unified Approach*, Prentice Hall, New York, 2002.

[PAJU97] K. Pajuski and J. Savusalo, "Wideband CDMA Test System," *Proc. IEEE Int. Conf. on Personal Indoor and Mobile Radio Communications*, PIMRC'97, Helsinki, Finland, 1-4 September 1997, pp. 669–672.

[PAPADI02] P. Papadimitratos and Z. J. Haas, "Secure Routing for Mobile Ad Hoc Networks," *SCS Communication Networks and Distributed Systems Modeling and Simulation Conference (CNDS 2002)*, San Antonio, TX, January 27–31,2002.

[PAREK93] A. Parekh and R. Gallager, "A Generalized Processor Sharing Approach to Flow Control the Single Node Case," *IEEE/ACM Transactions on Networking*, Vol. 3, No. 1, pp. 344–357, June 1993.

[PARK97] V. D. Park and M. S. Corson, "A Highly Adaptive Distributed Routing Algorithm for Mobile Wireless Networks," *Proceedings of the IEEE INFOCOM'97*, Kobe, Japan, pp. 1405–1413, 7–11 April 1997.

[PAXSON97] V. Paxson, "Measurements and Analysis of End-to-End Internet Dynamics," Ph.D. Thesis, Department of Computer Science, University of California, Berkeley, 1997.

[PEARL00] M. Pearlman, Z. Haas, P. Sholander, and S. Tabrizi, "On the Impact of Alternate Path Routing for Load Balancing in Mobile Ad Hoc Networks," *The First*

Annual Workshop on Mobile Ad Hoc Networking & Computing (MobiHoc 2000), Boston, MA, August 2000.

[PEARL99] M. R. Pearlman and Z. J. Haas, "Determining the Optimal Configuration for the Zone Routing Protocol," *IEEE Journal on Selected Areas in Communications*, August 1999 (Special Issue on Mobile and Wireless Networks).

[PERK00] C.E. Perkins, ed., *Ad Hoc Networking*, Addison- Wesley, Reading, MA, 2000.

[PERK02] C. E. Perkins, E. M. Royer, and S. R. Das, Ad hoc on-demand distance vector routing (AODV), Internet draft (draft-ietf-manet-aodv-13.txt), Feb 2003, Work in Progress.

[PERK94] C. E. Perkins and P. Bhagwat, "Highly Dynamic Destination-Sequenced Distance-Vector Routing (DSDV) for Mobile Computers," *Computer Communications Review*, pp. 234–244, October 1994.

[PERK95] C.E. Perkins and K. Luo, Using DHCP with computers that move, *Wireless Networks*, Vol. 1, No. 3, pp. 341–354, 1995.

[PERK99] C. E. Perkins and E. M. Royer, "Ad-hoc On-demand Distance Vector Routing," *Proceedings of the 2nd IEEE Workshop on Mobile Computing Systems and Applications (WMCSA '99)*, February 1999, pp. 90–100.

[PERLM99] R. Perlman, "Bridges, Routers, Switches, and Internetworking Protocols," *Interconnections*, 2d ed., Addison-Wesley Professional Computing Series, Boston, 1999.

[PETERS00] L. L. Peterson and B. S. Davie, *Computer Networks - A Systems Approach*," Morgan Kaufmann Publishers, San Francisco, 2000, ISBN 1-55860-368-9.

[PICKHO82] R. Pickholtz, D. Schilling, and L. Milstein, "Theory of Spread Spectrum Communication - a Tutorial," *IEEE Transactions on Communications*, Vol. COM-30, No. 5, pp. 855–884, May 1982,.

[PIRHO99] R. Pirhonen, T. Rautava, and J. Penttimen, "TDMA Convergence for Packet Data Services," *IEEE Personal Communications Magazine*, Vol. 6, No.3, pp. 68–73, June 1999.

[PIS99] Report to the President Information Technology: Transforming our Society 1999.

[PNNI-SPEC] PNNI SWG (Doug Dykeman ed.). PNNI Draft Specification. ATM Forum 94-0471R10, October 1995.

[POLLINI95] G.P. Pollini and K.S. Meier-Hellstern, "Efficient routing of information between interconnected cellular mobile switching centers," *IEEE/ACM Trans. on Networking*, Vol. 3, No. 6, pp. 765–774, 1995.

[PRAKAS98] R. Prakash, "Unidirectional Links Prove Costly in Wireless Ad-Hoc Networks," *Proceedings of the 3rd International Workshop on Discrete Algorithms and Methods for Mobile Computing and Communications - Dial M '99*, Seattle, pp. 15–22, August 20, 1998.

[PRASAD00] R. Prasad, W. Mohr, and W. Konhäuser, *Third Generation Mobile Communication Systems*, Artech House, London, Boston, 2000.

[QAYYUM01] A. Qayyum, L. Viennot, and A. Laouiti, "Multipoint Relaying: An Efficient Technique for Flooding in Mobile Wireless Networks," *35th Annual Hawaii International Conference on System Sciences (HICSS '2001)*, Maui, 2001.

[QMA98] Q. MA, "Routing Traffic with Quality-of-Service Guarantees in Integrated Services Networks," PhD Thesis, Department of Computer of Science, Carnegie Mellon University, 1998.

[QUALC92] An overview of the application of code division multiple access (cdma) to digital cellular systems and personal cellular networks. Qualcomm Inc., 1992.

[RAGHAV98] C. S. Raghavendra and S. Singh, PAMAS - Power-Aware Multi-Access Protocol with Signaling for Ad Hoc Networks, ACM Computer Communication Review, 1998.

[RAMAN96] S. Ramanathan and M. Steenstrup. "A survey of routing techniques for mobile communications networks," *ACM/Baltzer Mobile Networks and Applications*, Vol. 1, No. 2, pp. 89–103., 1996.

[RAMAN98] R. Ramanathan and M. Steenstrup, "Hierarchically-organized, multihop mobile wireless networks for quality-of-service support," *Mobile Networks and Applications*, Vol. 3, 1998.

[RANGELAN] Range LAN. http://www.proxim.com/products/rl2/7410.shtml.

[RANTA96] P. A. Ranta, A. Lappetelainen, and Z-C Honkasalo, "Interference cancellation by joint detection in random frequency hopping TDMA networks," *Proceedings of ICUPC96 conference*, Vol. 1, pp 428–432, 1996.

[RAPELI95] J. Rapeli, "UMTS: Target, System Concept, and Standardization in a Global Framework," *IEEE Personal Communications*, Vol. 2, pp. 20–28, February 1995.

[RAPP96] T.S. Rappaport, *Wireless Communications: Principles and Practice,* Prentice Hall, Englewood Cliff, NJ, 1996.

[RDMAR-ID] G. Aggelou and R. Tafazolli "Relative Distance Micro-discovery Ad Hoc Routing (RDMAR) Protocol," IETF Internet Draft, draft-ietf-manet-rdmar-00.txt, November 1999.

[REMON01] D. R. Bueno, K. Coreman, and R. Tafazolli, "Ad Hoc Networks," *Proceedings of the WWRF meeting*, Helsinki, Finland, May 10–11, 2001.

[REVEL97] D. Revel, C. Cowan, D. McNamee, C. Pu, and J. Walpole, "Predictable File Access Latency for Multimedia," *Proc. 5th International Workshop on Quality of Service (IWQOS'97)*, Columbia University, New York, pp. 401–404, 1997.

[RFC 1071] B. Draden, D. Borman, and C. Partridge, "Computing the Internet Checksum," RFC 1071, Sept. 1988.

[RFC 1190] C. Topolcic, "Experimental Internet Stream Protocol, version 2 (ST-II)," IETF RFC 1190, October 1990.

[RFC 1541] R. Droms, Dynamic host configuration protocol, Internet RFC 1541 (1993).

[RFC 1633] R. Braden, D. Clark, and S. Shenker, Integrated Services in the Internet Architecture - An Overview, IETF RFC1633, June 1994.

[RFC 2002] C. Perkins, ed. IPv4 Mobility Support. RFC 2002, October 1996.

[RFC 2205] R. Braden, L. Zhang, S. Berson, S. Herzog, and S. Jamin. "Resource ReSerVation Protocol (RSVP) — Version 1," Functional Specification. IETF RFC 2205, September 1997.

[RFC 2210] J. Wroclawski, "The Use of RSVP with IETF Integrated Services," RFC 2210, September 1997.

[RFC 2211] J. Wroclawski, "Specification of the Controlled-Load Network Element Service," RFC 2211, September 1997.

[RFC 2212] S. Shenker, C. Partridge, R. Guerin, "Specification of Guaranteed Quality of Service," RFC 2212, September 1997.

[RFC 2215] General Characterization Parameters for Integrated Service Network Elements, RFC 2215, INTERNET-DRAFT, September 1997

[RFC 2328] J. Moy. OSPF version 2, RFC 2328. IETF, April 1998.

[RFC 2386] A Framework for QoS-based Routing in the Internet, RFC 2386, IETF Internet Draft, August 1998.

[RFC 2453] G. Malkin. RIP version 2, RFC 2453. IETF, November 1998.

[RFC 2475] S. Blake, D. Black, M. Carlson, E. Davies, Z. Wang, and W. Weiss, "An Architecture for Differentiated Services," RFC 2475, December 1998.

[RFC 3140] D. Black, S. Brim, B. Carpenter, and F. L. Faucheur, Per-Hop Behavior Identification Codes, Internet IETF RFC3140, June 2001.

[RFC 791] J. Postel, Internet protocol, Internet RFC 791 (1981).

[RODOPLU99] V. Rodoplu and T.H.-Y. Meng, "Minimum energy mobile wireless networks," *IEEE J. Select. Areas Commun.*, Vol. 17, pp. 1333–1344,1999.

[ROEFS77] H. F. A. Roefs, "Binary Sequences for Spread-Spectrum Multiple-Access Communications," PhD thesis, University of Illionois, Urbana, Illinois, 1977.

[ROM90] R. Rom and M. Sidi, *Multiple Access Protocols: Performance and Analysis*, Springer-Verlag, New York, 1990.

[ROSE95] C. Rose and R. Yates, "Minimizing the average cost of paging under delay constraints," *Wireless Networks* , Vol. 1, No. 2, pp. 211–220, 1995.

[ROYER99] E. Royer and C.-K. Toh, "A Review of Current Routing Protocols for Mobile Ad-Hoc Networks" *IEEE Personal Communications*, April 1999.

[RUBENS98] D. Rubenstein, J. Kurose, and D. Towsley "Real-Time Reliable Multicast Using Proactive Forward Error Correction" , *Proceedings of NOSSDAV '98* , Cambridge, UK, July 1998.

[SANTINAV02] C. Santivanez, B. McDonald, I. Stavrakakis, and R. Ramanathan, "On the scalability of ad hoc routing protocols," *Proc. IEEE INFOCOMM 2002*, 2002.

[SARI01]   H. Sari, "A multimode CDMA with reduced intercell interference for broadband wireless networks," *IEEE Journal on Selected Areas in Communications*, Vol. 19, pp. 1316–1323, 2001.

[SARWA80]   D. V. Sarwate and M. B. Pursley, "Cross-correlation properties of pseudo-random and related sequences," *Proceedings of the IEEE*, Vol.68, pp. 593—619. IEEE, May 1980.

[SASAKI97]   A. Sasaki, "A perspective of Third Generation Mobile Systems in Japan," *IIR Conference Third Generation Mobile Systems, The Route Towards UMTS*, London, February 1997.

[SATYAN95]   M. Satyanarayanan, "Fundamental challenges of mobile computing," *Proc. ACM Symposium on Principles of Distributed Computing (PODC)*, 1995.

[SAUNDER99]   S. Saunders, *Antennas and Propagation for Wireless Communications Systems*," Wiley, UK, July 1999.

[SCHELEN97]   O. Schelen and S. Pink, "Sharing Resources through Advance Reservation Agents," *Proc. 5th International Workshop on Quality of Service (IWQOS'97)*, Columbia University, New York, pp. 265–276, 1997.

[SCHWAR80]   M. Schwartz, *Information, Transmission, Modulation, and Noise*, McGraw Hill, New York, 1980.

[SINGH98]   S. Singh and C. S. Raghavendra, "Power-Efficient MAC Protocol for Multihop Radio Networks," *Proco of IEEE PIRMC '98 conf*, Vol. 1, pp. 153–157, Sep. 1998.

[SINGH98a]   S. Singh, M. Woo, and C.S. Raghavendra, "Power-Aware Routing in Mobile Ad Hoc Networks," *Proc. 4th Annual ACM/IEEE IntI. Conf on Mobile Computing and Networking (MobiCom '98)*, Dallas, October 25–30, 1998, pp. 181–190.

[SINHA99]   P. Sinha, R. Sivakumar, and Bharghavan, "CEDAR: A Core-Extraction Distributed Ad Hoc Routing Algorithm," *Proceedings of IEEE INFOCOM '99*, New York, May 1999, pp. 202–209.

[SISAL97]   D. Sisalem, H. Schulzrinne, and F. Emanuel, "The Direct Adjustment Algorithm: A TCP-Friendly Adaptation Scheme," Technical report, GMD-FOKUS, Aug. 1997, available at http://www.fokus.gmd.de/usr/sisalem.

[SKOLD97]   J. Sköld, P. Schramm, P-O. Anderson, and M Gudmundson, "Cellular Evolution into Wideband Services," *Proceedings of VTC97*, Vol. 2, pp 485 –489, Phoenix, May 1997.

[SPRAG91]   J. D. Spragins, *Telecommunications protocols and design*," Addison-Wesley, Reading, MA, 1991.

[SRENG03]   V. Sreng, H. Yanikomeroglu, and D.D. Falconer, "Relayer selection strategies in cellular networks with peer-to-peer relaying," *IEEE Vehicular Technology Conference Fall 2003 (VTC'F03)*, October 2003, Orlando, FL.

[STAJNO99]   F. Stajno and R. J. Anderson, "The Resurrecting Duckling: Security Issues for Ad Hoc Wireless Networks," *Proceedings of Security Protocols 17th International Workshop*, Vol. 1796 of Lecture Notes in Computer Science, Cambridge, U.K., April 19–21, 1999, pp. 172–194.

[STEEN95]   M. E. Steenstrup, *Routing in Communications Networks*, Prentice-Hall, Cambridge, MA, 1995.

[STEPH99]   Stephenson, A. "QoS: The IP Solution, Delivering End-to-End Quality of Service for the Future of IP," *White Paper*, Lucent Technologies, December 1999.

[STERNAD01]   M. Sternad, A. Svennson, A. AhlÇn, T. Ottoson, N. Ericsson, S. Falahati, A.Ewerlid, and B. Mielczarek, "PCC Wireless IP - Optimizing Throughput and QoS over Fading Channels," *Nordic Radio Sympusium 2001*, 2001.

[STILIAD97]   D. Stiliadis and A. Varma, "A General Methodology for Designing Efficient Traffic Scheduling and Shaping Algorithm," *Proceedings of IEEE INFOCOM'97*, Kobe, Japan, April 1997.

[STOICA99]   I. Stoica and H. Zhang, "Providing Guaranteed Services Without Per Flow Management," *Proc. ACM Conf Applications, Technologies, Architectures, and Protocols for Computer Commun. (SIGCOMM '99)*, Cambridge, MA, August 1999, pp. 81–94.

[STOJME01]   I. Stojmenovic and X. Lin, "Loop-free hybrid single-path flooding routing algorithms with guaranteed delivery for wireless network," *IEEE Transactions on Parallel and Distributed Systems*, Vol. 12, No. 10, pp. 1023–1032, October 2001.

[STOJME01a] I. Stojmenovic, and X. Lin, "Power-aware localized routing in wireless networks," *IEEE Transactions on Parallel and Distributed Systems*, Vol. 12, pp. 1122–1133, 2001.

[STOJME02] I. Stojmenovic, "Location updates for efficient routing in ad hoc networks, *Handbook of Wireless Networks and Mobile Computing*, pp. 451–471, Wiley, New York, 2002.

[STRAYER92] W.T. Strayer, B. Dempsey, and A.Weaver, *XTP: The Xpress Transfer Protocol*, Addison Wesley, Reading, MA, 1992.

[SURE98] S. Suresh, M. Woo, and C. Raghavendra, "Power-Aware Routing in Mobile Ad Hoc Networks," *Proceedings of Mobicom 98 Conference*, Dallas, October 1998.

[SZE99] S. -Y. Ni, Y. -C. Tseng, and Y. -S. Chen, "The Broadcast Storm Problem in a Mobile Ad Hoc Network," *Proceedings of the fifth annual ACM/IEEE International conference on Mobile computing and networking*, August 15–19, 1999, Seattle, WA.

[TALUC97] F. Talucci, M. Gerla, and L. Fratta, "MACABI (MACA By Invitation): A Receiver Oriented Access Protocol for Wireless Multiple Networks," *PIMRC '97*, Helsinki, September 1–4, 1997.

[TANENB96] A.S. Tanenbaum, *Computer Networks*, 3d ed., Prentice Hall, Englewood Cliffs, NJ, 1996.

[TANG99] Z. Tang, and J.J. Garcia-Luna-Aceves., "Hop-reservation Multjple Access (HRMA) for Ad Hoc Networks," *Proceedings of IEEE INFOCOM '99*, 1999, pp. 194–201.

[TANG99a] Z. Tang, and J.J. Garcia-Luna-Aceves., "A Protocol for Topology Dependent Transmission Scheduling in Wireless Networks," *Proceedings of WCNC '99*, 1999, pp. 1333–1337.

[TEL-INTER02v36] Telecommunications$^{(r)}$ International, Vol. 36, No. 6, June 2002.

[TEL-INTER02v36a] Telecommunications$^{(r)}$ International, Vol. 36, No. 8, Aug 2002.

[TEL-INTER03v36] Telecommunications$^{(r)}$ International, Vol. 36, No. 10, October 2003.

[THOMSO97] K. Thomson, G. Miller, and R. Wilder, "Wide Area Traffic Patterns and Characteristics," *IEEE Network Magazine*, December 1997.

[TIA-IS95] Telecommunications Industry Assocation, Mobile station - base station compatibility standard for dual-mode wideband spread spectrum cellular system, TIA/EIA IS-95 (July 1993).

[TIMMER98] P. Timmers, "Business Models for Electronic Markets," *J. Electronic Markets*, Vol. 8, No. 2, 1998, pp. 3–8.

[TOBAGI90] F. Tobagi, "Fast Packet Switch Architectures for Broadband Integrated Networks," *Proc. IEEE*, Vol. 78, No. 1, pp. 133–167.

[TOH01] C.-K. Toh, "Maximum Battery Life Routing to Support Ubiquitous Mobile Computing in Wireless Ad Hoc Networks," *IEEE Commun. Mag.*, June 2001, pp. 138–147.

[TOH02] Toh, C.-K., *Ad Hoc Mobile Wireless Networks, Protocols and Systems*, Prentice Hall, New York, 2002.

[TOH96] C-K. Toh, "A Novel Distributed Routing Protocol to Support Ad Hoc Mobile Computing, *Proceedings of 15th IEEE Annual International Conference on Computers and Communications*, Phoenix, 1996, pp. 480–486.

[TSENG01] Y.-C. Tseng, , S.- Y. Ni, , and E.- Y. Shih, "Adaptive Approaches to Relieving Broadcast Storms in a Wireless Multihop Mobile Ad Hoc Network," *Proco 21st Int. Conf Distributed Computing Systems*, Phoenix, 2001, p. 481.

[UMTS-RPRT] UMTS-Forum, UMTS/IMT-2000 Spectrum, Report No. 6, 1999.

[URIE95] A. Urie, M. Streeton, and C. Mourot, "An Advanced TDMA Mobile Access System for UMTS," *IEEE Personal Communications*, Vol.12, No 1, pp 38–47, February 1995.

[VAIDYA00] N. H. Vaidya, P. Bahl, and S. Gupta, "Distributed Fair Scheduling in a Wireless LAN," *Proceedings of 6th Annual International Conference on Mobile Computing and Networking*, 2000.

[VAN01] K. V. Dam, S. Pitchers, and M. Barnard, "Body Area Networks: Towards a Wearable Future," *Proc. Wireless World Research Forum*, Munich, March 2001.

[VITERBI] VITERBI J. A., *CDMA: Principles of Spread Spectrum Communication*, Prentice Hall, Reading, NA, 1995.

[WALKE01] B. Walke, et al., "Self-Organizing Wireless Broadband Networks with Guaranteed Quality of Service," *Proceedings of the WWRF meeting*, Helsinki, Finland, May 10–11, 2001.

[WALKE01A] B. Walke, N. Esseling, J. Habetha, A. Hettich, A. Kadelka, S. Mangold, J. Peetz, and U. Vornefeld, "IP over Wireless Mobile ATM – Guaranteed Wireless QoS by HiperLAN2," *Proceedings of the IEEE*, Vol. 89, pp. 21–40, Jan. 2001.

[WALKE03] B. Walke, R. Pabst, and D. Schultz, A Mobile Broadband System based on Fixed Wireless Routers, Proceedings of International Conference on Communication Technology 2003, Vol. 2, pp. 1310–1317, Beijing.

[WALKE03a] B. Walke and R. Pabst, Relay-based Deployment Concepts for Wireless and Mobile Broadband Cellular Radio, White Paper, *Proceedings of the 9th WWRF Meeting*, Zurich, Switzerland, July 2003.

[WALKE85] B. Walke and G. Briechle, "A Local Cellular Radio Network for Digital Voice and Data Transmission at 60 GHz," *Proc. Intern. Conference on Cellular and Mobile Communications*, London, Nov. 1985, pp. 215–225.

[WANG01] Z. Wang, *Internet QoS: Architectures and Mechanisms for Quality of Service*," Morgan Kaufmann, March 2001.

[WANG91] Z. Wang and J. Crowcroft, "Shortest Path First with Emergency Exits," *ACM SIGCOMM 90*, September 1991.

[WANG96] Z. Wang and J. Crowcroft, "Quality of service routing for supporting multimedia applications," *IEEE J. Select. Areas Commun*, Vol. 14, pp. 1228–1234,1996.

[WANG97] F. Wang, B. Vetter, and S.F. Wu, "Secure Routing Protocols: Theory and Practice," *Tech. Rept.*, University of California at Davis, May 1997. Available from http:/ /shang.csc.ncsu.edu/papers.htm.

[WICKEL96] I.J. Wickelgren, "Local-area networks go wireless," *IEEE Spectrum*, Vol. 33, No. 9, pp. 34–40, 1996.

[WIELAND04] Ken Wieland, "I-mode in Europe still to make an impact," *Telecommunications Magazine Online*, pp. 23–25, August 2004.

[WILKIN95] T. Wilkinson, T. G. C. Phipps, and S. K. Barton, "A report on HIPERLAN standardization," *International Journal of Wireless Information Networks*, Vol. 2, pp. 99–120, April 1995.

[WOO01] K. Woo, C. Yu, H.Y. Youn, , and B. Lee, "Non-Blocking, Localized Routing Algorithm for Balanced Energy Consumption in Mobile Ad Hoc Networks," *Int'l Symp. on Modeling, Analysis and Simulation of Computer and Telecommunication Systems (MASCOTS 2001)*, August 2001, pp. 117–124.

[WOOD01] L. Wood, et al., "Effects on TCP of Routing Strategies in Satellite Constellations," *IEEE Communications Magazine*, March 2001, pp. 172–181.

[WU01] H. Wu, C. Qiao, S. De, and O. Tonguz. "Integrated Cellular and Ad Hoc Relaying Systems: iCAR." *IEEE J. Sel. Areas in Comm.*, Vol. 19, No. 10, pp. 2105–2115, Oct. 2001.

[WWRF] The "Wireless World Research Forum," http://www.wireless-world-research.org/general_info/

[XIAO00] H. Xiao, K.G. Seah, A. Lo, and K.C. Chua, "A Flexible Quality of Service Model for Mobile Adhoc Network," *Proceedings of IEEE VTC2000 - Spring*, Tokyo, May 2000.

[XIAO99] Internet QoS: the Big Picture Xipeng Xiao & Lionel M. Ni, IEEE Network, January 1999

[XIE93] H. Xie, S. Tabbane, and D.J. Goodman, "Dynamic location area management and performance analysis," *Proc. 43rd IEEE Vehicular Tech. Conf.*, pp. 536–539, 1993.

[XU00] Y. Xu, J. Heidemann, and D. Estrin, Adaptive energy-conserving routing for multihop ad hoc networks, Research Report 527 USC/Information Sciences Institute, October 2000.

[XU98] Z. Xu, S. Dai, and J.J. Garcia-Luna-Aceves, "Hierarchical routing using link vectors," *IEEE Infocom*, March 1998.

[XUE01] Y. Xue and B. Li, "A Location-aided Power-aware Routing Protocol in Mobile Ad Hoc Networks," *Proc. IEEE Symp. on Ad Hoc Mobile Wireless Networks / IEEE GLOBECOM 2001*, San Antonio, TX, November 25–29, 2001.

[YEE94] N. Yee, J. -P. M.G. Linnartz, and G. Fettweis, "Multi-carrier-cdma in indoor wireless network," *IEICE Transaction on Communications, Japan*, Vol. E77, No. B(7), pp. 900–904, July 1994.

[YOUNG79] W.R. Young, "Advanced mobile phone service: introduction, background, and objectives," *The Bell Systems Technical Journal*, Vol. 58, pp. 1–14, 1979).

[ZADEH01] Ali Nabi Zadeh and Bijan Jabbari, "Performance Analysis of Multihop Packet CDMA Cellular Networks," *In Proc. IEEE Globecom 2001*, Vol. 5, pp. 2875–2879, San Antonio, Texas, November 2001.

[ZAUMEN92] W. Zaumen and J. Carcia-Luna-Aceves, "Steady-State Response of Shortest-Path Routing Algorithms," *Proceedings of the IPCCC'92*, April 1992, pp.323–332.

[ZHAN95] H. Zhang, "Service Disciplines for Guaranteed Performance Service in Packet-Switching Networks," *Proceedings of the IEEE*, Vol. 83, No. 10, October 1995.

[ZHOU99] L. Zhou and Z,J. Haas, "Securing Ad Hoc Networks," *IEEE Network Magazine*, Vol. 13, No.6, Nov./ Dec. 1999.

[ZHU98] C. Zhu and M.S., A. Corson, "Five-Phase Reservation Protocol (FPRP) for Mobile Ad Hoc Networks," *Proceedings of IEEE INFOCOM '98*, San Francisco, CA, 1998, pp. 322–331.

[ZIMMER96] T.G. Zimmerman, "Personal Area Networks: near-field intrabody communication," *IBM Systems Journal*, Vol. 35, Nos. 3 & 4, 1996.

[ZIMMER96a] T.G. Zimmerman, "Wireless networked devices: a new paradigm for computing and communications," *IBM Systems Journal*, Vol. 38, No.4, 1999.

[ZRP-ID] Z. J. Haas, "The Zone Routing Protocol (ZRP) for Ad Hoc Networks," IETF, Internet Draft, draft-zone-routing-protocol-00.txt, November 1997.

[ZUSSM02] G. Zussman and A. Segall, Capacity Assignment in Bluetooth Scatternets - Analysis and Algorithms, Proc. Networking 2002, LNCS 2345.

Index

ABOUT THE AUTHOR

GEORGE AGGELOU is a professor of electronic engineering at the Institute of Technology, Greece and the Director of G-Alpha Telecomms, consultants who build next-generation networks for enterprises and industry. He received his Ph.D. from the University of Surrey. His doctoral research, which was fully sponsored from Lucent Technologies, Middlesex, UK, was on wireless ad hoc and third generation (3G) mobile networks. Prof. Aggelou has been actively involved in standardization efforts for wireless mobile networking for nearly 10 years, first as a researcher at IBM, T.J. Watson Research Center, in New York and later as Team Leader at Cisco in London. He is the cofounder of Mobile E-commerce Technologies Ltd., London. Prof. Aggelou is the recipient of the U.K. Prize for Research of Excellence 2000 for his research innovations on Mobile Ad Hoc Networking. Prof. Aggelou is an active participant in several IEEE technical committees and activities. He has served on the Grants Committee (UGC) of Hong Kong. He is also a member of the Research Grants Evaluation Committee (FP5 & FP6) of the European Union.

www.ingramcontent.com/pod-product-compliance
Lightning Source LLC
Chambersburg PA
CBHW060423220326
41598CB00021BA/2275